Gender and Environmental Education: Feminist and Other(ed) Perspectives

This timely book provides a starting point for critical analysis and discourse about the status of gendered perspectives in environmental education research.

Through bringing together selected writings of Annette Gough, it documents the evolving discussions of gender in environmental education research since the mid-1990s, from its origins in putting women on the agenda through to women's relationships with nature and ecofeminism, as well as writings that engage with queer theory, intersectionality, assemblages, new materialisms, posthumanism and the more-than-human. The book is both a collection of Annette Gough, and her collaborators, writings around these themes and her reflections on the transitions that have occurred in the field of environmental education related to gender since the late 1980s, as well as her deliberations on future directions.

An important new addition to the World Library of Educationalists, this book foregrounds women, their environmental perspectives, and feminist and other gendered research, which have been marginalised for too long in environmental education.

Annette Gough OAM is Professor Emerita of Science and Environmental Education in the School of Education at RMIT University. She has held senior appointments at RMIT and Deakin University and has been a visiting professor at universities in Canada, South Africa and Hong Kong, as well as being life fellow of the Australian Association for Environmental Education and the Victorian Association for Environmental Education.

World Library of Educationalists series

The *World Library of Educationalists* celebrates the important contributions to education made by leading experts in their individual fields of study. Each scholar has compiled a career-long collection of what they consider to be their finest pieces: extracts from books, journals, articles, major theoretical and practical contributions, and salient research findings.

For the first time ever the work of each contributor is presented in a single volume so readers can follow the themes and progress of their work and identify the contributions made to, and the development of, the fields themselves.

The distinguished careers of the selected experts span at least two decades and include Richard Aldrich, Stephen J. Ball, Elliot W. Eisner, John Elliott, Howard Gardner, John Gilbert, Ivor F. Goodson, David Hargreaves, David Labaree and E.C. Wragg. Each book in the series features a specially written introduction by the contributor giving an overview of their career, contextualizing their selection within the development of the field, and showing how their own thinking developed over time.

Researching Literate Lives
The Selected Works of Jerome C. Harste
Jerome C. Harste

The Sociology of Assessment: Comparative and Policy Perspectives
The Selected Works of Patricia Broadfoot
Patricia Broadfoot

Gender and Environmental Education: Feminist and other(ed) perspectives
The Selected Works of Annette Gough
Annette Gough

For more titles in this series visit www.routledge.com/World-Library-of-Educationalists/book-series/WORLDLIBEDU

Gender and Environmental Education: Feminist and Other(ed) Perspectives
The Selected Works of Annette Gough

Annette Gough

Routledge
Taylor & Francis Group

LONDON AND NEW YORK

Designed cover image: Cover photo by Julian Dolman

First published 2024
by Routledge
4 Park Square, Milton Park, Abingdon, Oxon OX14 4RN

and by Routledge
605 Third Avenue, New York, NY 10158

Routledge is an imprint of the Taylor & Francis Group, an informa business

British Library Cataloguing-in-Publication Data
A catalogue record for this book is available from the British Library

ISBN: 978-1-032-48820-2 (hbk)
ISBN: 978-1-032-48821-9 (pbk)
ISBN: 978-1-003-39093-0 (ebk)

DOI: 10.4324/9781003390930

Typeset in Galliard
by Apex CoVantage, LLC

This book is dedicated to my family, without whom this book would not exist: my parents, Pat and Ray Wilkinson, who supported my early education and interests; my life partner, Noel Gough, who has been my best friend, critic, co-parent and co-author for nearly 40 years; and especially my children, Kate and Simon, and my grandson, Liam, who will face the challenges of achieving a more just world for all life on Earth from the mess past generations have bequeathed them.

Contents

Figures and tables

Figures

Tables

Preface

Environmental education has been a focus of my career for nigh on 50 years, and I was humbled early in 2023 when my services to the field, and to tertiary education, were recognised with the awarding of an Order of Australia medal (OAM). Over these decades I have been a passionate advocate for the need for people to understand their environments, care for them and act for them, while recognising that there is no one true story of what is happening and what needs to be done to achieve environmental conservation goals or address climate change. Action and social transformation are needed, and formal schooling (and its associated teacher education) has been my focal point.

This book has its origins in my doctoral research, which was a feminist poststructuralist analysis of the beginnings of environmental education in Australia and internationally. As part of this research I examined the definitions of environmental education, the forming documents for the field, particularly those from UNESCO, and some of the people (the "founders") who helped frame these documents or who were instrumental in helping frame their interpretation and implementation in Australia and the United States of America in the 1970s through to the early 1990s.

While I started with a feminist analysis, this evolved to an analysis from the perspective of marginalised voices as I came to recognise that women's voices are just one of many marginalised people's voices silenced in the dominant environmental education discourses and that there were intersections in these voices (poor migrant women, for example).

The content of this book samples across my subsequent and related writings on feminisms, gender and environmental education. While my first article that drew attention to the absence of women from the groups that formulated the definitions of environmental education is not included here because the focus was more on language and Amero-Eurocentrism than gender (Greenall Gough 1993), the articles that built on and went beyond my doctoral thesis are collected here under three themes: putting women on the agenda, feminisms and nature in environmental education, and moving beyond feminisms and gender.

Within these main themes I explore the importance of feminist research in environmental education, and how thinking about this has changed over the past 25 years. Gender perspectives, and other othered perspectives which are the focus of social oppression, have long been marginalised in environmental education research and practice through being subsumed into the notion of "universalised people", the "norm". This volume highlights the importance of bringing gendered perspectives to the centre of discussions with a view to inspiring others to pursue such research. I hope you find it inspirational.

References

Greenall Gough, Annette. 1993. "Globalizing environmental education: What's language got to do with it?" *Journal of Experiential Education* 16(3): 32–39, 46. https:// doi.org/10.1177/105382599301600306

Acknowledgements

This book would not have been possible without the collaborations of my partners in print – Hilary Whitehouse, Judy Horacek, Connie Russell and Noel Gough (who is also my partner in life) – who share authorship of several of the articles included here.

Hilary and I have talked about writing a book like this together for over two decades. We haven't got there yet, but our collaborations in writing articles over the past 20 years are reflected in four of the chapters included here. Writing with Hilary is always a case of the sum being greater than the two parts. I just wish she lived closer (though visiting her in Cairns is always fun).

Judy's cartoons have been inspirational for me since Noel gave me her first book in 1992. When the call for abstracts for a special issue of *Environmental Education Research* on humour and environmental education was released, I immediately thought of Judy's cartoons and approached her about doing something together. The results of that duoethnography are included here.

Although not overtly included here as a co-author, Connie has been part of my writing adventures for over two decades. She edited the issue of the *Canadian Journal of Environmental Education*, in which my first article with Hilary, and the Camp Wilde article, appeared. She co-edited the section of the *International Handbook of Research of Environmental Education* that included my gender chapter, and we have co-edited special issues of *The Journal of Environmental Education* (together with Hilary). I wish she lived closer too!

What can I say about Noel's immeasurable contributions? Academically, he has been a critical friend, sounding board, provocateur, occasional collaborator in teaching, researching and writing over many years, and nurturer of my career. Part of my life for over half my life, through lots of ups and downs, this book reflects many of those experiences. Thank you for being there, my love. I look forward to our continuing adventures.

The small group of friends who joined us in Camp Wilde – Warren Sellers, Mary Aswell Doll, Peter Appelbaum and Sophia Appelbaum – made that writing experience an unexpected successful experience, as it stimulated others, including Joshua Russell and Jesse Bazzul, whom I have enjoyed working with in other contexts.

I would also like to acknowledge the contributions to my feminist thinking by colleagues at Deakin University (especially Jill Blackmore and Jane Kenway), and John Fien (who suggested long ago and far away that my doctoral research should look at "environmental education as a man-made subject", and I did).

I thank Judy Horacek for permission to include a number of her cartoons (in Chapter 9 and Chapter 14), the Australian Research Council for permission to reproduce Figure S1.1, the Australian Government Department of Education for permission to reproduce Table 3.1, Ryerson Image Centre for permission to reproduce Jo Spence's photo (Figure 11.1) and Deena Metzger for permission to reproduce her image (Figure 11.2).

Last, but not least, these articles and chapters were written on the unceded lands of the Woi wurrung and Boon wurrung peoples of the Kulin Nations; I respectfully acknowledge what their elders, past and present, have taught us about this land and sea.

The following chapters were reprinted from previous publications. I thank the publishers and editors for their support.

Chapter 2. "Recognising women in environmental education pedagogy and research: Toward an ecofeminist poststructuralist perspective". *Environmental Education Research, 5*(2), pp. 143–161. ©1999 Taylor and Francis Ltd.

Chapter 3. "The power and the promise of feminist research in environmental education". *Southern African Journal of Environmental Education, 19,* 28–39. Reprinted in accordance with Creative Commons Attribution license CC-BY-NC-SA.

Chapter 4. "The contribution of ecofeminist perspectives to sustainability in higher education". In P. Corcoran & A. Wals (Eds.), *Higher Education and the Challenge of Sustainability: Contestation, Critique, Practice, and Promise* (pp. 149–161). Dordrecht: Kluwer. ©2004 Reprinted by permission of SpringerNature.

Chapter 5. "Researching differently: Generating a gender agenda for research in environmental education". In R. B. Stevenson, M. Brody, J. Dillon, & A. Wals (Eds.), *International Handbook of Research on Environmental Education* (pp. 375–383). New York: Routledge.©2012 Taylor and Francis Group.

Chapter 6. Gough, A., & Whitehouse, H. "Centering gender on the agenda for environmental education research". *The Journal of Environmental Education, 50*(4–6), 332–347. ©2019 Taylor and Francis Ltd.

Chapter 7. Gough, A. & Whitehouse, H. "The 'nature' of environmental education research from a feminist poststructuralist standpoint". *Canadian Journal of Environmental Education, 8,* 31–43. ©2003 reprinted by permission of Canadian Journal of Environmental Education editors.

Chapter 8. Gough, A., & Whitehouse, H. "New vintages and new bottles: The 'nature' of environmental education from new material feminist and ecofeminist viewpoints". *The Journal of Environmental Education, 49*(4), 336–349. ©2018 Taylor and Francis Ltd.

Chapter 9. Gough, Annette and Hilary Whitehouse. "Challenging Amnesias: Feminist New Materialism/Ecofeminism/Women/Climate/Education". *Environmental Education Research 26*(9–10): 1420–1434. ©2020 Taylor and Francis Ltd.

Chapter 10. "Symbiopolitics, sustainability and science studies: How to engage with alien oceans". *Cultural Studies <=> Critical Methodologies, 20*(3), 272–282. ©2019 reprinted by permission of Sage Publications.

Chapter 11. "Listening to voices from the margins: Transforming environmental education". In J. Russell (Ed.), *Queer Ecopedagogies: Explorations in Nature, Sexuality, and Education* (pp. 161–181). ©2021 Reprinted by permission of SpringerNature.

Chapter 12. Gough, N., Gough A., Appelbaum, P., Appelbaum, S., Doll, M.A. & Sellers, W. Tales from "Camp Wilde: Queer(y)ing environmental education research". *Canadian Journal of Environmental Education, 8*, 44–66. ©2003 reprinted by permission of *Canadian Journal of Environmental Education* editors.

Chapter 13. Gough, Annette, and Judy Horacek. "The generativity of feminist and environmental cartoons for environmental education research and teaching". *Environmental Education Research 29* (4): 500–519. ©2022 Taylor and Francis Ltd.

Chapter 14. "Body/mine: A chaos narrative of cyborg subjectivities and liminal experiences". *Women's Studies, 34*(3–4): 249–264. ©2005 Taylor and Francis Ltd.

Chapter 15. "Education in the Anthropocene". *The Oxford Research Encyclopedia on Gender and Sexuality in Education*. ©2021 Reprinted by permission of Oxford Publishing Limited.

1 Reflections and refractions on gender and environmental education

Annette Gough

Abstract

This chapter provides a reflection on how I became engaged with feminist theorising in environmental education. Starting from a focus on "environmental education as a man-made subject", I explored and engaged with different feminist and ecofeminist theories before finding my voice for researching in this space. I explain how my position was very much influenced by Carolyn Merchant's partnership ethic and Sandra Harding's feminist standpoint epistemology and strong objectivity, as well as feminist poststructuralism research in education. I then discuss how the various articles and chapters contained in this volume came about, and how feminist research in environmental education has changed since the late 1980s.

Beginnings

When I started to investigate "environmental education as a man-made subject" in 1990 there had been very little written about environmental education from a feminist perspective. As I have discussed elsewhere, and as reflected in several of the chapters in this book, gender was not part of the agenda of environmental education in its early days. This should have been apparent to me as I was one of the few women at the 1975 Australian UNESCO Seminar (Linke 1977), the only female on the 15-member 1975 Curriculum Development Centre (CDC) Environmental Education Committee (as Secretary) and 1 of 2 women on the 11-member 1977 CDC Study Group on Environmental Education (as convenor). As noted in Chapter 2, according to the list of participants in final report of the Tbilisi conference on environmental education (UNESCO 1978, 83–99), there were only 55 females out of a total of 345 participants at the conference, a ratio of over 6:1 in favour of males (see Table 2.1). Thus, in contrast with today, there was definitely a dearth of women in environmental education deliberations at this time.

The ecofeminist movement emerged in the 1970s; Francoise d'Eaubonne first used the term "l'écoféminisme" and outlined her ecofeminism theory in her book *Le féminisme ou la mort* in 1974. Grounded in the activist social

DOI: 10.4324/9781003390930-1

movements of the time, d'Eaubonne (1974) describes ecofeminism as both an activist and academic/philosophical movement where "the convergence of ecology and feminism into a new social theory and political movement challenges gender relations, social institutions, economic systems, sciences, and views of our place as humans in the biosphere" (p. 28).

The ecofeminist movement grew with the feminist movements of the 1980s and 1990s, and as is illustrated in Figure 6.1, for example, references to ecofeminism in books peaked in 1996. While environmentalists have been calling for environmental education as a means of resolving environmental problems since the 1960s, for the main part, up until the late 1980s, feminists had not addressed environmental education as a strategy for achieving their goals. Women in outdoor education had started to draw attention to the different instructional needs of women in outdoor education programmes (e.g. Miranda and Yerkes 1982; Mitten 1985), and women in science education had drawn attention to the under-representation of women in science careers (e.g. Kahle 1985; Manthorpe 1982), but it was not until 1987 that feminist perspectives were introduced into the environmental literature. Giovanna Di Chiro (1987a, 1987b) argued for the close connection between the socially and politically constructed nature of environmental problems and the importance of developing a feminist perspective in environmental education for a more complete analysis of the problems. She (1987b, 15) wrote that, as environmental education is problem-solving focused,

> the feminist perspective offers a more complete analysis of the environmental issue and thereby a better understanding of the problem and its potential solutions. Such an analysis is a political one, in that it looks at how power relations (in, for example, gender, class, race) shape the world in which we live. It is political in that it asserts that the "polity" (human social world) determines and controls how this social world is and has been historically constructed and organised and hence, refutes the myth that the past and present state of the world is a "natural" and therefore justifiable progression.

In terms of Schuster and Van Dyne's (1984) stages of curriculum change, Di Chiro was taking a different approach to that adopted by the outdoor and science educators. While working from within a social justice agenda, she had leapt to the sixth stage of seeking a transformed "balanced" curriculum rather than drawing attention to the invisible women (stage 1), women as a disadvantaged subordinate group (stage 3) or women studied on their own terms (stage 4).

At the time of writing her article, and an extended version (1987a) of which was included in a Deakin monograph that also contained a chapter of mine (Greenall 1987), Di Chiro was a lecturer at Deakin University, where I had just commenced my doctoral studies (Gough 1994). It would be nice to say that her work influenced me, but at that time it did not – at that time my focus was more on the policy politics and history of environmental education. It was only when I changed my thesis focus in 1990 that Di Chiro's chapter and article took on importance.

When I began my searches for literature on feminist research in environmental education, what I found was both sparse and predominantly written by Australians or published in Australia. A similar finding was made by Di Chiro (1993) when, in 1991, she conducted an ERIC search using the descriptors "feminism" and "environmental education" which yielded only two articles, one her own (1987b) which was written and published in Australia, and the other by Ariel Salleh (1989), an Australian ecofeminist and social theorist. My own search in 1992 added two North American articles (Kremer et al. 1990–1991; Fawcett et al. 1991), one British article (Hallam and Pepper 1991) and more from Australia calls to consider women's perspectives when developing environmental education programmes (Brown and Switzer 1991a, 1991b, 1991c; Peck 1991, 1992). With the exception of the Di Chiro and Salleh articles, the preparation of the Australian literature on women and environmental education had been funded by, or was related to actions by, the Australian government. For example, the paper by Dianne Peck (1991, and reported in her 1992 article) was prepared as part of the Commonwealth Gender Equity in Curriculum Reform Project, and the other papers were prepared under the auspices of the Office of the Status of Women in the Commonwealth Department of Prime Minister and Cabinet (DPMC/OSW) or its National Women's Consultative Council (NWCC). However, little of this literature can be called feminist research in environmental education. Rather it was calls to consider women's perspectives when developing environmental education programmes (Brown and Switzer 1991a, 1991b, 1991c; Peck 1991, 1992) or calls for recognition of the link between women and ecology in an educational context (Di Chiro 1987a, 1987b; Hallam and Pepper 1991; Salleh 1989).

Positioning my voice

As my doctoral research was a poststructuralist feminist analysis of the foundations of environmental education, I immersed myself in feminist and ecofeminist literature. A lot of ecofeminist literature was being published in the 1990s (see Figure 6.1) so there was much to be reviewed from liberal, Marxist, socialist, radical and ecological positions. There was also much contestation between the different philosophies, in what Rosemarie Tong (1989, 238) calls a "kaleidoscopic" feminist thought, in which the "preliminary impression may be one of chaos and confusion, of dissension and disagreement, of fragmentation and splintering. But a closer look will always reveal new visions, new structures, new relationships . . . all of which will be different tomorrow than today". There were also debates between ecofeminists and social ecologists (such as Bookchin 1982) and deep ecologists (such as Devall and Sessions 1985).

Many of the ecofeminist arguments of this time centred on a belief that women had natural, cultural or ideological closeness to nature, which only sought to reinforce the binaries that my feminist poststructuralist analysis was arguing against. I was interested in the emerging convergent evolution in both feminism and ecofeminism towards the notion of paying attention to the differences among women, the socially constructed nature of both humanity and nature, and the appropriateness of deconstruction and other forms of poststructuralist analysis as a methodology for feminism and ecofeminism.

My background was in science education, and I was drawn to the writings of Sandra Harding and Carolyn Merchant. Merchant's (1980) *The Death of Nature: Women, Ecology and the Scientific Revolution* brought together my three interests of science, environment and feminism, and her socialist feminist positioning together with her writings around a partnership ethic, an environmental ethic for society "that treats humans (including male partners and female partners) as equals in personal, household, and political relations and humans as equal partners with (rather than controlled-by or dominant-over) nonhuman nature" (Merchant 1992, 188) felt right. She further argues that such an ethic treats nature and human nature as socially and historically constructed over time, transformed through human praxis, and rooted in an analysis of race, class and gender:

> Constructing nature as a partner allows for the possibility of a personal or intimate (but not necessarily spiritual) relationship with nature and for feelings of compassion for nonhumans as well as for people who are sexually, racially, or culturally different. It avoids gendering nature as a nurturing mother or a goddess and avoids the ecocentric dilemma that humans are only one of many equal parts of an ecological web and therefore morally equal to a bacterium or a mosquito.
>
> (1992, 188)

The feminist poststructuralist analysis part of my research required "attention to historical specificity in the production, for women, of subject positions and modes of femininity and their place in the overall network of social power relations" (Weedon 1987, 135). I found Sandra Harding's (1986, 1987, 1993a, 1993b) work on feminist standpoint theory and strong objectivity particularly constructive in developing my analysis. Harding (1993b, 56) argues that feminist standpoint theory

> sets out on a rigorous "logic of discovery" intended to maximise the objectivity of the results of research and thereby to produce knowledge that can be *for* marginalized people (and those who would know what the marginalized can know) rather than *for* the use only of dominant groups in their projects of administering and managing the lives of marginalized people.

These characteristics of Harding's feminist standpoint theory continue to inform my research:

- Standpoint approaches argue that all knowledge is "situated knowledge"
- Standpoint approaches intend to, and can, produce research that is "for women"

- Standpoint research is by its very focus overtly politically engaged in its conscious, intentional critical focus on the power relations that oppress women and other economically and politically vulnerable groups
- Standpoint approaches "study up": focusing on dominant institutions and their cultures and practices, not just groups less powerful than the researchers, which is usual practice
- Research that would identify the conceptual practices of power must start off with researchers' thought from women's lives instead of from those disciplinary or social policy frameworks that treat as natural women's oppression, domination and exploitation.
- Whatever strategy a researcher may use to start off inquiry from women's lives, a standpoint is an achieved and collective position, not an ascribed position or individual opinion.
- Standpoint-directed research is able to practice more effective methodological strategies and thus produce more objective accounts of nature and social relations than conventional research that attempts to achieve value neutrality.

One outcome from my analysis of the "founders" of environmental education in Australia and the United States of America was five principles for research and practice in environmental education:

- to draw attention to the racism and gender blindness in environmental education and to develop a willingness to listen to silenced voices and to provide opportunities for them to be heard;
- to foster working, individually and collectively, and equally, with other humans and with nonhuman nature rather than separating humans from nature;
- to recognise that knowledge is partial, multiple and contradictory;
- to develop understandings of the stories of which we are a part and our abilities to deconstruct them; and
- to recognise resistances to liberatory pedagogy in environmental education and to work with these resistances.

These principles have moved beyond the dominant notions of ecofeminism of the early 1990s and embraced the underlying concepts of Merchant's partnership ethic, as well as Harding's notions of feminist standpoint theory and strong objectivity. They also reflect understandings of what came to be known as feminist new materialism. For example, they are in accord with Stacy Alaimo's (1994) argument for an environmental feminism that stressed a political alliance between women and nature and one that would not slide into essentialism:

> focusing on the agency of women and nature can help keep environmentalism in the political arena and can oppose the appropriation of nature as resource by stressing nature as an actor and by breaking down the nature/culture divide, thus undermining the systems of domination.
>
> (p. 150)

More recently, Carol Taylor (2019, 39–40) has expanded on new material feminism and the more-than-human world, noting that

> New material feminism shares a social justice imperative with other modes of feminism. Like them, it is committed to finding ways to combat gender inequality, discrimination, and violence in education. More broadly, it shares with post-structuralism, post-colonialism, and intersectional studies a suspicion that the Enlightenment ideals of rationality, objectivity, and scientific progress have only delivered partial benefits for particular groups of people (mostly males, White, Western, able-bodied people) and that the narrative of "progress" it offers is also a partial affair designed to maintain the hegemony of those who benefit most from it. New material feminism, therefore, offers a radical set of tools for generating new understandings of subjectivity, relationality, and ethics, and suggests that these tools offer ways of fundamentally rethinking what we mean by – and how we do – social justice in a more-than-human world.

It is these principles and continuing reading and research drawing on feminism, poststructuralism, postcolonialism, intersectional studies, new materialism and ecofeminism that informed my writing about environmental education research and practices post thesis.

Writing feminism into environmental education

Following completion of my thesis I started writing articles – and a book (Gough 1997) – grounded in my doctoral research and other projects.

My first feminist articles (Gough 1999a, 1999b – see Chapters 2 and 3) were very much in the spirit of Schuster and Van Dyne's (1984) early stages of curriculum change – drawing attention to the absence of women in the discussions around the formulations of environmental education as a field and arguing for why their presence was important. This was also the case with Chapter 4 (Gough 2004). These articles were consistent with Valerie Brown and Margaret Switzer's (1991a, 16) argument that women are less likely to have scientific or economic training than men and, consequently, have less influence on the development of curriculum priorities. They also note that there is a need to compensate for the effects on environmental education research and teaching of the relative absence of women and women's interests from the professions of environmental science and economics: "This absence has meant that many questions on ecologically sustainable development from the fields of health, welfare, household management and social policy have neither been investigated nor included in environmental education" (1991a, 16). This observation led me to continue to aim to pursue gender equity research that transcended "the boundaries of race, ethnicity, class and socio-economic identities" (Krockover and Shepardson 1995, 223).

Sadly, women continue to be much, if not most, affected by environmental problems. As the Global Gender and Climate Alliance (2013) notes with respect to climate change,

- Women in developing countries are particularly vulnerable to climate change because they are highly dependent on natural resources for their livelihood.
- Women experience unequal access to resources and decision-making processes, with limited mobility in rural areas.
- Women make between 30 and 80 per cent of what men earn annually.
- Out of 140 countries surveyed by the World Bank, 103 impose legal differences on the basis of gender that may hinder women's economic opportunities.
- Women make up half of the agricultural workforce in the least developed countries.
- In developing countries they own between 10 and 20 per cent of the land.
- Two-thirds of the world's illiterate adults are women.
- Socio-economic norms can limit women from acquiring the information and skills necessary to escape or avoid hazards (e.g. swimming or climbing trees to escape rising water levels).
- Dress codes imposed on women can restrict their mobility in times of disaster, as can their responsibility for small children who cannot swim or run.
- A lack of sex-disaggregated data in all sectors often leads to an underestimation of women's roles and contributions, thus increasing gender-based vulnerability.

Ten years on the United Nations is still reporting that gender equality has not improved, and indeed it has probably gone backwards during the pandemic (Boecker 2023). Thus, we continually need to consciously put women on the environmental education agenda as the world is unlikely to read the United Nations' Sustainable Development Goal of gender parity by 2030 (United Nations 2015), and this is reflected in Chapters 5 and 6 (Gough 2013; Gough and Whitehouse 2019).

Beyond the women/nature binary

Ecofeminism suffered from a feminist backlash in the late 1990s, being then criticized as essentialist, elitist and ethnocentrist, "and effectively discarded" (Gaard 2011, 26). I had already positioned my writings at a distance from most forms of ecofeminism, although informed by them, but the binary association of women and nature persisted for many. Over more than a decade Hilary Whitehouse and I have attempted to destabilise traditional understandings and argued for poststructuralist and other positions as preferably approaches for environmental education research (Gough and Whitehouse 2003, 2018, 2020 – see Chapters 7, 8 and 9). Ecofeminists tended to be critical of poststructuralist approaches. For example, Salleh (1997, 9) argues "the tenets of deconstructive practice have been catechised and used as political rhetoric, resulting in an impractical nihilism when applied to everyday life". However, Hilary and I (Gough and Whitehouse 2003) saw poststructuralist positioning as enabling a rich diversity of viewpoints to be recognised and celebrated for what it reveals about social and environmental meanings and actions.

In our more recent articles (2018, 2020) we have argued that much of what is being conceptualised in feminist new materialism was also being recognised in ecofeminist writings of the 1990s and that this needs to be acknowledged by environmental educators. This position is consistent with Gaard's argument for an intersectional ecological-feminist approach that frames issues such as global gender justice and climate justice "in such a way that people can recognize common cause across the boundaries of race, class, gender, sexuality, species, age, ability, nation – and affords a basis for engaged theory, education, and activism" (2011, 44).

There has been a resurgence of interest in ecofeminism since its 2009 low point (see Figure 1.1), and much of this writing has reflected Gaard's argument. A broader understanding of ecofeminism, gender and environmental education research was also reflected in the two special issues of *The Journal of Environmental Education* on gender and environmental education that Hilary and I edited with Connie Russell. Here we were inspired by the international encouragement for centering gender on the environmental education agenda found in *The Future We Want*, the outcomes document adopted at the Rio + 20 United Nations Conference on Sustainable Development (2012), which reaffirmed the necessity for promoting "social equity, and protection of the environment, while enhancing gender equality and women's empowerment, and equal opportunities for all, and the protection, survival and development of children to their full potential, including through education" (paragraph 11). We called for "manuscripts that respond to the need for promotion of social equity and enhancing gender equality and women's empowerment within environmental education" (Gough et al. 2017, 5) and received enough acceptable submissions to warrant two issues of the journal. We were thrilled.

Going beyond the woman/nature binary, as reflected in Chapter 10, includes the discussion of how what counts as biological life is changing and that human and more-than-human life are very much entangled, together with discussing that gender should not be a silence in sustainability education because gender and more-than-human can enhance each other, with gender more particularly helping "to disrupt the somewhat flat equation of the more-than-human" (Probyn 2016, 113).

Troubling gender and nature

The remaining chapters span nearly 20 years, but all reflect my writings around troubling gender and nature. Section III opens with overt examples of queering environmental education (Chapters 11 and 12). Here queering is in the sense described by Mary Bryson and Suzanne de Castell (1993), who describe

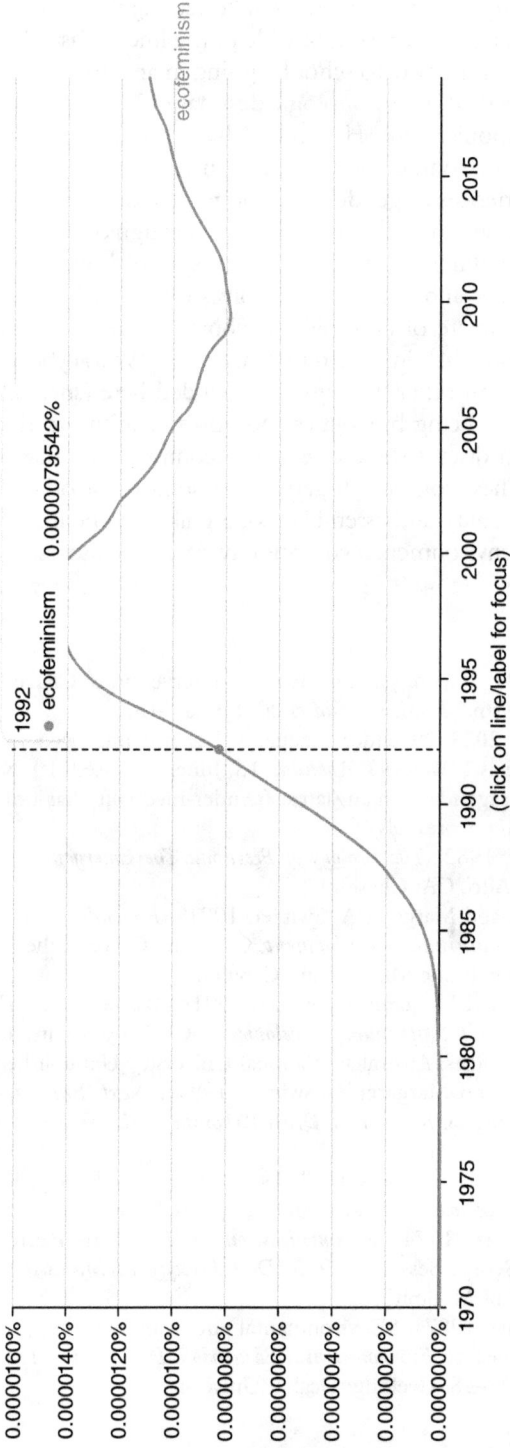

Figure 1.1 Ngram of use of "ecofeminism" in books update 1970–2019.

an actively queering pedagogy in terms of "queering its technics and scribbling graffiti over its texts, of colouring outside of the lines so as to deliberately take the wrong route on the way to school – going in an altogether different direction than that specified by a monologic destination" (p. 299). I would also call writing with cartoonist Judy Horacek (Chapter 13) as providing an opportunity to queer environmental education in a different way as we explored both of our experiences of gender, environment and environmental education through her cartoons and other experiences: through our duoethnography we certainly went in a different direction from a specified monologic destination.

Troubling gender and nature also involves exploring becoming more-than-human. I initially wrote of becoming a cyborg as a result of my breast cancer experiences (Chapter 14), informed by Haraway's (1985) cyborg manifesto, but this evolved, through other writings not included here (such as Gough 2015) to be discussions of being biopolitics and posthumanism, to the two examples included here that discuss the concept of becoming more-than-human (Chapters 10 and 15). These concepts, together with some of the others raised in Chapter 11 (intersectionality and assemblages), are likely to be important for future developments in environmental education research, as discussed in Chapter 16.

References

Alaimo, Stacy. 1994. "Cyborg and ecofeminist interventions: Challenges for an environmental feminism." *Feminist Studies 20* (1): 133–152.

Boecker, Brianna. 2023. "Gender inequality has not improved in a decade, new UN report reveals." *Women's Agenda*, 13 June. Accessed 10 November 2023. https://womensagenda.com.au/latest/gender-inequality-has-not-improved-in-a-decade-new-un-report-reveals/

Bookchin, Murray. 1982. *The Ecology of Freedom: The Emergence and Dissolution of Hierarchy*. Palo Alto, CA: Cheshire.

Brown, Valerie A., and Margaret A. Switzer. 1991a. *Engendering the Debate: Women and Ecologically Sustainable Development*. Canberra: Office of the Status of Women, Department of the Prime Minister and Cabinet.

Brown, Valerie A., and Margaret A. Switzer. 1991b. *Where have all the Women Gone? The Role of Gender in Sustainable Development*. Canberra: Centre for Resource and Environmental Studies, Australian National University. Unpublished paper.

Brown, Valerie A., and Margaret A. Switzer. 1991c. *Next: Engendering Sustainable Development Strategies. Response to Draft Reports of ESD Working Groups*. Canberra: Office of the Status of Women.

Bryson, Mary, and Suzanne de Castell. 1993. "Queer pedagogy: Praxis makes im/perfect." *Canadian Journal of Education 18* (3): 285–305.

d'Eaubonne, Francoise. 1974. *Le Féminisme ou La Mort*. Paris: Pierre Horay.

Devall, Bill, and George Sessions. 1985. *Deep Ecology: Living as if Nature Mattered*. Layton, UT: Gibbs M. Smith.

Di Chiro, Giovanna. 1987a. "Environmental education and the question of gender: A feminist critique." In *Environmental Education: Practice and Possibility*, edited by Ian Robottom, 23–48. Geelong: Deakin University.

Di Chiro, Giovanna. 1987b. "Applying a feminist critique to environmental education." *Australian Journal of Environmental Education 3* (1): 10–17.

Di Chiro, Giovanna. 1993. "Environmental education in the war years: A case study." In *Alternative Paradigms in Environmental Education Research,* edited by Rick Mrazek, 227–237. Troy, OH: North American Association for Environmental Education.

Fawcett, Leesa, Dian Marino, and Rebecca Raglon. 1991. "Playfully critical: Reframing environmental education." In *Confronting Environmental Challenges in a Changing World,* edited by John H. Baldwin, 250–254. Troy, OH: NAAEE.

Gaard, Greta. 2011. "Ecofeminism revisited: Rejecting essentialism and re-placing species in a material feminist environmentalism." *Feminist Formations 23* (2): 26–53.

Global Gender and Climate Alliance (GGCA). 2013. *Overview of Linkages between Gender and Climate Change.* Accessed 10 November 2023. https://issuu.com/undp/docs/pb1_asiapacific_capacity_final

Gough, Annette. 1994. "Fathoming the fathers in environmental education: A feminist poststructuralist analysis." PhD diss., Deakin University.

Gough, Annette. 1997. *Education and the Environment: Policy, Trends and the Problems of Marginalisation.* Australian Education Review Series No. 39. Melbourne: Australian Council for Educational Research.

Gough, Annette. 1999a. "Recognising women in environmental education pedagogy and research: Toward an ecofeminist poststructuralist perspective." *Environmental Education Research 5* (2): 143–161.

Gough, Annette. 1999b. "The power and the promise of feminist research in environmental education." *Southern African Journal of Environmental Education 19:* 28–39.

Gough, Annette. 2004. "The contribution of ecofeminist perspectives to sustainability in higher education." In *Higher Education and The Challenge of Sustainability: Contestation, Critique, Practice, and Promise,* edited by Peter B. Corcoran, and Arjen E.J. Wals, 149–161. Dordrecht, Netherlands: Kluwer Academic Press.

Gough, Annette. 2013. "Researching differently: Generating a gender agenda for research in environmental education." In *International Handbook of Research on Environmental Education,* edited by Robert B. Stevenson, Michael Brody, Justin Dillon, and Arjen Wals, 375–383. New York: Routledge.

Gough, Annette. 2015. "Resisting becoming a glomus body within Posthuman theorizing: Mondialisation and embodied agency in educational research." In *Posthumanism and Educational Research,* edited by Nathan Snaza, and John Weaver, 254–275. New York: Routledge.

Gough, Annette, Connie Russell, and Hilary Whitehouse. 2017. "Moving gender from margin to center in environmental education." *The Journal of Environmental Education 48* (1): 5–9.

Gough, Annette, and Hilary Whitehouse. 2003. "The 'nature' of environmental education research from a feminist poststructuralist standpoint." *Canadian Journal of Environmental Education 8:* 31–43.

Gough, Annette, and Hilary Whitehouse. 2018. "New vintages and new bottles: The 'nature' of environmental education from new material feminist and ecofeminist viewpoints." *The Journal of Environmental Education 49* (4): 336–349.

Gough, Annette, and Hilary Whitehouse. 2019. "Centering gender on the agenda for environmental education research." *The Journal of Environmental Education 50* (4–6): 332–347.

Gough, Annette, and Hilary Whitehouse. 2020. "Challenging amnesias: Feminist new materialism/ecofeminism/women/climate/education." *Environmental Education Research 26* (9–10): 1420–1434.

Greenall, Annette. 1987. "A political history of environmental education in Australia: Snakes and ladders." In *Environmental Education: Practice and Possibility*, edited by Ian Robottom, 3–21. Geelong: Deakin University.

Hallam, Nickie, and David Pepper. 1991. "Feminism, anarchy and ecology: Some connections." *Contemporary Issues in Geography and Education* 3 (2): 151–167.

Haraway, Donna. 1985. "A Manifesto for Cyborgs." *Socialist Review 80*: 65–107.

Harding, Sandra. 1986. *The Science Question in Feminism*. Ithaca: Cornell University Press.

Harding, Sandra, ed. 1987. *Feminism and Methodology: Social Science Issues*. Bloomington: Indiana University Press and Open University Press.

Harding, Sandra, ed. 1993a. *The "Racial" Economy of Science: Toward a Democratic Future*. Bloomington: Indiana University Press.

Harding, Sandra. 1993b. "Rethinking standpoint epistemology: 'What is strong objectivity?.'" In *Feminist Epistemologies*, edited by Linda Alcoff, and Elizabeth Potter, 49–82. New York/London: Routledge.

Kahle, Jane Butler. 1985. *Women in Science: A Report from the Field*. Philadelphia: Falmer Press.

Kremer, Kristine Benne, Gary W. Mullins, and Robert E. Roth. 1990–1991. "Women in science and environmental education: Need for an agenda." *The Journal of Environmental Education 22* (2): 4–6.

Krockover, Gerald H., and Daniel P. Shepardson. 1995. "Editorial: The missing links in gender equity research." *Journal of Research in Science Teaching 32* (3): 223–224.

Linke, Russell D. ed. 1977. *Education and the Human Environment: UNESCO Seminar Report*. Canberra: Curriculum Development Centre.

Manthorpe, Catherine A. 1982. "Men's science, women's science or science? Some issues related to the study of girls' science education." *Studies in Science Education 9* (1): 65–80.

Merchant, Carolyn. 1980. *The Death of Nature: Women, Ecology and the Scientific Revolution*. New York: Harper and Row.

Merchant, Carolyn. 1992. *Radical Ecology: The Search for a Livable World*. New York/London: Routledge.

Miranda, Wilma, and Rita Yerkes. 1982. "The need for research in outdoor education programmes for women." *Journal of Physical Education, Recreation & Dance 53* (4): 82–85.

Mitten, Denise. 1985. "A philosophical basis for a women's outdoor adventure program." *Journal of Experiential Education 8* (2): 20–24.

Peck, Diane. 1991. "Gender equity and environmental education." Unpublished draft paper.

Peck, Diane. 1992. "Environmental education – time to branch out." *The GEN*, January/February: 1, 4.

Probyn, Elspeth. 2016. *Eating the Ocean*. Durham: Duke University Press.

Salleh, Ariel. 1989. "Environmental consciousness and action: An Australian perspective." *Journal of Environmental Education 20* (2): 26–31.

Salleh, Ariel. 1997. *Ecofeminism as Politics: Nature, Marx and the Postmodern*. London: Zed Books.

Schuster, Marilyn, and Susan Van Dyne. 1984. "Placing Women in the Liberal Arts: Stages of Curriculum Transformation." *Harvard Educational Review* 54 (4): 413–429.

Taylor, Carol A. 2019. "Diffracting the curriculum: Putting "new" material feminist theory to work to reconfigure knowledge-making practices in undergraduate higher education." In *Theory and Method in Higher Education Research, Volume 5*, edited by Jeroen Huisman, and Malcolm Tight, 37–52. Bingley, UK: Emerald.

Tong, Rosemarie. 1989. *Feminist Thought: A Comprehensive Introduction*. Boulder, CO/San Francisco, CA: Westview Press.

UNESCO. 1978. *Intergovernmental Conference on Environmental Education: Tbilisi (USSR), 14–26 October 1977, Final Report*. Paris, UNESCO.

United Nations. 2012. *The Future We Want: Outcomes Document Adopted at Rio + 20*. Rio de Janeiro: United Nations. Accessed 10 November 2023. https://sustainabledevelopment.un.org/content/documents/733FutureWeWant.pdf

United Nations. 2015. *Transforming Our World: The 2030 Agenda for Sustainable Development*. Accessed 10 November 2023. https://sdgs.un.org/publications/transforming-our-world-2030-agenda-sustainable-development-17981

Weedon, Chris. 1987. *Feminist Practice and Poststructuralist Theory*. Oxford: Blackwell.

Section I

Putting women on the agenda

The five chapters included in this section draw on my writings, solo and with Hilary Whitehouse, which have argued for the importance of putting women on environment and environmental education agenda.

The first of these articles draws on some of the research from my doctoral thesis which looked at "environmental education as a man-made subject" through a feminist poststructuralist lens, and reports on research into the gaps and silences present in policies, pedagogy and research in environmental education from a feminist perspective. The documents analysed are UNESCO-UNEP discourses on environmental education including intergovernmental conference reports and the International Environmental Education Programme (IEEP) "green" series because these statements have been used to inform national- and school-level policies and programmes in environmental education in many places, and I saw it as important to critically examine the gendered world view implicit in these statements as a starting point for a discussion on how to destabilize these statements so that women are more recognised in environmental education pedagogy and research.

The second article built on the first article and explored the related issue of the potential of adopting feminist research methods and methodologies in environmental education research. This article was stimulated by my experiences of working with South African colleagues on a capacity building in environmental education research project and finding that gender was not on the research agenda in South Africa at that time (1998–1999): the focus was much more on equality writ large as a result of the relatively recent election of Nelson Mandela's government after years of apartheid. Thus, the article was designed to convince South African environmental educators to adopt different ways of thinking and perceiving in environmental education research by using feminist research strategies.

Chapter 4, which discusses the contribution of ecofeminist perspectives to sustainability in higher education, focuses on research into the absences of women's perspectives from sustainability policies, pedagogy and research, and it argues that ecofeminist pedagogies and research methodologies suggest new possibilities for the development of sustainability in higher education. In this

DOI: 10.4324/9781003390930-2

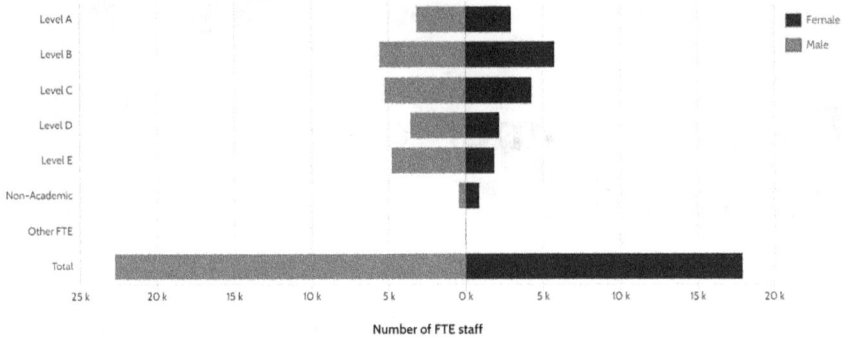

Figure S1.1 Number of FTE staff by gender by employment level, from the *State of Australian University Research 2018–19.* Reproduced with permission from the Australian Research Council.

chapter I draw attention to the imbalance in the distributions of women and men across Australian university positions (Table 4.1). More recent data from the Australian Research Council indicates that the proportion of women in higher-ranking positions is increasing, though still lower than men, except at lower ranks (see Figure 1.1).

As discussed in Chapter 1, the ecofeminist movement suffered from a feminist backlash in the late 1990s because of its perceived essentialism, elitism and ethnocentrism. However, the second decade of this century has seen a resurgence of interest in ecofeminism and feminist and intersectional perspectives in environmental education. Chapter 5 was written before the resurgence, but it provided a new opportunity to argue that environmental education research has rarely addressed areas of different women's experiences and knowledges, which means many useful insights have not been adequately pursued, but that using feminist research strategies to generate a gender agenda provides the basis for different ways of thinking and perceiving in environmental education research. In particular, the chapter explains why a feminist perspective is important in environmental education and what characterises feminist research; it also discusses what feminist research has been undertaken to date in environmental education and the potential for feminist research in environmental education.

The final chapter in this section comes from 50th anniversary special issue of *The Journal of Environmental Education*, in 2019. Hilary Whitehouse and I were invited to reflect on feminism and environmental education research since 1969, so we traced the history of feminist environmental education research across *The Journal of Environmental Education* and other environmental education research journals. We noted that, although there was some research in the 1990s and 2000s, until two special issues of *The Journal of Environmental Education* in 2016 and 2018, there had been a prolonged, even a deafening, silence around gender and eco/feminism in environmental education research. We argue that it is time for gender to be much higher on the agenda of environmental education researchers and of journals if we are to better achieve gender equality and more fully address the climate emergency within the field.

2 Recognising women in environmental education pedagogy and research

Annette Gough

Gough, A. (1999). Recognising women in environmental education pedagogy and research: Toward an ecofeminist poststructuralist perspective, *Environmental Education Research*, 5(2), pp. 143–161. doi: 10.1080/1350462990050202 ©1999 Taylor and Francis Ltd.

Summary

In the past, women have been overlooked in most environmental education programmes through being subsumed into the notion of 'universalised people'. However, women have a distinctive contribution to make to environmental education pedagogy and research which needs to be foregrounded. This chapter reports on research into the gaps and silences present in policies, pedagogy and research in environmental education from a feminist perspective. This research has been inspired by feminist critiques of critical pedagogy and the potentialities of feminist poststructuralist methodologies. In particular I focus on the silencing of marginalised perspectives in environmental education policy development, as well as in research conducted from the perspective of the dominant positivist research methodologies, and argue for the possibilities for new directions when poststructuralist pedagogies and research methodologies are used in environmental education.

Introduction

Women were noticeably in the minority at the international gatherings which formalised conceptions of environmental education. This absence of women can be seen as being related to the epistemological framework of environmental education being very much that of a man-made subject and to the content of the corresponding curriculum and research programmes tending to be determined by the male agenda. However, through recent environmental education statements, such as those emanating from the 1992 United Nations Conference on Environment and Development, through the development of feminist poststructuralist educational research strategies and other feminist critiques of science and society a significant place can be argued for a women's perspective in both pedagogy and research in environmental education.

DOI: 10.4324/9781003390930-3

This chapter is organised in three parts. Firstly, I present the findings of my research into the gender and language bases of UNESCO discourses on environmental education. Secondly, I discuss activities attempted to date to recognise women in environmental education pedagogy and research. Thirdly, I argue that women have a distinctive contribution to make to environmental education pedagogy and research and I present some pedagogical and research principles which, I believe, will both destabilise current approaches and provide a worthwhile and effective strategy for reconstituting environmental education as a more democratic human science.

Documenting the absence of women and other marginalised groups in environmental education discourses

Environmental education has been around now for 25 years, more or less. During this time the field has undergone many changes. It has also been associated with many other forms of education – such as science, social and outdoor – and it incorporates many elements of these. Its global history within UNESCO can be precisely charted because of the activities of the UNESCO-UNEP International Environmental Education Programme (IEEP). This programme had its origins in the recommendations from the 1972 United Nations Conference on the Human Environment held in Stockholm, Sweden. The outcomes of the various activities conducted through the IEEP can be read as attempts to universalise (make one for all) statements about environmental education. By making universal statements the IEEP could be seen as saying that there is only one problem and one solution, thereby masking any differences that may exist. However, this universalisation also can be read as an effect of colonisation of others by the English-speaking world, and of marginalisation of non-English-speaking views. Colonisation has a new meaning in this context. Here it is applied in the realm of ideas, texts, language and discourse rather than just in terms of geography. It is this possible reading of the ideas and language of the IEEP statements on environmental education that is explored in this article.

I have been working in the field for over two decades now, but I only recently started to look at the language of some of the statements and publications of UNESCO in the area of environmental education. This focus arose from my academic research interests in developing a critical perspective on the relationships between gender, science and environmental education. However, it has led me to look at issues of marginalisation of people and perspectives that are not Western, not English-speaking, and not male, in environmental education statements.

IEEP and other UNESCO statements have been used as the foundations of national and school-level policies and programmes in environmental education in many places. Thus the gendered worldview implicit in these statements should be critically examined as a starting point for a discussion on how to destabilise these statements so that women are more recognised in environmental education pedagogy and research.

UNESCO publications

A Western, Amero-Eurocentric, English-speaking and developed worldview has dominated the statements and discourses of environmental education for much, if not all, of the past two decades for a number of reasons. This dominant worldview originated in the scientific revolution of the seventeenth century ('the Age of Enlightenment') and is secular, empirical and mechanistic, and characterised by seeing the human species as apart from nature (and thus nature has no intrinsic value). Material progress is an essential part of this worldview, as is belief in technological innovation and a capitalist market-led economy. 'In essence, the dominant worldview has become so ingrained in our way of thinking, particularly in Western society, that it acts hegemonically to maintain itself as the dominant ideology' (Robottom & Hart, 1993, p. 29). Within environmental education, and education in general, this has led to a view of inquiry as analytic, reductionist and based on scientific neutrality, rationality, divisibility of knowledge and emphasis on quantitative measurement and observable phenomena (see, for example, Hart, 1990).

Most of the trend papers presented at the 1975 UNESCO Belgrade workshop were biased towards the developed world. The papers prepared as working documents for the Belgrade workshop were published by UNESCO (1977). This volume contains 16 papers, of which 15.5 were written by males (the Introduction was co-authored by a female). Of the 20 authors, 8 are from North America, 5 are from Europe, 4 represent UN agencies, 2 come from Africa and 1 is from South America. As one of the participants reported (Fensham, 1976, p. 4),

> Because of the overwhelming preponderance in published form of information about Environmental Education in countries like those of Europe and North America this bias was not surprising but it did tend to give the impression that the models for developing Environmental Education must also be those for the developed world.

The situation was little different at the 1977 UNESCO-UNEP Intergovernmental Conference on Environmental Education, held in Tbilisi, USSR. Here, according to the Final Report of the Conference (UNESCO, 1978, pp. 83–99), there were only 55 females out of a total of 345 participants, a ratio of over 6:1 (see Table 2.1).

It is not only the paucity of representation of women in the groups that formulated the foundational statements on environmental education that is significant. It is also likely that the women who were present were from scientific backgrounds and, thus, for example, were unlikely to see anything wrong with the generic 'man' in the statements.

Table 2.1 Gender ratios of participants at the UNESCO-UNEP Intergovernmental Conference on Environmental Education held at Tbilisi, USSR, October 1977

Category of Participant	Number of participants who were:		
	Male	*Female*	*Unknown (unable to be determined from information provided)*
Member States (delegate,	233	43	3
adviser, expert)	4	0	0
Non-Member States	15	1	0
UN Organisations	4	0	0
Intergovernmental	31	11	0
Organisations			
International Non-Government			
Organisations			
TOTAL	287	55	3

* Assembled from the Final Report of the Tbilisi Conference (UNESCO, 1978, pp. 83–99). The table omits the Conference Secretariat who officially took no part in the endorsing of the recommendations. The Secretariat included the Director-General of UNESCO and other UNESCO staff (26 males, 10 females (of whom only 1 was designated a section chief, 2 were in press and conference services and 7 were described as secretaries)), plus the interpretation, translation and typing services (27 males, 30 females).

UNESCO has continued to perpetuate this view by continuing to publish a Western perspective in the volumes of the Environmental Education series of the UNESCO-UNEP International Environmental Education Programme, even though, to adapt Fensham's words, these perspectives are largely irrelevant to most of UNESCO's countries. At the present time there are 30 volumes in the series, of which 29 are available. These have been published between 1983 and 1992 and prepared by authors whose national affiliations are shown in Table 2.2. Interestingly, in terms of the possible politics involved, Volume 19, *Analysis of Results of Environmental Education Pilot Projects*, seems never to have been released.

The gender of the authors and editors of these volumes is also very telling (see Table 2.3). The lack of participation by women is apparent.

The language in which the text is written is important. Western domination can be found in the languages used in the UNESCO-UNEP series, as Table 2.4 illustrates. The dominance of the use of English is perhaps explicable in terms of it being the universal language for United Nations documents; however, UNESCO is supposed to be a multilingual organisation. The use of English nevertheless conveys particular messages and limits access to the volumes to those who are very proficient in English. Gayatri Spivak (1987) argues that colonised races and peoples have been forced to articulate their experiences in the language of their oppressors. Ngugi Wa Thiong'O (1986, p. 4) further develops this point: 'The choice of language and the use to which it is put is central to a people's definition of themselves in relation to their natural and social environment, indeed in relation to the entire universe'.

Table 2.2 Authorship of UNESCO-UNEP Environmental Education series volumes

Country/Continent of author	No. of volumes authored	Comment
Africa	1	edited in USA
Australia	1	edited in USA
Europe	9	none edited in USA
India	3	all edited in USA
Jamaica	1	edited in USA
Philippines	1	edited in USA
United States of America	10	either not edited or edited in USA
UNESCO	3	
Unknown (because of unavailability)	1	all written in Europe, not edited in USA

Table 2.3 Gender of authors and editors of UNESCO-UNEP Environmental Education series volumes

	Male author	Male editor	Female author	Female editor	Unstated (UNESCO or other institution as author)	Unknown*
No. of volumes (total of 30 published)	44	8	13	3	5	4

* Either because gender was unable to be determined from name, e.g. 'Pansy', or because volumes were unavailable.

Table 2.4 Languages of UNESCO-UNEP Environmental Education series volumes

Language used in volume	No. of volumes in that language*
Arabic	12
English	28†
French	14 or 15·
Spanish	3 or 13#

* Of the 30 volumes, all except 2 have been published in English with some also being published in other language(s).

† Of these volumes, Volume 19 is claimed to be available in English, but it has never been sighted and it is listed as 'unavailable' in the UNESCO-UNEP newsletter, *Connect* (XIV(3): 6–8), and in recent correspondence.

· Volume 29 states that 15 volumes are in French, whereas Volume 30 states that 14 volumes are in French, as does Volume 21, which was published in 1992 (Volume 25 is the uncertain one).

UNESCO itself seems uncertain about the number of volumes published in Spanish. The listing of documents in the IEEP series given at the beginning of Number 29 only notes 3 Spanish volumes (all published in 1983), *Connect* (XIV(3): 6–8), lists 13 volumes published in Spanish. Both listings were published in 1989. The volumes published in 1990 and 1992 list 13 volumes as being available in Spanish.

Much work has already been undertaken on 'man made language'. For example, Spender (1990, p. 3) argues that women, 'having learnt the language of a patriarchal society we have also learnt to classify and manage the world in accordance with patriarchal order and to preclude many possibilities for alternative ways of making sense of the world'. That these documents have been written by males with a particular worldview means that alternative worldviews, such as those of women, are precluded.

Of particular concern in this article is the universalised nature of the statements made in these volumes given the diversity of cultures, environments, languages, religions, stages of 'development' and politics within the world, as well as differing stages of colonisation and post-colonisation. How can universal statements, generally made by Western English-speaking males, meet the needs and interests of such a diversity? And what can be done to recognise women's voices?

Authorship of IEEP volumes

UNESCO, of course, makes the usual disclaimer about the opinions being those of the authors and not necessarily coinciding with any official views of UNESCO, but it seems more likely that the views do at least coincide with those who direct the UNESCO-UNEP International Environmental Education Programme given that the authors were commissioned by them to prepare the volumes. Also, the contents of the various volumes have then been promoted through lead articles in the IEEP newsletter, *Connect*. That there is a particular acceptable worldview from the UNESCO and IEEP is also reflected in the authors (and their countries) they selected compared with the ones that were overlooked. That such politics are played, and voices are silenced, is supported by the above quote from Fensham (1976, p. 4) which he elaborated in a more recent interview (as cited in Greenall Gough, 1993, p. 15):

> because, even with the best will in the world, the liberals will find themselves confronted by groups who are much more radical than they are, and yet the liberals will always be in charge, because only liberals get put in charge of things.

The domination of the authorship of the IEEP volumes by Europeans and Americans is noticeable, especially when it is noted that all the volumes not written by Americans, Europeans or UNESCO have been subsequently edited by Americans. It could be contested that the editing of the volumes from outside Europe and the USA influenced the content of the texts. However, that such editing was carried out could nevertheless have resulted in the colonisation of the language of the texts by an American perspective. By colonisation of the language I mean changing the original author's language (and values) to suit the American editor's intentions and meanings rather than allowing the author's voice to be heard. This certainly seems to be an intent in some of the volumes. For example, Hungerford et al. (1989, p. i), who are Americans, universalise their work in terms of being:

> [A]n ideal around which a team of educational planners can make intelligent decisions about what their own curriculum should look like. Even though the curriculum outlined here may exceed the constraints placed upon a given school or nation, all of the major components should probably be represented in one way or another.

Another example comes from Marcinkowski et al. (1990, p. 1) where it is claimed:

> When implemented as intended, these guidelines will, in fact, result in teachers who are sufficiently competent and skilled to offer instruction in environmental education that will clearly contribute to the development of environmentally literate students.

Such statements raise questions about what makes these particular prototypes for an environmental education curriculum appropriate for places other than where they have been developed and whether the major components and guidelines they have identified are also appropriate. In making their universal statements the prototypes do not take into account that there are different perspectives, goals and strategies for change in other countries and that these are grounded in the different social and political contexts of each of these countries (Sangari, 1987). From a non-American perspective these types of materials are very much in the vein of what Wole Soyinka (in Slemon & Tiffin, 1989, p. ix) calls 'a second epoch of colonisation', or how 'Western theoretical practice applies itself, even with the best intentions, to the cultural productions of the non-Western world':

> We . . . have been blandly invited to submit ourselves to a second epoch of colonisation – this time by a universal-humanoid abstraction defined and conducted by individuals whose theories and prescriptions are derived from the apprehension of their worlds and their history, their social neuroses and their value systems.

Many authors are discussing 'the material, often devastating, consequences of a centuries-long imposition of Euro-American conceptual patterns onto a world that is at once "out there" and yet thoroughly assimilable to the psychic grasp of Western cognition' (Slemon & Tiffin, 1989, p. ix). Of particular relevance to environmental educators is the work by Vandana Shiva (1989) on women, ecology and development. She documents how the effects of the imposition of Western scientific knowledge and attitudes to economic development in India have destroyed life and threatened survival during the first epoch of colonisation.

Just as important for environmental education is the work currently being undertaken on the 'second epoch of colonization' which focuses on the realm of ideas, texts, language and discourse. This work builds on the recognition that knowledge is socially constructed and that language is a key participant in producing the reality people wish to present (Hawthorn in Spivak, 1990, p. 17):

> All that we can know is what we say about the world – our talk, our sentences, our discourse, our texts. There's nothing outside these texts, no extra texts. There's nothing prior to these texts, no pretexts, there are just more texts. Indeed this claim itself is just another text.

This statement encapsulates one of my major concerns with the texts of the IEEP. By promulgating a Western, Eurocentric, English-speaking worldview through the authorship of these texts IEEP are supporting the imposition of the associated conceptual patterns, and corralling of social meanings, rather than allowing other voices to be heard. Once in print these views become legitimated and difficult to contest, particularly if they are published in a language other than that which is natural for the reader. As an environmental educator I am concerned that practitioners and policy makers then focus on the pedagogy of implementing these texts rather than looking at the language and the gendered worldview that imbue the IEEP texts. If we do this then our actions are limited by the existing discourses and their assumptions, such as nature having no intrinsic value but only a utilitarian value.

Interestingly, it is not only in the realm of environmental education that UNESCO seems to be silencing 'Other' non-Western, non-male voices. Zia Sardar (1993) notes a similar phenomenon in the field of future studies where UNESCO has, with one exception, only sampled Western literature in the compilation of an authorative bibliography in its supposedly global reference tool, *UNESCO Future Scan*.

Putting women on the environmental education agenda

Since the earliest days of proclaiming an ecological crisis, environmentalists have been calling for environmental education as a means of resolving environmental problems as they variously see them. However, for the main part, feminists, and particularly ecofeminists, have not addressed environmental education as a strategy for achieving their goals. Thus, the literature on feminist research in environmental education is both recent and sparse, and, until recently, has been almost totally Australian. A similar finding was made by Giovanna Di Chiro (1993, p. 228) when she conducted an ERIC search using the descriptors 'feminism' and 'environmental education' which yielded only two articles, one her own (1987a) which was written and published in Australia, and the other by Ariel Salleh (1989), an Australian ecofeminist and social theorist. My own search adds two North American articles (Kremer et al., 1990–1991; Fawcett et al., 1991), one British article

(Hallam & Pepper, 1991), and two recent articles in *Environmental Education Research* (Hampel et al.,1996; Pawlowski, 1996; Storey et al., 1998) have drawn attention to gender differences in environmental knowledge, concerns and behaviours. The remaining literature which specifically relates gender to environmental education is Australian (Brown & Switzer, 1991a, 1991b, 1991c; Peck, 1991, 1992; Brown & Broom, 1992, Department of Prime Minister and Cabinet/Office of the Status of Women, 1992; NWCC, 1992; Gough, 1994, 1997b; Greenall Gough, 1993; Barron, 1995; White-house & Taylor, 1996).

There are also statements from the United Nations Conference on Environment and Development (UNCED) where the activities for promoting education, public awareness and training include fostering opportunities for women and eliminating gender stereotyping in curricula (*Agenda 21*, UNCED, 1992: see, for example, Paragraphs 24.2(e) and 36.5 (m)).

However, lack of reciprocity is an issue between the 'Global Action for Women Towards Sustainable and Equitable Development' and the 'Promoting Education, Public Awareness and Training' chapters in *Agenda 21* (UNCED, 1992, Chapters 24 and 36 respectively). The 'Women' chapter has as its overall goal, achieving active involvement of women in economic and political decision making, with emphasis on women's participation in national and international ecosystem management and control of environmental degradation. One of its objectives for national governments (UNCED, 1992, Paragraph 24.2(e)) is

> To assess, review, revise and implement, where appropriate, curricula and other educational material, with a view to promoting the dissemination to both men and women of gender-relevant knowledge and valuation of women's roles through formal and non-formal education.

The activities for governments related to such objectives are broadly concerned with achieving equality of opportunity for women (such as through eliminating illiteracy): increasing proportions of women as decision makers in implementing policies and programmes for sustainable development; and recognising women as equal members of households with respect to both workloads and finance. Consumer awareness is particularly mentioned, as are 'programmes to eliminate persistent negative images, stereotypes, attitudes and prejudices against women through changes in socialization patterns, the media, advertising, and formal and non-formal education' (UNCED, 1992, Paragraph 24.3(i)). Here, women's knowledge is being recognised and valued as something different rather than assuming that women will achieve equality simply through equal opportunity, although there are some elements of a liberal view present.

Unfortunately these views are not matched in the 'Education' chapter. Here women are generally included with all sectors of society, although specific mention is made of the high illiteracy levels among women which need to be addressed (UNCED, 1992, Paragraph 36.4(a)) in the objectives. In the activities, women are mentioned in the following terms (UNCED, 1992, Paragraph 36.5(m)): 'Governments and educational authorities should foster opportunities for women in non-traditional fields and eliminate gender stereotyping in curricula'. No mention is made of recognising and valuing women's knowledge, and the perspective seems once more to be that of liberal feminism, although the experience and understanding of sustainable development of indigenous peoples are affirmed as playing a part in education and training (UNCED, 1992, Paragraph 36.5(n)).

The emphasis in many of the Australian (government-sponsored) documents on women, the environment and education can be generally characterised in terms of taking account of women's interests. For example, in their discussion paper on women and ecologically sustainable development, Brown and Switzer (1991a, p. 11) see education as 'the primary avenue through which society transmits social values and behaviour patterns on environmental management to both the current and the future generation'. They also note that 'women are much less likely than men to have access to the scientific training which would assist in explicating environmental issues'. They thus include 'environmental education which includes women's interests' as one of five inter-related policy principles and strategies 'needed to respond to the risks and responsibilities of Australian women with respect to the environment' (1991a, p. 14). They suggest that the goal of including women's interests in environmental education could be achieved through:

- ensuring that there is overall emphasis in research and education on the impact of 'female' industries on the environment;
- including human and social development, the needs of human communities, and their interactions with the natural environment in curricula; and
- providing practical training in conflict management and the negotiated settlement of environmental disputes. (1991a, p. 16)

Brown and Switzer (1991a, p. 16) note that women are less likely to have scientific or economic training than men and, consequently, have less influence in developing curricula which give high priority to issues of importance to women, such as reduction of toxic wastes and information on safety standards. They also note that there is a need to compensate for the effects on research and teaching of the relative absence of women and women's interests from the professions of environmental science and economics: 'this absence has meant that many questions on ecologically sustainable development from the fields of health, welfare, household management and social policy have neither been investigated nor included in environmental education' (Brown & Switzer, 1991a, p. 16). Their orientation can be read as being more consistent with liberal feminism (where women just need to be given equal opportunities with men) rather than the more radical forms of feminism, or even ecofeminism. The term *ecofeminisme* was coined by the French writer Francoise d'Eaubonne in her book *Le feminisme ou la mort* (d'Eaubonne, 1980, pp. 213–252, with an English translation of one section included in Marks and de Courtivron, 1980, pp. 64–67) 'to represent women's potential for bringing about an eco-logical revolution to ensure human survival on the planet. Such an ecological revolution would entail new gender relations between women and men and between humans and nature' (Merchant, 1990, p. 100). Nowadays the term is used 'to describe both the diverse range of women's efforts to save the Earth and the transformations of feminism in the West that have resulted from the new view of women and nature' (Diamond & Orenstein, 1990, p. ix).

While many forms of feminism and ecofeminism have much to contribute to discussions about the environment, I particularly argue for the relevance of feminist poststructuralism (or poststructuralist feminism) as an orientation for future research and pedagogy in environmental education. According to Chris Weedon (1987, pp. 40–41), feminist poststructuralism is 'a theory which decentres the rational, self-present subject of humanism, seeing subjectivity and consciousness, as socially produced in language, as a site of struggle and potential change' (Weedon, 1987, p. 41). Feminist poststructuralism has rein-forced the notion that there is no such subject as the universalised woman, and that there is no 'one true story' (Harding, 1986) for environmental education. This chapter thus concludes with a discussion of possible new directions in environmental education content and research which take into account sub-jectivity that acknowledges women as gendered, classed, raced and aged indi-viduals, among other characteristics. These include some research principles which could guide such research and curriculum development in environmen-tal education in the future. Such a perspective is sorely needed if women are to be recognised in environmental education pedagogy and research.

Towards a feminist poststructuralist perspective

If it is accepted that there are links between language and power in the dis-courses of environmental education, then it would seem important to exam-ine the multiplicity of meanings in these discourses. Such an examination would provide less partial and distorted descriptions and explanations using language which stresses context and interaction and democratic models of order. A methodology for doing this arises out of feminist poststructuralist analysis which is 'a mode of knowledge production which uses poststructural-ist theories of language, subjectivity, social processes and institutions to under-stand existing power relations and to identify areas and strategies for change' (Weedon, 1987, pp. 40–41).

Cleo Cherryholmes (1988, p. 177) provides further explication: 'Poststruc-tural analysis points beyond structure, utility, and instrumentality. Our ability to shape and design the social world can be enhanced, I hope, if we outline, examine, analyze, interpret, criticize, and evaluate the texts and discourse-practices that surround us'. While my original interest grew out of feminist poststructuralist analysis, and continues to be informed by its discourses, the following discussion is informed by the broader constructions (and decon-structions) of poststructuralist analysis per se.

Working from the frames suggested by Weedon (1987) for feminist post-structuralist analysis, and by Cherryholmes (1988) and Davies (1994) for poststructuralist analysis, to date I have drafted four guiding principles which I am exploring in my own work. These principles, which are still being developed and which are all grounded in an opposition 'to the longing for "one true story" that has been the psychic motor for Western science' (Harding, 1986, p. 193), include:

- to recognise that knowledge is partial, multiple and contradictory;
- to draw attention to the racism and gender blindness in environmental education;
- to develop a willingness to listen to silenced voices and to provide opportunities for them to be heard; and
- to develop understandings of the stories of which we are a part and our abilities to deconstruct them.

Listening to the voice of the marginalised in environmental education

Only a few authors have questioned the colonial domination implicit in the environmental education discourses which have been published. As mentioned previously, Fensham (1976) is the one who has drawn attention to the Other voices. Leopoldo Chiappo (1978), a Peruvian, and Daniel Vidart (1978), a Colombian, are other critics. However, other voices that should have been more vocal, such as the report on *Environmental Education in Asia and the Pacific* (UNESCO Regional Office for Education in Asia and the Pacific, 1981), have been silent, choosing instead to adopt the master narrative or dominant discourse. Chiappo (1978, p. 456) questions whether the inhabit-ants of the needy South can 'accept as valid the way of seeing and interpreting ecological facts adopted by the countries of the super-industrialized, wealthy North'. He (1978, p. 456) asserts that it is 'necessary to reveal the ideology that underlies the attitude of dominance', and then asks, 'What are the central issues of environmental education?'. His answers differ greatly from the gener-ally accepted ones. For him, the fundamental issues of environmental educa-tion are the awakening of critical awareness of the social and political factors of the environmental problem and the development of a new ethic of liberation: 'Failure to tackle these two issues may reduce environmental education to a purely pedagogical and informative exercise' (1978, p. 458). He also draws attention to the silences in the Tbilisi Declaration: 'it is true only for what it says, not for what it does not say' (1978, p. 463). For example, by omit-ting the word 'economic' from its reference to 'the new international order' (in Recommendation 3, UNESCO 1978, p. 27), Chiappo argues that 'it has left out the essential, since the issue is essentially an economic one' (1978,

p. 464). Interestingly, the much debated (Fensham, 1976) 'A. Environmental Situation' section of The Belgrade Charter did refer to 'The recent United Nations Declaration for a New International Economic Order'. The change is therefore remarkable and opens up questions of the politics being played at the Tbilisi conference.

Chiappo (1978, p. 457) asserts that the perceptions of environmental problems which underlie the dominant conceptions of environmental education are not as the North (or what elsewhere in this article has been called West) portrays them: 'The present crisis is not due to lack of resources, but to the unjust exploitation and distribution of resources. It is the result of wastage and profit-seeking. The industrial mentality is what must be challenged'. Both Merchant (1980, 1992) and Shiva (1989) argue similarly, linking modern Western-style scientific knowledge and economic development with the death of nature. For example, Shiva (1989, p. xiv) argues:

> The Age of Enlightenment, and the theory of progress to which it gave rise, was centred on the sacredness of two categories: modern scientific knowledge and economic development. Somewhere along the way, the unbridled pursuit of progress, guided by science and development, began to destroy life without any assessment of how fast and how much of the diversity of life on this planet is disappearing. The act of living and of celebrating and conserving life in all its diversity – in people and in nature – seems to have been sacrificed to progress, and the sanctity of life been substituted by the sanctity of science and development.

From my own experiences and readings, the industrial development mentality still dominates much environmental education, and many of the environmental problems are framed in terms of a reductionist 'us' against 'them' discourse, where, for example, the cause of environmental deterioration and the exhaustion of the planet is attributed to population growth in the Third World.

However, there is also an increasing questioning of both the effects of science and technology and of the worldview implicit in the knowledge that frames both science and economic development. This critique is forthcoming from a number of sources, but particularly from Asian as well as Western scientists, feminists and postmodernists. Haraway (1989), for example, documents how national interests are reflected in the research orientations of primatologists: whereas American and European primatologists are obsessed with studying sex and war among primate groups, Japanese primatologists focus on the construction of a specifically Japanese scientific cultural identity where 'the Japanese monkeys became part of a complex cultural story of a domestic science' (1989, p. 244). According to Haraway (1989, p. 263), Indian primatology has a different orientation again, representing 'a post-colonial nation with a sophisticated national primatology and the political and technical ability to restrain western biomedical and military hegemony over its own inhabitants, human and animal'.

Given this growing recognition that there is no one way of looking at the world, no 'one true story', but a multiplicity of stories then we should be looking at a multiplicity of strategies for policies, pedagogies and research in environmental education. These strategies should be ones that are not universal and part of the dominant discourse but ones which are from the lives of the colonised and marginalised.

Linda Hutcheon (1990, p. 176) asks, 'How do we construct a discourse, which displaces the effects of the colonizing gaze while we are still under its influence?'. The task is to dismantle colonialism's system, expose how it has silenced and oppressed its subjects and find ways for their voices to be heard. As Edward Said (1985, p. 91) argues, the problems are concerned with

> how the production of knowledge best serves communal, as opposed to factional, ends, how knowledge that is nondominative and noncoercive can be produced in a setting that is deeply inscribed with the politics, the considerations, the positions, and the strategies of power.

Some answers to these problems are being suggested. For example, Homi Bhabha (1985, 1986, 1994) has asserted that the colonised is constructed within a disabling master discourse of colonialism which specifies a degenerate native population to justify its conquest and subsequent rule. Gayatri Spivak (1987, 1990) argues that colonised races and peoples have intimate experience of the politics of oppression and repression and have been forced to articulate their experiences in the language of their oppressors. In environmental education we need to be working to disrupt the oppression of the native voices and listening to people expressing themselves in their own languages. This we will not find in the IEEP texts, unless we are Western, Eurocentric and English-speaking.

Postcolonialism and feminism have developed as parallel discourses which have much in common, and some writers are starting to draw the two together (e.g. Spivak, 1987, 1990). Feminist and postcolonial discourses both seek to reinstate the marginalised in the face of the dominant (the former coloniser), and both are oriented to the future, 'positing societies in which social and political hegemonic shifts have occurred' (Ashcroft et al., 1989, p. 177). Postcolonialism provides a possible approach for it 'challenges how imperial centers of power construct themselves through the discourse of master narratives and totalizing systems; they contest monolithic authority wielded through representations of "brute institutional relations" and the claims of universality' (Giroux, 1992, p. 20).

Feminist critique provides another, yet similar, approach from the perspective that 'Women in many societies have been relegated to the position of "Other", marginalized and, in a metaphorical sense, "colonized", forced to pursue guerrilla warfare against imperial domination from positions deeply imbedded in, yet fundamentally alienated from, that *imperium*' (Ashcroft et al., 1989, p. 174). The experiences of India's Chipko movement (see Merchant, 1992; Shiva, 1989), and the Kenyan Greenbelt Movement (see Merchant, 1992), are examples of women pursuing ecological guerrilla warfare as attempts to maintain or achieve sustainability. Merchant (1992, p. 200, 206) argues that 'many of the problems facing Third World women today are the historical result of colonial relations between the First and Third Worlds', but that Third World women 'are making the impacts of colonialism and industrial capitalism on the environment and on their own lives visible'.

In this there is both a challenge and a dilemma (or two or three) for environmental educators and researchers. The politics of difference goes beyond being 'simply oppositional in contesting the mainstream (or *male*stream) for inclusion', and beyond being 'transgressive in the avant-gardist sense of shocking conventional bourgeois audiences' (West, 1990, p. 19). It aligns itself 'with demoralized, depoliticized and disorganized people in order to empower and enable social action and, if possible, to enlist collective insurgency for the expansion of freedom, democracy and individuality' (West, 1990, p. 19). We need to be aware that there are ways other than those to which we are accustomed of looking at the environment and its problems. And we need to be aware that those ways and the voices that accompany them have to date been silenced in the IEEP texts.

Some, too, are starting to relate the discourses of colonisation to those of environmentalism, particularly ecofeminism. Since the seventeenth century, European colonisation in the Americas, Africa, Asia, Australia and the Pacific has resulted in a colonial ecological revolution that has disrupted native ecologies and native peoples' modes of subsistence. For example, Carolyn Merchant (1992, p. 201) notes, 'Third World women have borne the brunt of environmental crises resulting from colonial marginalisation and ecologically sustainable development'. The Indian physicist and environmentalist Vandana Shiva (1989) links the violation of nature with the violation and marginalisation of women in the Third World. There have also been some attempts to relate these discourses of ecofeminism to environmental education – see for example the work of Giovanna Di Chiro (1987a, 1987b) and Valerie Brown and Meg Switzer (1991a, 1991b, 1991c) – but such efforts have so far been rare.

The role of women in achieving sustainable development was recognised in the Rio Declaration (1992): 'Women have a vital role in environmental management and development. Their full participation is therefore essential to achieve sustainable development'. However, women's access to education can be limited. For example, the literacy rate for women in Pakistan as a whole is estimated at only 23 per cent, and in villages as nil (del Nevo, 1993), and gender as well as caste affects women's access to education. Although silenced as well as colonised and marginalised through this process, women do have a vital role in environmental education: they are often responsible for the health and nutrition of their families as well as their other duties, and their involvement is essential if we are to overcome the barriers to successful implementation of environmental education. As a way of overcoming the barriers perhaps, as Sandra Harding (1991, p. 268, emphasis as in original) suggests, we should be moving from

> *including* others' lives and thoughts in research and scholarly projects to *starting from* their lives to ask research questions, develop theoretical concepts, design research, collect data, and interpret findings . . . that would provide less partial and distorted accounts of nature and social relations.

Ashis Nandy (1986, p. xv) argues similarly that we must choose the slave's standpoint, not only because the slave is oppressed but also because the slave represents a higher-order cognition which perforce includes the master as human, whereas the master's cognition has to exclude the slave except as a 'thing'. Vandana Shiva (1989, p. 53) also argues that liberation should begin from the colonised and end with the coloniser. In environmental education we are concerned with both the liberation of nature and of people, thus we should consider starting from others' lives, both human and non-human (see, for example, Devall & Sessions, 1985). Although, it is important to note that some ecofeminists (see, for example, Warren, 1990) dispute the perspectives of deep ecologists such as Devall and Sessions, so this is not a simple solution, but rather something to be discussed in a future article.

The dominant accounts of the environment and its problems, and the modern Western development views of science that dominate the IEEP publications do not encourage starting from others' lives. Rather, their authors know the one true story and they are concerned with implementing their message. Yet, as I have argued here, their message is flawed. It is grounded in the same view of science that has precipitated the environmental crisis – so shouldn't environmental educators look elsewhere, start from others' lives, to look for solutions?

Conclusion

The environmental education publications of UNESCO and the IEEP are an attempt by a privileged few to teach other nations how to live. We, as environmental educators, and particularly as researchers, should instead be listening to the silenced voices speaking in their own languages, and encouraging our students to do the same. The colonialism and marginalisation implicit in the dominant discourses, such as the IEEP documents, are barriers to the successful implementation of environmental education through teacher education. The documents are based on worldviews, and written in languages, that are quite alien to non-Western, non-English-speaking and non-male people. And this 'Other' category is very varied, with the differences within being as great as or greater than the differences from the West (Inayatullah, 1993). Elizabeth Minnich's (1989, p. 286) comments about the absurdity of expecting a Black woman to learn the same things as a white man studying alongside her are appropriate in this context:

> We have seen Black people admitted to institutions that continue to offer the same curriculum they offered when Black people were excluded . . . an overwhelmingly if not exclusively white curriculum . . . (But) . . . [t] he full absurdity of assuming that a Black woman, studying a curriculum that is by and about white men is having the same experience, is learning the same things as a white man studying alongside of her is still not fully evident to some educators.

The language of the publications is also important. A.S. Byatt, in her award-winning novel *Possession*, summarises this position well when she writes 'it's the language that matters, isn't it, it's what went on in her mind' (1991, p. 55). Although, in the case of environmental education statements, it is very much what went on in his mind. As has been discussed in this chapter, the domination of the discourses of environmental education by the English language and by Amero-Eurocentric authors, and the potential this has for privileging only certain male, English-speaking voices in those discourses, raises many concerns about environmental education per se. If we are to critically confront the environmental crisis then we must listen to more voices than these. What is clear is that we must contest statements such as 'environmental education does have a substantive structure' (Hungerford et al., 1983, p. 2) when it is clear that such statements are made by white, Western, Amero-Eurocentric, English-speaking males. These males have also been associated with the IEEP texts (see, for example, Wilke et al., 1987; Hungerford et al., 1988, 1989; Marcinkowski et al., 1990) and invoked their structure for environmental education in these texts (Trudi Volk is a rare female voice in this group, but she speaks English, the language of the coloniser, in a white Western developed world voice, and it could be argued that she has been coopted by the patriarchal discourses).

It seems to be time to move beyond the silencing effect of colonialism to a stage where sufficient space can be created so that 'the colonised can be written back into history' (Parry, 1987, p. 39), and written into the discourses of environmental education. Like the African SONED (M'Mwereria, 1993), we need to be rejecting co-option, exclusion and marginalisation of non-Western, non-English-speaking and non-male peoples and affirming alternatives. Rather than studying other peoples as objects (as is often done in school projects), or including their lives and thoughts in projects (as is starting to happen in, for example, Huckle, 1988 and Williamson-Fien, 1993), we need to start from others' lives to develop less partial and less distorted accounts of nature and social relations.

Environmental education is not the only area affected in this way; we all should be looking at texts for who is saying what and for what purpose, looking for the gaps and silences in those texts, and asking questions about whether the discourses which are of interest to us start from others' lives or perpetuate the dominant discourses. A feminist poststructuralist perspective provides some promise for change.

Note on contributor

Annette Gough is a Senior Lecturer in environmental and science education in the Faculty of Education, Deakin University, Australia. This paper, together with one on the founders of environmental education which also published in Environmental Education Research (Gough, 1997a), is based on aspects of her doctoral research in which she analysed the founders and foundations of environmental education from a feminist poststructuralist perspective. Her other research interests include feminist epistemologies, postcolonialism, critical inquiry, curriculum policy and curriculum reform in both science education and environmental education. Correspondence: Faculty of Education, Deakin University (Burwood Campus), 221 Burwood Highway, Burwood, Victoria, Australia 3125.

References

ASHCROFT, B., GRIFFITHS, G. & TIFFIN, H. (1989) *The Empire Writes Back: Theory and Practice in Post-Colonial Literatures* (New York/London, Routledge).

BARRON, D. (1995) Gendering environmental education reform: Identifying the constitutive power of environmental discourses, *Australian Journal of Environmental Education, 11*, pp. 107–120.

BHABHA, H.K. (1985) Signs taken for wonders: Questions of ambivalence and authority under a tree outside Delhi, May 1817, *Critical Inquiry, 12*(1), pp. 148–172.

BHABHA, H.K. (1986) The other question: Difference, discrimination and the discourse of colonialism, in BARKER, F. et al. (Eds) *Literature, Politics and Theory* (New York, Methuen).

BHABHA, H.K. (1994) *The Location of Culture* (London/New York, Routledge).

BROWN, V.A. & BROOM, D.H. (1992) *Engendering Sustainable Development Occasional Paper No. 1* (Canberra, National Women's Consultative Council).

BROWN, V.A. & SWITZER, M.A. (1991a) *Engendering the Debate: Women and Ecologically Sustainable Development* (Canberra, Office of the Status of Women, Department of the Prime Minister and Cabinet).

BROWN, V.A. & SWITZER, M.A. (1991b) *Where have all the Women Gone? The Role of Gender in Sustainable Development* (Canberra, Australia, Centre for Resource and Environmental Studies, Australian National University), unpublished paper.

BROWN, V.A. & SWITZER, M.A. (1991c) *Next: Engendering Sustainable Development Strategies. Response to Draft Reports of ESD Working Groups* (Canberra, Office of the Status of Women, Department of the Prime Minister and Cabinet).

BYATT, A.S. (1991) *Possession* (London, Vintage).

CHERRYHOLMES, C.H. (1988) *Power and Criticism: Poststructural Investigations in Education* (New York/London, Teachers College Press).

CHIAPPO, L. (1978) Environmental education and the third world, *Prospects, VIII*(4), pp. 456–465.

DAVIES, B. (1994) *Poststructuralist Theory and Classroom Practice* (Geelong, Deakin University Press).

D'EAUBONNE, F. (1980) Feminism or death, in MARKS, E. & DE COURTIV-RON, I. (Eds) *New French Feminisms: An Anthology* (Amherst, University of Massachusetts Press).

DEL NEVO, M. (1993) Learning to heal, *New Internationalist, 243*, p. 3.

DEPARTMENT OF PRIME MINISTER AND CABINET. OFFICE OF THE STATUS OF WOMEN (DPMC/OSW). (1992) *Women and the Environment* (Canberra, Australian Government Publishing Service).

DEVALL, B. & SESSIONS, G. (1985) *Deep Ecology: Living as if Nature Mattered* (Layton, Gibbs M. Smith).

DI CHIRO, G. (1987a) Applying a feminist critique to environmental education, *Australian Journal of Environmental Education, 3*(1), pp. 10–17.

DI CHIRO, G. (1987b) Environmental education and the question of gender: A feminist critique, in ROBOTTOM, I. (Ed) *Environmental Education: Practice and Possibility* (Geelong, Deakin University Press).

DI CHIRO, G. (1993) Environmental education in the war years: A case study, in MRAZEK, R. (Ed) *Alternative Paradigms in Environmental Education Research* (Troy, North American Association for Environmental Education).

DIAMOND, I. & ORENSTEIN, G.F. (Eds) (1990) *Reweaving the World: The Emergence of Ecofeminism* (San Francisco, Sierra Club Books).

FAWCETT, L., MARINO, D. & RAGLON, R. (1991) Playfully critical: Reframing environmental education, in BALDWIN, J.H. (Ed) *Confronting Environmental Challenges in a Changing World. Selected Papers from the Twentieth Annual Conference of the North American Association for Environmental Education (NAAEE)* (Troy, NAAEE).

FENSHAM, P.J. (1976) *A Report on the Belgrade Workshop on Environmental Education* (Canberra, Curriculum Development Centre).

GIROUX, H.A. (1992) *Border Crossings: Cultural Workers and the Politics of Education* (New York/London, Routledge).

GOUGH, A. (1994) Fathoming the fathers in environmental education: A feminist poststructuralist analysis. Unpublished doctoral dissertation. Deakin University, Geelong, Australia.

GOUGH, A. (1997a) Founders of environmental education: Narratives of the Australian environmental education movement, *Environmental Education Research, 3*(1), pp. 43–57.

GOUGH, A. (1997b) *Education and the Environment: Policy, Trends and the Problems of Marginalisation* (Melbourne, Australian Council for Educational Research).

GREENALL GOUGH, A. (1993) *Founders in Environmental Education* (Geelong, Deakin University Press).

HALLAM, N. & PEPPER, D. (1991) Feminism, anarchy and ecology: Some connections, *Contemporary Issues in Geography and Education, 3*(2), pp. 151–167.

HAMPEL, B., HOLDSWORTH, R. & BOLDERO, J. (1996) The impact of parental work experience and education on environmental knowledge, concern and behaviour among adolescents, *Environmental Education Research, 2*(3), pp. 287–300.

HARAWAY, D.J. (1989) *Primate Visions: Gender, Race and Nature in the World of Modern Science* (New York, Routledge).

HARDING, S. (1986) *The Science Question in Feminism* (Ithaca, NY, Cornell University Press).

HARDING, S. (1991) *Whose Science? Whose Knowledge? Thinking from Women's Lives* (Ithaca, NY, Cornell University Press).

HART, P. (1990) Environmental education in Canada: Contemporary issues and future possibilities, *Australian Journal of Environmental Education*, 6, pp. 45–65.

HUCKLE, J. (1988) *What We Consume Unit 5: Brazil, WWF-UK Project* (Richmond, UK, Richmond Publishing).

HUNGERFORD, H.R., PEYTON, R.B. & WILKE, R.J. (1983) Yes, EE does have definition and structure, *Journal of Environmental Education*, *14*(3), pp. 1–2.

HUNGERFORD, H.R., VOLK, T.L., DIXON, B.G., MARCINKOWSKI, T.J. & SIA, A.P.C. (1988) *An Environmental Education Approach to the Training of Elementary Teachers: A Teacher Education Programme*, UNESCO-UNEP International Environmental Education Programme Environmental Education Series No. 27 (Paris, UNESCO).

HUNGERFORD, H.R., VOLK, T.L. & RAMSEY, J.M. (1989) *A Prototype Environmental Education Curriculum for the Middle School*, UNESCO-UNEP International Environmental Education Programme Environmental Education Series No. 29 (Paris, UNESCO).

HUTCHEON, L. (1990) Circling the downspout of Empire, in ADAM, I. & TRIFFIN, H. (Eds) *Past the Last Post: Theorizing Post-Colonialism and Post-Modernism* (Calgary, University of Calgary Press).

INAYATULLAH, S. (1993) A response to Zia Sardar's 'colonizing the future', *Futures*, 25(2), pp. 190–195.

KREMER, K.B., MULLINS, G.W. & ROTH, R.E. (1990–1991) Women in science and environmental education: Need for an agenda, *Journal of Environmental Education*, 22(2), pp. 4–6.

MARCINKOWSKI, T.J., VOLK, T.L. & HUNGERFORD, H.R. (1990) *An Environmental Education Approach to the Training of Middle Level Teachers: A Prototype Programme*, UNESCO-UNEP International Environmental Education Programme Environmental Education Series No. 30 (Paris, UNESCO).

MERCHANT, C. (1980) *The Death of Nature: Women, Ecology and the Scientific Revolution* (New York, Harper & Row).

MERCHANT, C. (1990) Ecofeminism and feminist theory, in DIAMOND, I. & ORENSTEIN, G.F. (Eds) *Reweaving the World: The Emergence of Ecofeminism* (San Francisco, CA, Sierra Club Books).

MERCHANT, C. (1992) *Radical Ecology: The Search for a Livable World* (New York/ London, Routledge).

MINNICH, E.K. (1989) From the circle of the elite to the world of the whole: Education, equality and excellence, in Pearson, C.S., Shavlik, D.L. & Touchton, J.G. (Eds) *Educating the Majority: Women Challenge Tradition in Higher Education* (New York, Macmillan).

M'MWERERIA, G.K. (1993) Southern networks for environment and development (SONED) – Africa region, *The Independent Sectors' Network*, 26, p. 2.

NANDY, A. (1986) *The Intimate Enemy* (Delhi, Oxford University Press).

NATIONAL WOMEN'S CONSULTATIVE COUNCIL (NWCC). (1992) *A Question of Balance: Australian Women's Priorities for Environmental Action* (Canberra, National Women's Consultative Council).

NGUGI WA THIONG'O. (1986) *Decolonizing the Mind: The Politics of Language in African Literature* (London/Nairobi, Heinemann).

PARRY, B. (1987) Problems in current theories of colonial discourse, *Oxford Literary Review*, 9(1–2).

PAWLOWSKI, A. (1996) Perception of environmental problems by young people in Poland, *Environmental Education Research*, 2(3), pp. 287–300.

PECK, D. (1991) Gender equity and environmental education, unpublished draft paper.

PECK, D. (1992) Environmental education – time to branch out, *The GEN*, January/ February, pp. 1,4.

THE RIO DECLARATION ON ENVIRONMENT AND DEVELOPMENT. (1992) Reprinted in *Earth Ethics*, 4(1), pp. 9–10.

ROBOTTOM, I. & HART, P. (1993) *Research in Environmental Education: Engaging the Debate* (Geelong, Deakin University Press).

SAID, E.W. (1985) Orientalism reconsidered, *Cultural Critique*, 1, pp. 89–107.

SALLEH, A.K. (1989) Environmental consciousness and action: An Australian perspective, *Journal of Environmental Education*, 20(2), pp. 26–31.

SANGARI, K.K. (1987) The politics of the possible, *Cultural Critique*, 7, pp. 157–186.

SARDAR, Z. (1993) Colonizing the future: The 'other' dimension of futures studies, *Futures*, 25(2), pp. 179–187.

Shiva, V. (1989) *Staying Alive: Women, Ecology and Development* (London, Zed Books).

SLEMON, S. & TIFFIN, H. (1989) Introduction, *Kunapipi*, XI(1), pp. ix–xxiii.

SPENDER, D. (1990) *Man Made Language*. Revised edition (London, Pandora).

SPIVAK, G.C. (1987) *In Other Worlds: Essays in Cultural Politics* (London, Methuen).

SPIVAK, G.C. (1990) *The Post-Colonial Critic: Interviews, Strategies and Dialogues* (New York/London, Routledge).

UNESCO. (1977) *Tbilisi Declaration*. As reprinted in Connect, III(1), pp. 1–9.

UNESCO. (1978) *Intergovernmental Conference on Environmental Education: Tbilisi (USSR), 14–26 October 1977*, Final Report (Paris, UNESCO).

UNESCO REGIONAL OFFICE FOR EDUCATION IN ASIA AND THE PACIFIC. (1981) *Environmental Education in Asia and the Pacific*, Bulletin Number 22 (Bangkok, UNESCO).

UNITED NATIONS CONFERENCE ON ENVIRONMENT AND DEVELOPMENT (UNCED). (1992) *Agenda 21* (Rio de Janeiro, UNCED). Final, advanced version as adopted by the Plenary on 14 June 1992.

VIDART, D. (1978) Environmental education – theory and practice, *Prospects VIII*(4), pp. 466–479.

WARREN, K.J. (1990) The power and the promise of ecological feminism, *Environmental Ethics* 12(2), pp. 125–146.

WEEDON, C. (1987) *Feminist Practice and Poststructuralist Theory* (Oxford, Blackwell).

WEST, C. (1990) The new cultural politics of difference, in FERGUSON, R., GEVER, M., MINH-HA, TRINH T. & WEST, C. (Eds) *Out There: Marginalization and Contemporary Cultures* (New York, The New Museum of Contemporary Art & Cambridge, The MIT Press).

WHITEHOUSE, H. & TAYLOR, S. (1996) A gender inclusive curriculum model for environmental studies, *Australian Journal of Environmental Education*, 12, pp. 77–83.

WILKE, R.J., PEYTON, R.B. & HUNGERFORD, H.R. (1987) *Strategies for the Training of Teachers in Environmental Education*, UNESCO-UNEP International Environmental Education Programme Environmental Education Series No. 25 (Paris, UNESCO).

WILLIAMSON-FIEN, J. (1993) *Women's Voices* (Windsor, Global Learning Centre (Qld) Incorporated).

3 The power and the promise of feminist research in environmental education

Annette Gough

Gough, A. 1999. The power and the promise of feminist research in environmental educa-tion. *Southern African Journal of Environmental Education, 19*, 28–39. Reprinted in accordance with Creative Commons Attribution license CC-BY-NC-SA.

Summary

I have been arguing for recognition of the absence and need for inclusion of women's perspectives in environmental education research and pedagogy for some time (see, for example, Greenall Gough 1993; Gough 1997b, 1999). In this chapter I explore the related issue of the potential of adopting feminist research methods and methodologies in environmental education research. This exploration includes a discussion of the important of developing a feminist perspective, the characteristics of feminist educational research and a review of feminist research in environmental education. The chapter concludes with a discussion of feminist poststructuralist research as a powerful and promising approach for future research in environmental education.

Introduction

The title of this chapter is a play on a significant article in ecofeminist lit-erature, Karen Warren's (1990) "The power and the promise of ecological feminism". I chose this title in order to be playful in the spirit of Patti Lather (1991), but as I started to put the chapter together, the playfulness became even more meaningful as it gave me the opportunity to relate Karen War-ren's work to that of another significant author. I found the beginning of a quote from Foucault in Patricia Duncker's (1996) recent fiction *Hallucinat-ing Foucault* that seems to encapsulate (if Foucault can ever encapsulate!) my approach:

> There *are* times in life when the question of knowing if one can think differently than one thinks, and perceive differently than one sees, is absolutely necessary if one is to go on looking and reflecting at all.
>
> (Foucault 1990, p. 8, my emphasis)

DOI: 10.4324/9781003390930-4

The need to change (decentre) the perspectives from which we think and perceive (and act!) is one of the foundations of environmental education. Hence I believe it is important to explore the potential of Foucault's writings, and others, for their relevance for environmental education. It should be noted that I quote Foucault in the context of this chapter while recognising that several feminist researchers have drawn attention to the tensions between Foucault and feminism (from both positive and negative perspectives – see, for example, Diamond & Quinby 1988; Gore 1993; McNay 1992; Ramazaglou 1993). The power and relevance of Foucault's writing, "a type of thought strikingly attractive in its combination of extreme orderliness and brilliant intuitive insight, providing as it did an entirely new and excitingly different point of view on familiar scenery" (O'Farrell 1997b, p. 1), is too great to ignore him – as the recent *Foucault: The Legacy* (O'Farrell 1997a) indicates. This is a 780-page testament by 72 authors to the legacy of Foucault in fields as diverse as history, art, architecture, philosophy, psychoanalysis, feminism, medicine, government, management, public relations, environment, 'Third World', education and health.

The combination of the Foucault quote and Karen Warren's publications is also opportune because contained within a collection edited by Karen Warren (1994) is an article by Phillip Payne criticising Karen Warren's article "The power and the promise of ecological feminism" for the inclusion of first-person narrative (Payne 1994). From my perspective, first-person narrative is one of the powerful aspects of feminist research – "the personal is political" (Hanisch 1970 in Humm 1989, p. 162) – as it provides the opportunity to consciously reflect and come to think and perceive differently. This phrase "stresses the psychological basis of patriarchal oppression . . . [it] makes a direct relation between sociality and subjectivity so that to know the politics of women's situation is to know women's personal lives" (Humm 1989, p. 162). Thus in this chapter I argue for different ways of thinking and perceiving in environmental education research – by using feminist research strategies.

Why is a feminist perspective in environmental education important?

Developing a feminist perspective in environmental education is important because the vast majority of work in environmental education to date has been concerned with universalised subjects rather than recognising multiple subjectivities. It is time that we started to generate different ways of knowing and seeing environments in order that we might understand human relationships with them better. As Brown and Switzer (1991, p. iv) argue,

> Women and men contribute to maintaining environmental, economic and social sustainability in distinctive ways. For women these contributions are made through:
>
> • their public roles as the majority of the workforce in the health, education, welfare and service industries;

- their private roles as care-givers, farm managers, educators of children, and the principal purchasers of food and consumer goods; and
- the many public (paid) and private (unpaid) arenas where women have a major responsibility for the management of change and the transmission of social values.

Such a perspective is not intended to essentialise women as caretakers of the earth's household, obsessed with green cleaners, nor to cast women as symbols of nature. Rather, the intention in developing a feminist perspective in environmental education is to recognise the complexity of human roles and relationships with respect to environments, and that there are multiple subjectivities and multiple ways of knowing and interacting with environments which cannot be encapsulated within the notion of universalised subjects.

Most importantly, by pursuing a feminist perspective in environmental education we will be able to construct "less partial, less distorted" (Harding 1991) accounts of environments. Such a pursuit is also consistent with feminist praxis – a term which recognises "a continuing feminist commitment to a political position in which 'knowledge' is not simply defined as 'knowledge what 'but also as 'knowledge for' (Stanley 1990, p. 15). It indicates rejection of the theory/research divide, "seeing these as united manual and intellectual activities which are symbiotically related (for all theorising requires 'research' of some form or another)" (Stanley 1990, p. 15). Thirdly, it centres interest on methodological/epistemological concerns: " 'how' and 'what' are indissolubly interconnected and . . . the shape and nature of the 'what' will be a product of the 'how' of its investigation" (Stanley 1990, p. 15). A central concern of feminist praxis is thus the reconstituting of knowledge. As Dale Spender (1985, p. 5) argues, "at the core of feminist ideas is the crucial insight that there is no one truth, no one authority, no one objective method which leads to the production of pure knowledge". I therefore argue that through feminist research in environmental education we will be able to reconstitute knowledge about environments, and for environments.

Evidence that men and women do think differently about environmental issues is apparent in the data collected by Brown (1995). Table 3.1 compares Australian women's priority concerns about the environment with a survey of issues prioritised in the scientific literature.

The different role of women with respect to the environment is also recognised in *Agenda 21*. However, lack of reciprocity is an issue between the 'Global Action for Women Towards Sustainable and Equitable Development' and the 'Promoting Education, Public Awareness and Training' chapters in *Agenda 21* (UNCED 1992, Chapters 24 and 36 respectively). The 'Women' chapter has as its overall goal achieving active involvement of women in economic and political decision making, with emphasis on women's participation in national and international ecosystem management and control of

Table 3.1 Ranking of environmental issues by Australian women compared with the scientific-government agenda

Women's ranking	Scientific/government agenda ranking
1. Toxic wastes and waste management	(4)
2. Nuclear wastes and accidents	(7)
3. Loss of animal and plant species	(10)
4. Poverty and its environmental effects	(8)
5. Land degradation and deforestation	(2)
6. Energy use and consumerism	(5)
7. War and militarism	(9)
8. Human population growth	(3)
9. Climate change	(1)
10. Misuse of technology	(6)

environmental degradation. One of its objectives for national governments (UNCED 1992, Paragraph 24.2(e)) is

> To assess, review, revise and implement, where appropriate, curricula and other educational material, with a view to promoting the dissemination to both men and women of gender-relevant knowledge and valuation of women's roles through formal and non-formal education.

Other objectives addressed topics such as increasing the proportion of women decision makers, eliminating obstacles to women's full participation in sustainable development, achieving equality of access to opportunities for education, health and so on for women, equal rights in family planning and prohibiting violence against women. The activities for governments related to such objectives are broadly concerned with achieving equality of opportunity for women (such as by eliminating illiteracy). increasing proportions of women as decision makers in implementing policies and programmes for sustainable development, and recognising women as equal members of households both with respect to workloads and finance. Consumer awareness is particularly mentioned, as are

> programmes to eliminate persistent negative images, stereotypes, attitudes and prejudices against women through changes in socialization patterns, the media, advertising, and formal and non-formal education.
> (UNCED 1992, Paragraph 24.3(i))

Here, women's contributions to society are being recognised and valued as something different, rather than assuming that women will achieve equality simply through equal opportunity, although there are some elements of a liberal feminist view[1] present.

Unfortunately these views are not matched in the 'Education' chapter of *Agenda 21*. Here women are generally included along with all sectors of society, although specific mention is made of the high illiteracy levels among women which need to be addressed (UNCED 1992, Paragraph 36.4(a)) in the objectives. In the activities, women are mentioned in the following terms (UNCED 1992, Paragraph 36.5(m)): "Governments and educational authorities should foster opportunities for women in non-traditional fields and eliminate gender stereotyping in curricula". No mention is made of recognising and valuing women's roles in achieving sustainable development, and the perspective seems once more to be that of liberal feminism, although indigenous peoples' experience and understanding of sustainable development are affirmed as playing a part in education and training (UNCED 1992, Para. 36.5(n)).

There is a vast body of literature relating women and the environment, much too much to be listed here. However, it is significant to note that the Australian Government has published a statement on *Women and the Environment* (Department of Prime Minister and Cabinet/Office of the Status of Women 1992), and I do recommend Merchant (1996) and Salleh (1997) for an overview that does not approach the relationship of women and the environment from a perspective of women as goddesses.

What is feminist research?

According to Lather (1991, p. 17), "to do feminist research is to put the social construction of gender at the centre of one's inquiry . . . feminist researchers see gender as a basic organising principle which profoundly shapes and/or mediates the concrete conditions of our lives".

Feminist research is thus openly ideological, aiming to correct both the invisibility of female experience and its distortion.

Feminist educational research methods take many forms, including statistical, interview, ethnographic, survey, cross-cultural, oral history, content analysis, case studies and action research (as described in Reinharz 1992). As Reinharz (1992, p. 4) argues, "feminists have used all existing methods and have invented some new ones as well", and Stanley (1990, p. 12) makesa similar point "that there is no one set of methods or techniques . . . which should be seen as distinctly feminist. Feminists should use any and every means available".

According to Reinharz (1992, p. 240) in her most useful text on feminist methods in social research, "feminist methodology is the sum of feminist research methods". In making this statement, Reinharz epitomises a problem identified by Harding (1987, p. 2) "that social scientists tend to think about methodological issues primarily in terms of methods of inquiry". Indeed, 'method' and 'methodology' are terms that are frequently intertwined, used interchangeably or confused in feminist research scholarship, and contestation abounds as to whether or not there is a feminist research method or methodology. For example, Harding (1987, p. 1) argues against the idea of a distinctive feminist method of research "on the grounds that preoccupation with method mystifies what have been the most interesting aspects of feminist research processes", in particular the differences between method, methodology and epistemology. Harding (1987, p. 2) goes on to argue that

> it is new methodologies and new epistemologies that are requiring these new uses of familiar research techniques. If what is meant by a 'method of research' is just this most concrete sense of the term, it would undervalue the transformations feminist analyses require to characterize these in terms only of the discovery of distinctive methods of research.

The confusion between method (techniques for gathering evidence), methodology (a theory and analysis of how research should proceed) and epistemology (issues about an adequate theory or justificatory strategy) is not the sole province of feminist research. Such confusions abound in non-feminist research, too. In both feminist and non-feminist research 'method' is often used to refer to all aspects of research, thus making discussions about distinctiveness, particularly in regard to feminist research, difficult.

Although Reinharz does not distinguish techniques for gathering evidence (method) from theory and analysis of how research should proceed (methodology), which she should do, she does encapsulate an important aspect of feminist research when she writes that

> feminist research is driven by its subject matter, rather than by its methods . . . feminist research will use any method available and any cluster of methods needed to answer the questions it sets for itself.
> (1992, p. 213)

However, what makes feminist research distinctive, according to Harding (1987), is that it opens up

- new empirical and theoretical resources (women's experiences),
- new purposes of social science research (for women), and
- new subject matter of inquiry (locating the researcher in the same critical plane as the overt subject matter).

Although early feminist research was largely positivistic, recent methodologies have been more concerned with

> generating and refining . . . more interactive, contextualised methods in the search for pattern and meaning rather than for prediction and control . . . Hence feminist empirical work is multi-paradigmatic.
> (Lather 1991, p. 18)

Feminist research methodologies now include the whole range from post-positivistic concerns with prediction through interpretive, constructivist, phenomenological and ethnographic concerns to understand, to emancipatory methodologies such as critical, participatory and action research, and postmodern concerns such as poststructuralism and deconstruction. Reinharz (1992, p. 240) proposes ten characteristics of feminist research (which she calls methodology, but I prefer to call approach):

1. Feminism is a perspective, not a research method.
2. Feminists use a multiplicity of research methods.
3. Feminist research involves an ongoing criticism of nonfeminist scholarship [To this I would add criticism of feminist scholarship too!].
4. Feminist research is guided by feminist theory.
5. Feminist research may be transdisciplinary.
6. Feminist research aims to create social change.
7. Feminist research strives to represent human diversity.
8. Feminist research frequently includes the researcher as a person.
9. Feminist research frequently attempts to develop special relations with the people studied (in interactive research).
10. Feminist research frequently defines a special relation with the reader.

These characteristics share much in common with environmental education research, particularly the use of a multiplicity of methods (by some researchers), adopting a transdisciplinary focus, and aiming to create social change.

What feminist research has already been undertaken in environmental education?

The literature on feminist research in environmental education is both recent and sparse and, until recently, has been almost totally Australian. A similar finding was made by Giovanna Di Chiro (1993, p. 228) when she conducted an ERIC search using the descriptors 'feminism' and 'environmental education' which yielded only two articles, one her own (1987) which was written and published in Australia, and the other by Ariel Salleh (1989), an Australian ecofeminist and social theorist. My own search added two North American articles (Kremer et al. 1990–1991; Fawcett et al. 1991), one British article (Hallam & Pepper 1991), and three recent articles in *Environmental Education Research* (Hampel et al. 1996 [again Australians!], Pawlowski 1996 [Polish], Storey et al. 1998 [UK/Brazil]) have drawn attention to gender differences in environmental knowledge, concerns and behaviours. The remaining literature which specifically relates gender to environmental education is Australian (including Barron 1995; Brown & Switzer 1991; Department of Prime Minister and Cabinet/Office of the Status of Women 1992; Gough 1994; Gough 1997b; Greenall Gough 1993; NWCC 1992; Peck 1992; Whitehouse & Taylor 1996).

There are also statements from the United Nations Conference on Environment and Development (UNCED) where the activities for promoting education, public awareness and training include fostering opportunities for women and eliminating gender stereotyping in curricula (UNCED *Agenda 21* 1992, see, for example, Paragraphs 24.2(e) and 36.5 (m)).

Giovanna Di Chiro (1987) places a feminist perspective on environmental education within a socially critical framework. She grounds her ecofeminist perspective in radical and socialist feminism and asserts that (1987, p. 40):

> A feminist perspective [on] environmental education offers a more complete analysis of environmental problems and therefore a better understanding of those problems and their potential solutions. Such an analysis is political, in that it examines how power relations (in, for example, gender, class, race) shape the world in which we live; it asserts that the 'polity' (human social world) determines and controls how this social world is and has been historically constructed and organised, and hence refutes the myth that the past and present state of the world is a 'natural' and therefore justifiable progression. Moreover, environmental education's analysis of socioenvironmental problems is political in that it believes that if human social relations create the problems that can also change and improve them.

In particular, she argues (1987, p. 41) that the environmental problem is socially constructed and should be viewed as a social problem, that environmental education should engage in a feminist critique of environmental problems, and that it should engage in self-criticism "in order to understand how it is responsible as an educational enterprise for maintaining certain 'un-environmental' values and ideologies". While others (such as Salleh 1989; Fawcett et al. 1991; Peck 1992) also argue for the first two points to varying degrees, Di Chiro seems to be alone in asserting the need for environmental education to be self-critical as well as socially critical.

Ariel Salleh (1989, p. 27) describes an attempt to enact such an approach in a case study of a group of upper-working-class women coming together "to see what might be done about household waste recycling in their local community". Few of the group had completed high school (only one had a university degree); all were over 30 and either were or had been married with children. All were already involved in some kind of ecologically sound practice at home. Opening up the issue with the technique of consciousness-raising as a catalyst for moving from personal to political concerns, Salleh (1989, p. 30) found that

> it was the tensions growing out of the consciousness-raising process itself that undermined the possibility of their participation in an environmental program . . . Their workforce and personal marginality were so severe that they lacked the necessary human support and self-assurance to transform their critical stance into a collaborative praxis.

The paucity of feminist research in environmental education could be considered surprising when related fields, such as science education, have a history of feminist research (see, e.g., Parker, Rennie, & Fraser, 1996). Feminist scholarship in science started about the same time as the ecofeminist movement2 (in the early 1970s), but, perhaps because of the greater social status of science or because of some of the more extreme3 writings of some ecofeminists, feminist research in science education has received a much higher profile and generated many more studies than in environmental education:

> The less education, the more likelihood that respondents would choose education or individual actions as the principal action for women, and the less likely they would suggest changing social frameworks or political action. More of the respondents who had not proceeded beyond school level gave individual or personal answers, than women from the other three levels of education.

Salleh (1989, p. 30) suggests that for the group she studied to become more effectual "they would need their self-image bolstered by more social affirmation . . . but beyond this are issues of access to such resources as status, time, skill, and political models", and concludes by hypothesising that

> if such women gained more experience alongside men in the rough and tumble "real work world", they might discover the simultaneous empowerment and disenchantment that makes for an ecofeminist praxis.

Leesa Fawcett, Dian Marino and Rebecca Raglon (1991) take an approach with some similarities to Di Chiro's in their focus on reconnecting humans and nature in a reframing of environmental education. They argue for a transformative vision of ecofeminism,[2] which draws on radical feminists' focus on biological differences between women and men, spiritualism, and personal connections to nature, together with socialist feminists' view of human nature and nature as socially and historically constructed. They combine this vision with their notion of social change motivated by materialism – where "the powers behind social change (spiritualism, personal experience, and materialism) are not mutually exclusive" (1991, p. 251). Building from this vision, they suggest an approach to environmental education that works with resistance – "we try to understand how resistance can maintain the status quo and identify where transformations can occur" – and explores ideas of personal and social change, resistance, difference, and powerlessness (1991, p. 251). This includes exploring human-animal continuities while engaging in making new meanings. However, they are not looking for a totalising replacement to humans' fractured relations with nature: "Perhaps an affection for some of the chaos around us would be a better goal" (1991, p. 250) because they see paradox and contradiction as being useful for not "reducing the world" (1991, p. 252).

In developing her gender equity and environmental education guidelines Peck (1992) took Brown and Switzer's (1991) recommendations (cited earlier in this paper) into consideration, but she adopted a much more critical approach to society. Although her brief could have been interpreted from the liberal feminist perspective of simply identifying gender equity issues in environmental education which provide equal opportunities for girls and boys, she chose to adopt a critical approach. Peck convergently evolved her perspective in apparent ignorance of the research and writings of Di Chiro (1987), Salleh (1989), and Fawcett et al. (1991), and she has developed the implications of a feminist perspective on environmental education in greater detail.

As her starting point Peck (1992, p. 1) defines what she means by environmental education, equating it with environmental literacy, as the need for all people "to be aware of and concerned about the environment, and to have the knowledge, skills, values, attitudes, motivation and commitment which will enable them to participate in the care and conservation of the environment". She notes that girls in particular appear to be very interested in this area. She also comments that almost no work has been done on identifying gender equity issues in environmental education and argues that this is an urgent need.

That there is a paucity of feminist research in environmental education could be considered surprising when related fields such as outdoor education, and, in particular, science education have a history of feminist research (see, for example, Parker et al. 1996). Though, as I have argued elsewhere – see, for example, Greenall Gough 1993; Gough 1994; Gough 1997a – throughout its history environmental education has been dominated by universalised (masculine) perspectives. The history of feminist scholarship in science (see, for example, Keller & Longino 1996) actually started about the same time as the ecofeminist movement, but, perhaps because of the greater social status of science or because of some of the more extreme writings of some ecofeminists, feminist research in science education has received a much higher profile and generated many more studies than in environmental education.

What is the power and promise of feminist research in environmental education?

A major reason for wishing to explore the power and the promise of feminist research in environmental education is that the most interesting educational research publications I have read in recent times have been written by feminist researchers. The issues and approaches that are being explored in their work are often very relevant to the concerns of environmental educators and they provide some hope for new directions and possible successes. Feminist research has been an absence in environmental education research for too long.

While my personal disposition is towards critical and poststructuralist feminist research, all feminist research methodologies (and methods!) can be applied in environmental education research contexts. We can do many types of research which put the social construction of gender at the centre of the inquiry, whether we are seeking to predict, understand, emancipate or deconstruct, and we need more stories from women's lives relating to environments that we can use in environmental education (along the lines of Jane Williamson-Fien's (1993) *Women's Voices*).

In the remainder of this chapter I argue for poststructuralist research as the most promising approach for achieving the power and promise of feminist research in environmental education.

The power and the promise of doing feminist poststructuralist research in environmental education is that it calls into play a deconstructionist impulse which provokes consideration of the gendered positions made available to students and of understandings of gender identity. This approach is consistent with Bronwyn Davies (1994, p. 78), who suggests that a pedagogy informed by poststructuralist theory might begin

> with turning its deconstructive gaze on the fundamental binarisms of pedagogy itself: teacher/student, adult/child, internal/external, society/individual, reality/fiction, knower/known, nature/culture, objective/subjective [because] each of these underpin or hold together both what we understand as pedagogy and the discourses through which pedagogy is done.

Such an examination of the binaries of environmental education practices could also form the basis for a research agenda.

Another aspect of the appeal of feminist poststructuralist research as an approach for environmental education is summarised by Ben Agger. He (1992, p. 118) argues:

> Another primary aim of feminist cultural criticism is to decenter men from their dominance of various official canons and genres. Equally as troubling as the omission of women from the canon and from criticism is the installation of men as those who speak for women – universal subjects of world history. A good deal of poststructural feminist criticism has focused on the issue of the voices in which culture is expressed, the standpoints from which knowledge is claimed.

Decentring the male perspectives that dominate environmental education discourses is a challenge for the future. However, such decentring will not be an easy task. As Grace (1994, p. 19) argues, "men's interests, women's interests and the common interest are all difficult contingent alliances, and the ideological claim to them forms part of continuing negotiations of power". Nevertheless, there is a need to tell 'less partial, less distorted' stories in our research. As I noted earlier, by studying women's experiences we open up new empirical and theoretical resources, provide a new purpose for social science (for women rather than men) and a new subject matter of inquiry (while studying women is not new, it is new to study them "from the perspective of their own experiences so that women can understand themselves and the world" (Harding 1987, p. 8)).

Empowerment, and the search for more empowering ways of knowing, is a central concern of both critical and poststructuralist theorising; it is also a goal that is shared by many environmental educators. Emancipatory (empowerment) research involves "analyzing ideas about the causes of powerlessness, recognizing systemic oppressive forces, and acting both individually and collectively to change the conditions of our lives" (Lather 1991, p. 4), and Ellsworth (1989, p. 306), for example, argues that "critical pedagogies employing this strategy prescribe various theoretical and practical means for sharing, giving or redistributing power to students". However, empowerment is not something done to or for someone; it is a process one undertakes for oneself in the development of a new relationship within their own particular contexts. We need some research in environmental education which focuses on how women have been empowered, and are empowering themselves, rather than looking at universalised subjects (traditional research has focused on men's experiences and "asked only the questions about social life that appear problematic from within the social experiences that are characteristic for men" (Harding 1987, p. 6)).

Because of its political nature and concerns with empowerment, it is important to look at power and knowledge in the context of environmental education. This can be a focus of both critical and poststructuralist research. Power and knowledge as they are exercised through discourses are central aspects of poststructuralist theory:

> discourses . . . are ways of constituting knowledge, together with the social practices, forms of subjectivity and power relations which inhere in such knowledges and the relations between them.
>
> (Weedon 1987, p. 108)

The texts, myths and meanings of our culture and our relationships with nature need to be deconstructed in order that we know the stories of which we are a part. Such deconstruction and critical analysis will help practitioners and students to recognise whose interests are being served at particular moments in environmental issues. It will help them to understand that

> it does make a difference who says what and when. When people speak from the opposite sides of power relations, the perspective from the lives of the less powerful can provide a more objective view than the perspective from the lives of the more powerful.
>
> (Harding 1991, pp. 269–270)

The challenge is to encourage the development of alternative stories (discourses) by drawing attention to whose knowledge is legitimated and valorised in the power/knowledge structures of the dominant discourses. In particular, women's knowledge needs to be recognised and valued.

The dominant discourses in environmental education treat the subject of knowledge as homogeneous and unitary. Thus, in the behaviourist/individualist model which dominates much of these discourses, there is an emphasis on individuals having 'the right behaviour' and the knowledge of how to 'get it right'. This implies a power relationship where some take it as their role to set out what those 'right behaviours' are. However, it is no longer possible to find a universal subject: as subjects/agents of knowledge we are all part of multiple, heterogeneous and contradictory or incoherent positionings of race, class, gender and ethnicity, and there is no one right way of knowing or behaving.

Such multiple subjectivities are constantly achieved through relations with others (both real and imagined) which are themselves made possible through discourse. Accepting that the subjects of knowledge are multiple rather than homogeneous, unitary and universal has implications for curriculum, pedagogy and research in environmental education. Exploring and developing such possibilities for environmental education is a challenge for the future, and it is one that must include the power and promise of feminist research.

While poststructuralist research in environmental education is a challenge, I believe it also offers much promise which is consistent with the stated goals of environmental education. The dominant discourses of environmental education recognise that the environment and environmental problems are complex, not simple. For example, the guiding principles of environmental education from Tbilisi (UNESCO 1978, p. 27) refer to the multidisciplinary nature of the environment – "consider the environment in its totality – natural and built, technological and social (economic, political, technological, cultural-historical, moral, aesthetic)" – and "the complexity of environmental problems". While not overt nor necessarily intended, such a perspective could involve multiple readings or interpretations of the environment which are consistent with adopting poststructuralist pedagogy and research approaches.

An anonymous reviewer of an earlier draft of this paper commented:

Just because the authors of [the Tbilisi] report advocate that multiple environments should be considered in EE, [this] does not imply [that] any particular perspective should be adopted in viewing those environments. Such a consideration can take an externalized, objectivist or positivist stance toward examining these different environments. Many approaches to EE that claim to be based on these principles in fact have interpreted these principles from such a behavourist or positivist perspective.

If the Tbilisi statements can be (and have been) appropriated in this way by behaviourists to be consistent with their stance, it seems to me that it is quite legitimate for me to read them as offering the potential for multiple readings consistent with feminist poststructuralist research approaches. The multiple readings I am suggesting are not only of nature but also of the individuals and groups that are concerned with the particular environment or environmental issue: there is a need to develop local or situated knowledges which disrupt oppressions. We should be listening to multiple stories in the spirit of a partnership ethic (and its precepts) rather than following and an egocentric or homocentric ethic (Merchant 1996).

Whatever the issue or environment, there are multilevel meanings of narratives and texts, and multiple stories which can be told. There is not 'one true story' about the environment. The knowledges involved in dealing with environments are multiple, involving both humans (where each human is a multiple subject) and nonhuman nature (which also has a multiple subjectivity), and must be considered as such. Thus poststructuralist pedagogy and research is also consistent with a partnership ethic (and feminist research) in that it is concerned with listening to the voices of the marginalised as well as those of the dominant discourses.

Poststructuralist research is also concerned with deconstructing power/ knowledge relationships, which is also a goal of a partnership ethic in environmental education. As in critical research, it is important to analyse who has the power and what can be done to dismantle or subvert it through developing counter-hegemonic and oppositional discourses. However, in contrast with a critical research approach, this power can be other than economic. Poststructural research is concerned with providing opportunities for more than the dominant voices to be heard. We need to know the stories of which we are a part and to develop local knowledges.

Poststructuralist research that is consistent with a partnership ethic is also concerned with the liberation of nature and people. The goal is "to work toward a socially-just, environmentally sustainable world" (Merchant 1996, p. 222). It is time to stop trying "from the outside, to dictate to others, to tell them where their truth is and how to find it" (Foucault 1990, p. 9). By engaging in feminist and poststructuralist research in environmental education, we will be able to come closer to achieving this goal because we will have less partial and less distorted stories.

Acknowledgement

I would like to thank Eureta Janse van Rensburg for the encouragement to submit this paper to SAJEE and the anonymous reviewers of my paper for their constructive comments. I hope I have done them justice.

Note on contributor

Annette Gough is a Senior Lecturer in environmental and science education in the Faculty of Education, Deakin University, Australia. She has been active in the environmental education movement in Australia for 25 years and was recognised for her contributions to the field in 1992 through the awarding of a life fellowship of the Australian Association for Environmental Education. One of her major areas of interest is the intersection of feminism and environmental education in a research context. Her other research interests include feminist epistemologies, postcolonialism, critical inquiry, curriculum policy and curriculum reform in both science education and environmental education. Correspondence: Faculty of Education, Deakin University (Melbourne Campus), 221 Burwood Highway, Burwood, Victoria, Australia 3125.

Notes

1 The term 'liberal feminism' is often used to characterise the dominant form of feminism up to the 1960's. Its current form, inspired by the works of Simone de Beauvoir (The Second Sex 1972) and Betty Friedan (The Feminine Mystique 1963), "emanates from the classical liberal tradition that idealizes a society in which individuals are provided maximal freedom to pursue their own interest . . . [and] endorses a highly individualistic conception of human nature" (Warren 1987, 8). This conception locates our uniqueness as humans in our capacity for rationality and/or the use of language (Jaggar 1983) and "when reason is defined as the ability to comprehend the rational principles of morality, then the value of individual autonomy is stressed" (Tong 1989, II). For liberal feminists, the attainment of knowledge is an individual project and their epistemological goal is "to formulate value-neutral, intersubjectively verifiable, and universal rules that enable any rational agent to attain knowledge 'under a veil of ignorance' "(Warren 1987, 9).

   Historically, liberal feminists have argued that women do not differ from men as rational agents, and that it is only their exclusion from educational and economic opportunities which has prevented women from realising their potential (Jaggar 1983). However, according to Tong (1989, 11), the current status of liberal feminist thought is difficult to determine because liberalism "is in the process of reconceptualizing, reconsidering and restructuring itself". Critiques of liberal feminism focus on the alleged tendencies to accept male values as human values; to over-emphasise the importance of individual freedom over that of the common good; to adhere to normative dualism, and to valorize a gender neutral humanism over a gender-specific feminism (Jaggar 1983; Tong 1989).

2 The term 'ecofeminism' was coined in 1974 by Francois d'Eaubonne "who called upon women to lead an ecological revolution to save the planet. Such an ecological revolution would entail new gender relations between women and men and between humans and animals" (Merchant 1996, 5).

References

Agger, B. 1992. *Cultural Studies as Critical Theory*. Falmer Press, London.

Barron, D. 1995. Gendering environmental education reform: Identifying the constitutive power of environmental discourses. *Australian Journal of Environmental Education*, 11, 107–120.

Brown, V.A. 1995. Women who want the earth: Managing the environment is a gender issue. Paper presented at the Conference "Towards Beijing – Women, Environment and Development in the Asian and Pacific Regions". Victoria University of Technology, Melbourne, 9–11 February 1995.

Brown, V.A. & Switzwer, M.A. 1991. *Engendering the Debate: Women and Ecologically Sustainable Development.* Office of the Status of Women, Department of the Prime Minister and Cabinet, Canberra.

Davies, B. 1994. *Poststructuralist Theory and Classroom Practice.* Deakin University, Geelong.

De Beauvoir, S. 1972. *The Second Sex.* Penguin, Harmondsworth.

Department of Prime Minister and Cabinet. Office of the Status of Women (DPMC/ OSW). 1992. *Women and the Environment.* Australian Government Publishing Service, Canberra.

Di Chiro, G. 1987. Environmental education and the question of gender: A feminist critique. In I. Robottom (Ed.), *Environmental Education: Practice and Possibility.* Deakin University, Geelong, pp. 23–48.

Di Chiro, G. 1993. Environmental education in the war years: A case study. In R. Mrazek (Ed.), *Alternative Paradigms in Environmental Education Research.* North American Association for Environmental Education, Troy, pp. 227–237.

Diamond, I. & Quinby, L. (Eds.). 1988. *Feminism and Foucault: Reflections on Resistance.* Northeastern University Press, Boston.

Duncker, P. 1996. *Hallucinating Foucault.* Picador, London.

Ellsworth, E. 1989. Why doesn't this feel empowering? Working through the repressive myths of critical pedagogy. *Harvard Educational Review,* 59(3), 297–324.

Fawcett, L., Marino, D. & Raglon, R. 1991. Playfully critical: Reframing environmental education. In J.H. Baldwin (Ed.), *Confronting Environmental Challenges in a Changing World. Selected Papers from the Twentieth Annual Conference of the North American Association for Environmental Education (NAAEE).* NAAEE, Troy, pp. 250–254.

Foucault, M. 1990. *The Use of Pleasure (Volume 2 of the History of Sexuality).* Vintage, New York.

Friedan, B. 1963. *The Feminine Mystique.* W.W. Norton, New York.

Gore, J.M. 1993. *The Struggle for Pedagogies: Critical and Feminist Discourses as Regimes of Truth.* Routledge, New York.

Gough, A. 1994. Fathoming the fathers in environmental education: A feminist poststructuralist analysis. Unpublished doctoral dissertation. Deakin University, Geelong, Australia.

Gough, A. 1997a. Founders of environmental education: Narratives of the Australian environmental education movement. *Environmental Education Research,* 3(1), 43–57.

Gough, A. 1997b. *Education and the Environment: Policy, Trends and the Problems of Marginalisation.* Australian Council for Educational Research, Melbourne.

Gough, A. 1999. Recognising women in environmental education pedagogy and research: Toward an ecofeminist poststructuralist perspective. *Environmental Education Research,* 5(2), 143–161.

Grace, F. 1994. Do theories of the state need feminism? *Social Alternatives,* 12(4), 17–20.

Greenall Gough, A. 1993. *Founders in Environmental Education.* Deakin University, Geelong.

Hallam, N. & Pepper, D. 1991. Feminism, anarchy and ecology: Some connections. *Contemporary Issues in Geography and Education,* 3(2), 151–167.

Hampel, B., Holdsworth, R. & Boldero, J. 1996. The impact of parental work experience and education on environmental knowledge, concern and behaviour among adolescents. *Environmental Education Research,* 2(3), 287–300.

Hanisch, C. 1970. The personal is political. In S. Firestone & A. Koedt (Eds.), *Notes from the Second Year: Women's Liberation Major Writings of the Radical Feminists.* Radical Feminism, New York, pp. 76–78.

Harding, S. (Ed.). 1987. *Feminism and Methodology.* Indiana University Press, Bloomington.

Harding, S. 1991. *Whose Science? Whose Knowledge? Thinking for Women's Lives.* Cornell University Press, Ithaca.

Humm, M. 1989. *The Dictionary of Feminist Theory.* Harvester Wheatsheaf, Hemel Hempstead.

Jaggar, A.M. 1983. *Feminist Politics and Human Nature.* Rowman and Allanheld, Totowa.

Keller, E.F. & Longino, H. (Eds.). 1996. *Feminism and Science.* Oxford University Press, Oxford.

Kremer, K.B., Mullins, G.W. & Roth, R.E. 1990–1991. Women in science and environmental education: Need for an agenda. *Journal of Environmental Education,* 22(2), 4–6.

Lather, P. 1991. *Getting Smart: Feminist Research and Pedagogy with/in the Postmodern.* Routledge, New York/London.

McNay, L. 1992. *Foucault and Feminism: Power, Gender and the Self.* Polity Press, Cambridge.

Merchant, C. 1996. *Earthcare: Women and the Environment.* Routledge, New York.

NWCC. National Women's Consultative Council. 1992. *A Question of Balance: Australian Women's Priorities for Environmental Action.* NWCC, Canberra.

O'Farrell, C. (Ed.). 1997a. *Foucault: The Legacy.* Queensland University of Technology, Brisbane.

O'Farrell, C. 1997b. Introduction. In C. O'Farrell (Ed.), *Foucault: The Legacy.* Queensland University of Technology, Brisbane, pp. 1–9.

Parker, L.H., Rennie, L.J. & Fraser, B.J. (Eds.). 1996. *Gender, Science and Mathematics: Shortening the Shadow.* Kluwer, Dordrecht.

Pawlowski, A. 1996. Perception of environmental problems by young people in Poland. *Environmental Education Research,* 2(3), 279–300. doi:10.1080/1350462960020302

Payne, P. 1994. Restructuring the discursive moral subject in ecological feminism. In K.J. Warren (Ed.), *Ecological Feminism.* Routledge, New York, pp. 139–157.

Peck, D. 1992. Environmental education – time to branch out. *The GEN,* January/February, 1 & 4.

Ramazaglou, C. (Ed.). 1993. *Up Against Foucault: Explorations of Some Tensions Between Foucault and Feminism.* Routledge, London/New York.

Reinharz, S. 1992. *Feminist Methods in Social Research.* Oxford University Press, New York/Oxford.

Salleh, A. 1997. *Ecofeminism as Politics: Nature, Marx and the Postmodern.* Zed Books, London.

Salleh, A.K. 1989. Environmental consciousness and action: An Australian perspective. *Journal of Environmental Education,* 20(2), 26–31.

Spender, D. 1985. *For the Record: The Making and Meaning of Feminist Knowledge.* Women's Press, London.

Stanley, L. (Ed.). 1990. *Feminist Praxis: Research, Theory and Epistemology in Feminist Sociology.* Routledge, New York/London.

Storey, C., Da Cruz, J.G. & Camargo, R.F. 1998. Women in action: A community development project in Amazonas. *Environmental Education Research,* 4(2), 187–199.

Tong, R. 1989. *Feminist Thought: A Comprehensive Introduction.* Westview Press, Boulder/San Francisco.

UNESCO. 1978. *Intergovernmental Conference on Environmental Education: Tbilisi (USSR), 14–26 October 1977. Final Report.* UNESCO, Paris.

United Nations Conference on Environment and Development (UNCED). 1992. *Agenda 21.* UNCED, Rio de Janeiro. Final, advanced version as adopted by the Plenary on 14 June 1992.

Warren, K.J. 1987. Feminism and ecology: Making connections. *Environmental Ethics,* 9(1), 3–20.

Warren, K.J. 1990. The power and the promise of ecological feminism. *Environmental Ethics, 12*(2), 125–146.

Warren, K.J. (Ed.). 1994. *Ecological Feminism.* Routledge, New York.

Weedon, C. 1987. *Feminist Practice and Poststructuralist Theory.* Blackwell, Oxford.

Whitehouse, H. & Taylor, S. 1996. A gender inclusive curriculum model for environmental studies. *Australian Journal of Environmental Education, 12,* 77–83.

Williamson-Fien, J. 1993. *Women's Voices.* Global Learning Centre, Queensland.

4 The contribution of ecofeminist perspectives to sustainability in higher education

Annette Gough

Gough, A. (2004). The contribution of ecofeminist perspectives to sustainability in higher education. In P.B. Corcoran & A.E.J. Wals (Eds.), *Higher Education and The Challenge of Sustainability: Contestation, Critique, Practice, and Promise* (pp. 149–161). Dordrecht, Netherlands: Kluwer Academic Press. ©2004 Reprinted by permission of SpringerNature.

Abstract

In the past, in much of higher education, women, as students and academics, have struggled for recognition, and women have been overlooked in most sustainability programmes through being subsumed into the notion of "universalised people". However, women have a distinctive contribution to make to sustainability policy, pedagogy and research that needs to be foregrounded. This chapter discusses research into the absences of women's perspectives from sustainability policies, pedagogy and research and argues that ecofeminist pedagogies and research methodologies suggest new possibilities for the development of sustainability in higher education. I argue for the power and the promise of adopting such a perspective for the higher education context. At the technical level actions should include increasing proportions of women as decision makers in implementing policies and programs for sustainable development, recognising women as equal members in workplaces both with respect to workloads and finance, and including women's perspectives in the content of higher education courses. However, it is equally important that ecofeminist perspectives inform research and pedagogies in higher education. In developing my arguments I focus on changing institutional and philosophical actions to make women's and men's lives in higher education more democratic, academic knowledge less partial, and sustainability achievable.

Introduction

Higher education institutions are the gatekeepers of knowledge production, accreditation, legitimation and dissemination. What they choose to include, exclude or denigrate can make all the difference to the cognitive and operational capacities of their students as future citizens (e.g. Maher & Tetreault, 2001; Odora Hoppers, 2001).

DOI: 10.4324/9781003390930-5

In the past, in much of higher education women, as students and academics, have struggled for recognition (Christian-Smith & Kellor, 1999; Currie et al., 2002; Davies et al., 1994; Heilbrun, 2002; Kelly, 1985; Merrill, 1999; Morley, 1999), and women[1] have been overlooked in most sustainability programmes through being subsumed into the notion of "universalized people" (Braidotti et al., 1994; Buckingham-Hatfield, 2002; Gough, 1999a, 1999b, Salleh, 1997). However, women have a distinctive contribution to make to sustainability policy, pedagogy and research that needs to be foregrounded. This chapter discusses research into the absences of women's perspectives from sustainability policies, pedagogy and research and argues that ecofeminist pedagogies and research methodologies suggest new possibilities for the development of sustainability in higher education.

The ecofeminist movement has developed in parallel with the environment and environmental education movements since the 1970s, but there has been little dialogue between it and the other two. Chapter 24 of *Agenda 21* (UNCED, 1992) had as its overall goal, achieving active involvement of women in economic and political decision making, with emphasis on women's participation in national and international ecosystem management and control of environmental degradation. This perspective has been overlooked to date in most forms and sectors of education, so in this chapter I argue for the power and the promise of adopting such a perspective for the higher education context. At the technical level actions should include increasing proportions of women as decision makers in implementing policies and programmes for sustainable development, recognising women as equal members in workplaces both with respect to workloads and finance, and including women's perspectives in the content of higher education courses. However, it is equally important that ecofeminist perspectives inform research and pedagogies in higher education. In developing my arguments I focus on changing institutional and philosophical actions to make women's and men's lives in higher education more democratic, academic knowledge less partial and sustainability achievable. As Uma Narayan and Sandra Harding argue:

> It is worth recollecting that the deepest forms of sexism and androcentrism – the ones most difficult even to identify, let alone to eradicate, have not been those visible in the intentional actions of individuals (which is not to excuse such overt or covert sexism and androcentrism). It has not been sexist or androcentric motivations or prejudices of individuals – their false beliefs and bad attitudes – that has given women the most trouble. Rather, it has been the institutional, societal, and civilizational or philosophic forms of sexism and androcentrism that have exerted the most powerful effects on women's and men's lives – the forms least visible to us in our daily lives.
>
> (Narayan & Harding, 2000, pp. vii–viii)

Gender equality and sustainability

The Rio Declaration (1992, Principle 21) stated: "Women have a vital role in environmental management and development. Their full participation is therefore essential to achieve sustainable development". The recognition of women in the Rio Declaration was the result of lobbying by groups like the Women's Environment and Development Organisation (WEDO) in New York. As WEDO co-chair and former US Congress member Bella Abzug explains, "although we women are the vast majority of grassroots activists, very few of us are in positions of power, setting the priorities and making decisions on issues to be tackled nationally and internationally" (1991, p. 2). So their goals are political: "to encourage women's global solidarity and empowerment, to expand and deepen women's networks, to educate and inform, and to create a local, national and global capacity to act" (Abzug, 1991, p. 2). The Johannesburg Declaration on Sustainable Development (United Nations, 2002) expanded on strategies to achieve women's full participation in sustainable development and included strong commitments to ensuring women's empowerment, emancipation and gender equality within all activities related to enacting the implementation plan from the Summit.

Many texts in the past decade and more have argued that there is a close relationship between women and sustainable development (e.g. Braidotti et al., 1994; Merchant, 1992; Salleh, 1997; Shiva, 1989), and most recognise that, while gender equality is a prerequisite for sustainable development, the relationship is not a simplistic one. Recently Susan Buckingham-Hatfield (2002) explored ways in which women have, and have not, become more involved in environmental decision making since UNCED. She concluded that health care, education and economic status are prerequisites for women to be meaningfully involved in environmental decision making, and that structural inequalities are currently overwhelming progress towards sustainable development. Eradication of poverty is recognised by the UN as being essential for sustainable development, as is the importance of universal basic education. However, although women's literacy levels have increased in the past decade, "economic inequality has seen little change and gendered health difference remain significant" (Buckingham-Hatfield, 2002, p. 233).

Women's place in higher education

Historically universities have been uni-vocal places exclusively for white males of the ruling classes. When individual members of minority and marginal groups in society have been able to access the structures of knowledge and power represented by the academy, they have had to defend both their presence and expertise on a regular basis. Many authors have documented the university's long association with the development and legitimation of men's knowledge and scholarship, and the tenuous relationships between women and their legitimate claims to knowledge (e.g. Christian-Smith & Kellor,

1999; Currie et al., 2002; Davies et al., 1994; Gillett, 1981; Heilbrun, 2002; Kearney & Ronning, 1996; Kelly, 1985; Luke & Gore, 1992, Morley, 1999).

Although universities had been in existence for centuries, it was only in the latter half of the nineteenth century that women were able to enter the academy as students. Margaret Gillett (1981) wrote of the experiences of women seeking admission to McGill University in Canada:

> She could not defend herself, she could not argue for nor claim intellectual equality because she could not know what it was that she did not know. She could not even be a party to deciding whether or not she ought to know. Thus, even though she was high on a pedestal, the politics of ignorance kept her in her subordinate place.
>
> (in Kelly, 1985, p. 4)

Although women are still confronting issues of the politics of ignorance in many places around the world,[2] at the university level in English-speaking nations this politics was being substantially challenged in the late nineteenth century. The first woman to graduate from a university in the British "Empire" was in Canada in 1875. In 1877 the first woman graduated in New Zealand, and in 1883 the first woman graduated from the University of Melbourne, the first in Australia. Although women had been permitted to enrol in all courses (except in medicine) at the University of London from 1878, they were not admitted as full members of either Oxford or Cambridge universities until 1920 and 1948 respectively. The universities took the women's fees, allowed them to sit the usual examinations but they would not formally confer degrees upon them: "Conservatives feared that the full admission of women to Oxford and Cambridge would erode the 'distinctively manly spirit' on which their fame rested" (Kelly, 1985, p. 2). Women's experiences at these universities during this period are eloquently portrayed by Virginia Woolf ([1929] 1977) in *A Room of One's Own*, and more recently by Ian McEwan (2002). The content of university courses studied by women is also an issue:

> In the early days of university education for women, opinions were divided between those who thought women should conform to the same educational patterns as men and those who felt strongly that there should be a differential educational pattern – women should study subjects requiring so-called "feminine aptitudes". . . Had the women not pursued a similar course to that taken by the men, it is doubtful whether their qualifications would have been accepted in the community. From that time forward, the University has maintained a common curriculum in all subjects.
>
> (Blackwood, in Kelly, 1985, p. vii)

The early women graduates were proud to have entered the halls of academia on men's terms, but this unquestioning acceptance of men determining what it was worthwhile and necessary to know, and constructing and administering rules which perpetuated their own power, was challenged by feminists of the late twentieth century who drew attention to the socially constructed nature of knowledge and curriculum, and the need to change its form and focus to recognise the "politics of knowledge". For example, Sandra Harding notes,

> Feminists have argued that traditional epistemologies, whether inten- tionally or unintentionally, systematically exclude the possibility that women could be 'knowers' or *agents of knowledge*; they claim that the voice of science is a masculine one; that history is written from only the point of view of men (of the dominant race and class); that the subject of a traditional sociological sentence is always assumed to be a man. They have proposed alternative theories that legitimate women as knowers.
>
> (Harding, 1987, p. 3, emphasis in original)

More recently, Harding (1991, 1998; Narayan & Harding, 2000) has argued that gender is not the only influence on the production of knowledge and that class, race, sexual orientation, culture, ethnicity, age and religion are also significant. These issues are taken up later in this chapter.

Another aspect of women's place in higher education is their rank in the institutions. As Table 4.1 illustrates, women are still confronted by a "glass ceil- ing" and are few in number at academic leadership (associate professor and professor) levels (13.8% of female academics compared with 36% of male aca- demics, and only 22% of all Australian academics are female[3]).

If higher education is to contribute effectively to sustainability, a dual agenda of advancing women as academic leaders and renewal of higher educa- tion institutions is essential. As Sheryl Bond (1996, p. 51) argues, women are needed in senior decision-making positions, not just for numerical balance but

> to diversify and strengthen the leadership of the academy [because] women . . . by virtue of their own life experience, embrace the experi- ence and knowledge of women, as well as men . . . sufficient numbers

Table 4.1 Distributions of women and men across ranks: Australian universities 1996 (from Anderson et al., 1997, Table 6.13)

	Female		*Male*		*Total*	
Professor/Prof Fellow	115	4.5	1504	16.6	1619	13.9
Assoc Prof/Rdr/Snr Res Fellow	237	9.3	1756	19.4	1993	17.1
Snr Lect/Fellow/Snr Res Fellow	687	26.9	3020	33.3	3707	31.9
Lecturer/Research Fellow	1466	57.4	2590	28.6	4056	34.9
Other	47	1.8	201	2.2	248	2.1
Total	2552	100	9071	100	11623	100

of women are needed to generate change in institutional policies and practices, in who is admitted to the academy, what is taught, the methodologies of teaching and learning, as well as the definition of questions which excite the mind and warrant investigation.

(Bond, 1996, p. 51)

There is a gender dimension in all academic activity – research, teaching and administration – which goes beyond a disciplinary discussion and traverses all fields in academia. There also appears to be a correlation between increasing the number of women in academe and higher education institutions focusing on gender, ethnicity and culture (Maher & Tetreault, 2001). Thus it is vital that these dimensions (and others) are addressed if we are to achieve meaningful social change and achieve sustainability.

An ecofeminist perspective

The French writer Francoise d'Eaubonne coined the term *ecofeminisme* in her 1974 book *Le féminisme ou la Mort*, in which she "called upon women to lead an ecological revolution to save the planet. Such an ecological revolution would entail new gender relations between women and men and between humans and nature" (Merchant, 1992, p. 184). There are many paths into ecofeminism – from different forms of feminism (radical, Marxist, socialist, liberal), from environmentalism, from the study of political theory and history and from exposure to nature-based religion. Indeed, just as there is not one feminism, there is not one ecofeminism. However, the origins and trajectories of ecofeminism are contested, and much ecofeminist energy has been consumed in arguing the various ideological positions. What is clear is that ecofeminism is not a monolithic, homogeneous ideology (which causes difficulties for some scholars), but it nevertheless provides a very different notion of what constitutes political change: "while ecofeminism recognises the severity of the crisis, it also recognises that the methods we choose in dealing with the problems must be life affirming, consensual and non-violent" (Diamond & Orenstein, 1990, p. xii). This progressive and critical social theory orientation of ecofeminism[4] (which it shares with the feminist and environment movements) is important in reshaping the basic socio-economic relations and the underlying values of society and institutions.

Since the term "ecofeminism" was coined in 1974 and named as a grassroots women-initiated environment movement, its meaning has expanded from just being concerned with ecological feminism to now recognising "that there are important connections between how one treats women, people of color, and the underclass on one hand and how one treats the nonhuman natural environment on the other" (Warren, 1997a, p. xi). This enriched vision of ecofeminism is inspiring the research and other work of scholars in a variety of academic and vocational fields including anthropology, biology, chemical engineering, communication studies, education, environmental studies, literature, political science, recreation and leisure studies, and sociology.

There are many reasons for taking the philosophical basis of the ecofeminist movement seriously. As Warren (1997b, pp. 13–14) argues, there are significant empirical data which suggest that:

1. the historical and causal significance of ways in which environmental destruction disproportionally affects women and children;
2. the epistemological significance of the "invisibility of women", especially women who know (e.g. about trees), for policies which affect both women's livelihood and ecological sustainability;
3. the methodological significance of omitting, neglecting, or overlooking issues about gender, race, class, and age in framing environmental policies and theories;
4. the conceptual significance of mainstream assumptions, e.g. about rationality and the environment, which may inadvertently, unconsciously, and unintentionally sanction or perpetuate environmental activities, with disproportionately adverse effects on women, children, people of color, and the poor;
5. the political and practical significance of women-initiated protests and grassroots organising activities for both women and the natural environment;
6. the ethical significance of empirical data for theories and theorising about women, people of color, children, and nature;
7. the theoretical significance of ecofeminist insights for any politics, policy, or philosophy; and
8. the linguistic and symbolic significance of language used to conceptualise and describe women and nonhuman nature.

These data should form the basis for reforming the policies and practices of teaching, research and administration in higher education if we are to achieve sustainability. However, before moving on, I believe that it is important to confront the misreading of ecofeminism as "essentialist", that is, that women are closer to nonhuman nature.[5] Many ecofeminists (including those mentioned in footnote 5) have argued that ecofeminism is not essentialist. For example, in one of the clearer explanations, Braidotti et al. argue that we must recognise that

> women and nature are simultaneously subjugated, and that this subjugation takes historically and culturally specific forms. If women take themselves seriously as social agents and as constitutive factors in this process, their praxis to end this double subjugation can be rooted not so much in women's equation with nature but in taking responsibility for their own lives and environment
>
> (Braidotti et al., 1994, p. 75).

In this way we pursue a critical social theory within specific cultural, temporal and physical positions as a body of ideas, which avoids essentialism – but we must be vigilant and not reify "women's ways of knowing" (Gilligan, 1982).

Higher education, sustainability and feminist research and pedagogies[6]

Elsewhere I have discussed the significance of the absence of women from the international gatherings that formalised conceptions of environmental education, and the need to recognise them in environmental education research and pedagogy (Gough, 1999a), so I will not repeat those arguments here. Although my personal disposition is towards critical and poststructuralist feminist research and pedagogy, all feminist pedagogies and research methodologies can be applied in the development of sustainability in higher education. We can do many types of research and teaching which puts the social construction of gender at the centre of the inquiry, and we need more stories from women's lives relating to environments that we can use in the development of sustainability in higher education. Like Donna Haraway,

> I want to argue for a doctrine and practice of objectivity that privileges contestation, deconstruction, passionate construction, webbed connections, and hope for transformation of systems of knowledge and ways of seeing. But not just any partial perspective will do; we must be hostile to easy relativisms and holisms built out of summing and subsuming parts.
> (Haraway, 1991, pp. 191–192)

From an ecofeminist perspective there are many challenges for higher education institutions' research, teaching and administration if they are to have a role in sustainability. For example, Brown and Switzer (1991, p. 16) note that women are less likely to have scientific or economic training than men and, consequently, have less influence in developing curricula which give high priority to issues of importance to women, such as reduction of toxic wastes and information on safety standards. They also note that there is a need to compensate for the effects on research and teaching of the relative absence of women and women's interests from the professions of environmental science and economics: "this absence has meant that many questions on ecologically sustainable development from the fields of health, welfare, household management and social policy have neither been investigated nor included in environmental education" (Brown & Switzer, 1991, p. 16).

Feminist research and pedagogy in environmental education is generally still at the stage of attempting to "add women" to traditional analyses rather than engaging in the more distinctive features of feminist research and pedagogy. In many ways environmental education is still mainly at Stage 1 of Kreinberg and Lewis' (1996) model of curriculum change for science education (adapted for environmental education) with only a few moving to Stage 2 and beyond:

Stage 1: Absence of women in environmental education not noticed;
Stage 2: The search for the missing women in environmental education;
Stage 3: Why are there so few women in environmental education?;
Stage 4: Studying women's experience in environmental education;

Stage 5: Challenging the paradigms of what environmental education research and pedagogy are;

Stage 6: A transformed, gender-balanced environmental education curriculum and research agenda.

What makes feminist research distinctive, according to Harding (1987), is that it moves beyond "adding women" and opens up new empirical and theoretical resources (women's experiences); new purposes of social science research (for women); and new subject matter of inquiry (locating the researcher in the same critical plane as the overt subject matter).

This distinctiveness is more than a concern with research methods; it is a concern with epistemologies and ontologies as well as methodologies, and this has significant implications for both research and pedagogy in education for sustainability as we move towards Kreinberg and Lewis' stages 5 and 6.

New empirical and theoretical resources (women's experiences)

Traditionally, environmental education has analysed only male experiences or has constructed universalised subjects that are not distinguished as male or female. Yet there is no universal *man*, only culturally, racially, socio-economically (and so on) different men and women with fragmented identities. Environmental education has not addressed areas of women's experiences and knowledge which are equally resources for environmental education research processes. "One distinctive feature of feminist research is that it generates its problematics from the perspective of women's experiences" and "uses these experiences as a significant indicator of the reality against which hypotheses are tested" (Harding, 1987, p. 7).

> All of us live in social relations that naturalize, or make appear intuitive, social arrangements that are in fact optional; they have been created and made to appear natural by the power of the dominant groups. Thus, it is not necessary to have any *particular* form of human experience in order to learn how to generate less partial and distorted belief from the perspective of women's lives. It is 'only' necessary to learn how to overcome – to get a critical, objective perspective on – the 'spontaneous consciousness' created by thought that begins in one's dominant social location.
> (Harding, 1991, p. 287, emphases as in original)

Education for sustainability needs to adopt this feature for its research and pedagogy to address the current partial and distorted understandings of ourselves and the world around us that are produced by the traditional approaches to environmental education.

New purposes of social science research (for women)

"In the best of feminist research, the purposes of research and analysis are not separable from the origins of research problems" (Harding, 1987, p. 8).

Therefore, if the focus of education for sustainability begins with what is problematic from the perspective of women's experience then it is designed for women (as well as for the environment).

New subject matter of inquiry (locating the researcher in the same critical plane as the overt subject matter)

The best feminist analysis "insists that the inquirer her/himself be placed in the same critical plane as the overt subject matter", that is, "the class, race, culture, and gender assumptions, beliefs, and behaviors of the researcher her/himself must be placed with the frame of the picture that she/he attempts to paint" (Harding, 1987, p. 9). It is also important to critically examine "the differential ground for the scholar's, the teacher's, and the students' knowledge and authority, in order to put them all into relation with one another" (Maher & Tetreault, 2001, p. 209). Thus, rather than appearing as an anonymous voice, the researcher and teacher is a real individual with specific and locatable interests. In the case of education for sustainability, this means that the whole focus for the research and pedagogy in higher education will need to change.

Feminist research provides an opportunity to re-vision the world – "to know it differently than we have ever known it; not to pass on a tradition but to break its hold on us" (Rich, 1990, p. 484 in Crotty, 1998, p. 182) – by transforming common methodologies and methods and working against the "limiting of human possibility through culturally imposed stereotypes, lifestyles, roles and relationships" (Crotty, 1998, p. 182). The adoption of the three distinctive features of feminist research outlined above will lead to better ("less partial, less distorted") education for sustainability research stories and pedagogical practices that will benefit the field as a whole. Already some academics are moving in this direction. For example, several of the authors in Kearney and Ronning (1996) have adopted some of these approaches, particularly the university curriculum in development studies, which includes a gender dimension in Brazil (D'Avilo-Neto et al., 1996), and Lotz-Sisitka and Burt (2002) discuss their experiences in writing environmental education research texts. Other examples can be found for environmental studies and science in the work of Whitehouse and Taylor (1996) and Gough (2001).

The features of feminist research and an ecofeminist perspective outlined above resonate particularly well with recent work on indigenous knowledge systems and higher education being conducted in South Africa by, for example, Catherine Odora Hoppers (2001). She discusses the subjugation of indigenous knowledge systems by the past regimes and the need for "the reconstruction of knowledge, the critical scrutiny of existing paradigms and the epistemological foundations of existing academic practice, and the identification of the limitations they impose on creativity" (Odora Hoppers, 2001, p. 73) if South African society is to find sustainable and inclusive ways forward. As I have argued throughout this chapter, an ecofeminist perspective on sustainability includes not only women but also recognises that class, race, sexual orientation, culture, ethnicity, age and religion are significant. We need to support these new perspectives through higher education.

Conclusion

Because institutional sexism and androcentrism "have exerted the most powerful effects on women's and men's lives" (Narayan & Harding, 2000, p. viii), gender issues in higher education warrant urgent attention if we are to achieve sustainability. This will involve more than "adding women"; cultural attitudes and practices are key impediments to women's equal access to education and positions in higher education, and to getting sustainability onto the higher education agenda.

Kearney (1996) suggests a strategy for change based on four key premises: the promotion of feminine leadership in the academy, the mainstreaming of the gender dimension in the curriculum, continued research on barriers to women's equality, and acknowledgement of the gender dimension of (sustainable) development. I have discussed each of these issues here to varying degrees.

However, we also need to address the impediments posed by the new managerialism that pervades higher education, with its focus on corporate mission statements, goals, monitoring procedures and performance measures. According to Morley (1999, p. 28), the new managerialism promotes three "Es": economy, efficiency and effectiveness. Two other "Es" are noticeably absent – equity and environment – yet these are the areas of most concern for this chapter (and book). Social justice values are perceived as irrelevant to management theories based on marketisation, consumerism, individual rights and choice, and "equity is off the agenda" (Ball, 1994, p. 125). The challenge for us is to not only get the other "Es"– equity and environment – onto the agenda again but to have them prioritised as part of corporate mission statements, goals, monitoring procedures and performance measures. This reform will require cultural and ideological change too, not simply structural or technical tweaking.

Finally, it may be useful to think of the reforms in terms of a critical and conserving agenda that reasserts traditional university values from feminist and sustainability perspectives. Currie et al. (2002) include in these values: democratic collegiality, professional autonomy and integrity, critical dissent and academic freedom, and the public interest value of universities; however, not all of these are equally relevant in developing ecofeminist perspectives in higher education. Democracy is an active vehicle for social justice, and this is central in our concerns around ecofeminism and sustainability. So too is academic freedom as a key legitimating function of the university. Too often the new managerialist university displays a lack of respect for staff and their intellectual skills. It is the public interest value of higher education institutions that is paramount as part of the university's role "in developing active and critical-thinking citizens. . . [with] tolerance and compassion. . . [who] work for social justice. . . [and] the social transformation of society" (Currie et al.,

2002, p. 190). The South Africa government has already recognised that there is a role for universities in their social reconstruction agenda – including "the promotion of a critical citizenry" and addressing "its particular development challenges" (South African Council on Higher Education Size and Shape Task Team, 2000, p. 25, as cited in Currie et al., 2002, p. 190). While this is not as consistent with an ecofeminist and sustainability framework as it could be, it is a step in the right direction – towards developing an ecofeminist perspective to sustainability in higher education.

Biography

Dr Annette Gough is Associate Professor in Science and Environmental Education in the Faculty of Education at Deakin University, Melbourne, Australia. Here she lectures and researches in environmental education, with a particular research interest in feminist and poststructuralist methodologies. She is also an adjunct professor at the University of Victoria, BC, Canada, and a visiting professor at Rhodes University, South Africa. She was the first female president of the Australian Association for Environmental Education (1984–1986) and is a life fellow of the Association. She has published extensively on environmental education in many international publications, including her book *Education and the Environment: Policy, Trends and the Problems of Marginalisation* (1997, ACER Press). She was managing editor of the *Australian Journal of Environmental Education* and is an advisory editor for the *Canadian Journal of Environmental Education*, the *Southern African Journal of Environmental Education* and the *Journal of Biological Education*.

Notes

1 While I recognise that women are one of many marginalised groups in society, and discuss the importance of indigenous knowledge systems later in this chapter, the main emphasis in this chapter is on the contribution of ecofeminist perspectives to sustainability in higher education.

2 For example, 66% of the world's illiterates are women (Kearney, 1996, p. 1), and in some areas, such as Pakistani villages, women are 100% illiterate (del Nevo, 1993).

3 Similar figures apply in other English speaking countries – see Kearney and Ronning (1996) and Merrill (1999).

4 Ariel Salleh (1997, p. 69) notes that a number of prominent ecofeminists, such as Carolyn Merchant (1992) and Vandana Shiva (1989), have been influenced by Marx's work.

5 Accusations of essentialism are an obvious concern of most ecofeminists: the indexes of texts such as Braidotti et al. (1994), Salleh (1997) and Warren (1994, 1997c) all contain entries for "essentialism".

6 In this section I use the terms "environmental education" and "education for sustainability" interchangeably. This is because the majority of work to date in this field has been known as environmental education rather than education for sustainability. However, I believe that an important component of the shift from environmental education to education for sustainability as a terminology is the incorporation of ecofeminist perspectives into the latter where they have definitely been absent from the former (see Gough, 1999a, 1999b). I found aspects of Bill Scott's (2002) discussion of sustainability and learning useful in this regard, although I found his silence on the gender dimension alarming.

References

Abzug, B. (1991). Women want an equal say in UNCED. *The Network '92*, 9(2).

Anderson, D., Arthur, R. & Stokes, T. (1997). *Qualifications of Australian Academics Sources and Levels 1978–1996*. Canberra, ACT: Evaluations and Investigations Program, Department of Employment, Education, Training and Youth Affairs. Retrieved October 13, 2002, from: www.detya.gov.au/archive/highered/eippubs/eip97-11/front.htm

Ball, S. (1994). *Education Reform – A Critical and Post-Structuralist Approach*. Milton Keynes: Open University Press.

Bond, S.L. (1996). The experience of feminine leadership in the academy. In M.-L. Kearney & A.H. Ronning (Eds.), *Women and the University Curriculum: Towards Equality, Democracy and Peace* (pp. 35–52). London: Jessica Kingsley Publishers & Paris: UNESCO.

Braidotti, R., Charkiewicz, E., Hausler, S. & Wieringa, S. (1994). *Women, the Environment and Sustainable Development: Towards a Theoretical Analysis*. London: Zed Books/INSTRAW.

Brown, V.A. & Switzer, M.A. (1991). *Engendering the Debate: Women and Ecologically Sustainable Development*. Canberra, ACT: Office of the Status of Women, Department of the Prime Minister and Cabinet.

Buckingham-Hatfield, S. (2002). Gender equality: A prerequisite for sustainable development. *Geography*, 18(3), 227–233.

Christian-Smith, L.K. & Kellor, K.S. (Eds.). (1999). *Everyday Knowledge and Uncommon Truths: Women of the Academy*. Boulder, CO: Westview Press.

Crotty, M. (1998). *The Foundations of Social Research: Meaning and Perspective in the Research Process*. St Leonards, NSW: Allen & Unwin.

Currie, J., Thiele, B. & Harris, P. (2002). *Gendered Universities in Globalized Economies: Power, Careers, and Sacrifices*. Lanham: Lexington Books.

Davies, S., Lubelska, C. & Quinn, J. (Eds.). (1994). *Changing the Subject: Women in Higher Education*. London: Taylor & Francis.

D'Avilo-Neto, M.I., Baptista, C.A. & Calicchio, R. (1996). Women and development: Perspectives and challenges within the university curriculum. In M.-L. Kearney & A.H. Ronning (Eds.), *Women and the University Curriculum: Towards Equality, Democracy and Peace* (pp. 69–90). London: Jessica Kingsley Publishers & Paris: UNESCO.

del Nevo, M. (1993, May). Learning to heal. *New Internationalist*, 243, 3.

Diamond, I. & Orenstein, G.F. (Eds.). (1990). *Reweaving the World: The Emergence of Ecofeminism*. San Francisco: Sierra Club Books.

Gillett, M. (1981). *We Walked Very Warily: A History of Women at McGill*. Montreal: Goden Press.

Gilligan, C. (1982). *In a Different Voice*. Cambridge, MA: Harvard University Press.

Gough, A. (1999a). Recognising women in environmental education pedagogy and research: Towards an ecofeminist poststructuralist perspective. *Environmental Education Research*, 5(2), 143–161.

Gough, A. (1999b). The power and the promise of feminist research in environmental education. *Southern African Journal of Environmental Education*, 19, 28–39.

Gough, A. (2001). Pedagogies of science (in)formed by global perspectives: Encouraging strong objectivity in classrooms. In J.A. Weaver, M. Morris & P. Appelbaum (Eds.), *(Post) Modern Science (Education): Propositions and Alternative Paths* (pp. 275–300). New York: Peter Lang.

Haraway, D.J. (1991). *Simians, Cyborgs, and Women.* London: Free Association Books.

Harding, S. (1987). Is there a feminist method? In S. Harding (Ed.), *Feminism and Methodology* (p. 114). Bloomington: Indiana University Press.

Harding, S. (1991). *Whose Science? Whose Knowledge? Thinking for Women's Lives.* Ithaca, NY: Cornell University Press.

Harding, S. (1998). *Is Science Multicultural? Postcolonialisms, Feminisms, and Epistemologies.* Bloomington: Indiana University Press.

Heilbrun, C.G. (2002). *When Men were the Only Models We had.* Philadelphia, PA: University of Pennsylvania Press.

Kearney, M.-L. (1996). Women, higher education and development. In M.-L. Kearney & A.H. Ronning (Eds.), *Women and the University Curriculum: Towards Equality, Democracy and Peace* (pp. 1–33). London: Jessica Kingsley Publishers & Paris: UNESCO.

Kearney, M.-L. & Ronning, A.H. (Eds.). (1996). *Women and the University Curriculum: Towards Equality, Democracy and Peace.* London: Jessica Kingsley Publishers & Paris: UNESCO.

Kelly, F. (1985). *Degrees of Liberation: A Short History of Women in the University of Melbourne.* Parkville, Victoria: The Women Graduates Centenary Committee of the University of Melbourne.

Kreinberg, N. & Lewis, S. (1996). The politics and practice of equity: Experiences from both sides of the Pacific. In L.H. Parker, L.J. Rennie & B.J. Fraser (Eds.), *Gender, Science and Mathematics: Shortening the Shadow* (pp. 177–202). Dordrecht, The Netherlands: Kluwer Academic Publishers.

Lotz-Sisitka, H. & Burt, J. (2002). Being brave: Writing environmental education research texts. *Canadian Journal of Environmental Education,* 7(1), 132–151.

Luke, C. & Gore, J. (1992). Women in the academy: Strategy, struggle, survival. In C. Luke & J. Gore (Eds.), *Feminisms and Critical Pedagogy* (pp. 192–210). New York: Routledge.

Maher, F.A. & Tetreault, M.K.T. (2001). *The Feminist Classroom: Dynamics of Gender, Race, and Privilege.* Lanham: Rowman & Littlefield.

McEwan, I. (2002). *Atonement.* London: Vintage.

Merchant, C. (1992). *Radical Ecology: The Search for a Livable World.* New York: Routledge.

Merrill, B. (1999). *Gender, Change and Identity: Mature Women Students in Universities.* Aldershot: Ashgate.

Morley, L. (1999). *Organising Feminisms: The Micropolitics of the Academy.* New York: St Martin's Press.

Narayan, U. & Harding, S. (2000). Introduction. In U. Narayen & S. Harding (Eds.), *Decentering the Center: Philosophy for a Multicultural, Postcolonial, and Feminist World* (pp. vii–xvi). Bloomington: Indiana University Press.

Odora Hoppers, C.A. (2001). Indigenous knowledge systems and academic institutions in South Africa. *Perspectives in Education,* 19(1), 73–85.

Rich, A. (1990). When we dead awaken: Writing as re-vision. In D. Bartholomae & A Petrosky (Eds.), *Ways of Reading: An Anthology for Writers* (pp. 482–496). Boston: Bedford Books of St Martin's Press.

The Rio Declaration on Environment and Development. (1992). Reprinted in *Earth Ethics,* 4(1), 9–10.

Salleh, A. (1997). *Ecofeminism as Politics: Nature, Marx and the Postmodern.* London: Zed Books.

Scott, W.A.H. (2002, April 25). *Sustainability and Learning: What Role for the Curriculum? University of Bath Inaugural Lecture.* Retrieved October 13, 2002, from: www.bath.ac.uk/cree/scott.htm

Shiva, V. (1989). *Staying Alive: Women, Ecology and Development.* London: Zed Books.

South African Council on Higher Education Size and Shape Task Team. (2000). *Towards a New Higher Education Landscape.* Pretoria: Council on Higher Education.

United Nations. (2002, August 26–September 4). *Report of the World Summit on Sustainable Development.* Johannesburg South Africa. A/CONF.199/20*. Retrieved April 7, 2003, from: www.johannesburgsummit.org/html/documents/summit_docs/131302_wssd_report_reissued.pdf

United Nations Conference on Environment and Development (UNCED). (1992). *Agenda 21.* Rio de Janeiro, Brazil: UNCED. Final, advanced version as adopted by the Plenary on 14 June 1992.

Warren, K.J. (Ed.). (1994). *Ecological Feminism.* New York: Routledge.

Warren, K.J. (1997a). Introduction. In K.J. Warren (Ed.), *Ecofeminism: Women, Culture, Nature* (pp. xixvi). Bloomington: Indiana University Press.

Warren, K.J. (1997b). Taking empirical data seriously. In K.J. Warren (Ed.), *Ecofeminism: Women, Culture, Nature* (pp. 3–20). Bloomington: Indiana University Press.

Warren, K.J. (Ed.). (1997c). *Ecofeminism: Women, Culture, Nature* (pp. xi–xvi). Bloomington: Indiana University Press.

Whitehouse, H. & Taylor, S. (1996). A gender inclusive curriculum model for environmental studies. *Australian Journal of Environmental Education,* 12, 77–83.

Woolf, V. ([1929] 1977). *A Room of One's Own.* London: Collins.

5 Generating a gender agenda for environmental education

Annette Gough

Gough, A. (2013). Researching differently: Generating a gender agenda for research in environmental education. In R. B. Stevenson, M. Brody, J. Dillon, & A. Wals (Eds.), *International Handbook of Research of Environmental Education* (pp. 375–383). New York: Routledge for the American Educational Research Association. doi:10.4324/9780203813331.ch35

Abstract

This chapter argues that to date, environmental education research has rarely addressed areas of different women's experiences and knowledges, which means many useful insights have not been adequately pursued but that using feminist research strategies to generate a gender agenda provides the basis for different ways of thinking and perceiving in environmental education research. Thus, the emphasis here is on women's experiences and knowledge and the perspective these bring to environmental education research. In particular, the chapter explains why a feminist perspective is important in environmental education and what characterises feminist research, it also discusses what feminist research has been undertaken to date in environmental education, and the potential for feminist research in environmental education.

Introduction

The need to change the perspectives from which we think and act is one of the foundations of environmental education. As Foucault (1990) wrote, "There are times in life when the question of knowing if one can think differently than one thinks, and perceive differently than one sees, is absolutely necessary if one is to go on looking and reflecting at all" (p. 8). In this instance, using feminist research strategies to generate a gender agenda provides the basis for different ways of thinking and perceiving in environmental education research.

The major contributions of feminist research, in all its many forms, have been to raise the question of epistemological claims, such as who can be an agent of knowledge, what counts as knowledge, what constitutes and validates knowledge, and what the relationship should be between knowing and being. Feminist questions put the social construction of gender at the center of research

DOI: 10.4324/9781003390930-6

(Lather, 1991) for "what 'grounds' feminist standpoint theory is not women's experiences but the view from women's lives" (Harding, 1991, p. 269).

Feminism enables people to revision their world – "to know it differently than we have ever known it; not to pass on a tradition, but to break its hold on us" (Rich, 1972, p. 19). As Heilbrun (1999) noted,

> Women began to portray the new possibilities that, as a result of feminism, they found themselves confronting. They began to question . . . all strictures about women and about the institutions in which women now, in even greater numbers, and in a state of awakening, found themselves.
>
> (p. 8)

Ecological feminists have embraced personal and political action to "fully engage in the interweaving of humour, irony, grace, resistance, struggle and transformation" (Sandilands, 1999, p. 210) to envision a more democratic future for all. And, occasionally, environmental education researchers whose work is informed by (at least aspects of) feminism are publishing their work (see, e.g., Barrett, 2005; Bodzin et al., 2010; Carrier, 2007; Di Chiro, 1987; Fawcett, 2000, 2002; Fontes, 2002; Gough, 1997, 1999a, 1999b, 1999c, 2004; Gough & Gough, 2003; Gough & Whitehouse, 2003; Hallen, 2000; Lotz-Sisitka & Burt, 2002; Lousley, 1999; Malone, 1999; McKenzie, 2004, 2005; Newbery, 2003; Russell, 2003; Russell & Bell, 1996; Russell et al., 2002; Wane & Chandler, 2002). But, despite these in-roads, the subject of gender remains marginal to much environmental education research.

Many researchers still consider a "human" subjectivity homogeneous, ungendered, and unproblematic, whereas a vast edifice of sociological research reveals the opposite to be the case. Environmental education research remains bound with traditional epistemological frameworks of scientific research, which have, in Harding's (1987) words, "whether intentionally or unintentionally, systematically exclude(d) the possibility that women could be 'knowers' or *agents of knowledge*" (p. 3, emphasis in original). For example, writings on significant life experience research can be critiqued as remaining blind to gendered subjectivities (Gough, 1999b).

Addressing the balance is simply not a matter of "adding women" to traditional analyses. Rather, what is needed is a transformative process in which new empirical and theoretical resources are opened up to reveal new purposes and subjects for inquiry. And what needs to come under scrutiny is the implicit constitution of the assured, homogeneous, and universalized human subject of much environmental research. "Human" identity as constituted through positivist research regimes is not inclusive of all the different ways of being in the world.

Much past (and present) environmental education research has analyzed only male experiences or has constructed universalized subjects, which are not distinguished as male or female. Yet, there is no universal "Man" who acts as a powerful agent on an equally symbolic "Environment" – except perhaps in the imaginations of writers who reproduce these discourses. "Man" is not a term that is logically inclusive of women. Early formulations of environmental education, such as the IUCN (1970) definition, referred to as "the interrelatedness among man, his culture, and his biophysical surroundings" (as cited in Linke, 1980, pp. 26–27). Although more recent environmental education literature is gender neutral in its language, this too is a problem, as neutral voice is still interpreted as male by readers of both genders. As Cherryholmes (1993) argues, "texts that deny gender present themselves as generic. They pretend to speak the truth and truth is gender-neutral. Authoritative texts are distanced, objective, have a single voice (otherwise they would not be authoritative), are value-neutral, dispassionate and controlling" (p. 10). Perhaps the shackles of the past are proving difficult to shrug off, but the practice of creating gender-blind binaries is exclusive of lived experience.

In reality, we have culturally, racially, socioeconomically, and sexually (and so on) different people with fragmented identities whose experiences and understandings can only be constituted through the lenses of subjectivity. Given there is growing recognition that there is no one way of looking at the world, no "one true story"; rather, a multiplicity of stories, then we should look at a multiplicity of strategies for policies, pedagogies, and research in environmental education. These strategies should be strategies that are neither universal nor part of the dominant discourse, but strategies that are from the lives of the colonized and marginalized, including the lives of women.

However, to date, environmental education research has rarely addressed areas of different women's experiences and knowledges, which means many useful insights have not been adequately pursued. Environmental education research has ignored other aspects of human identities too, but these are beyond the scope of this chapter. Rather, the emphasis here is on women's experiences and knowledge and the perspective these bring to environmental education research.

Why is a feminist perspective in environmental education important?

Developing a feminist perspective in environmental education is important, because the vast majority of work in environmental education to date has been concerned with universalized subjects, rather than recognizing multiple subjectivities. It is time that we start generating different ways of knowing and seeing environments so we might understand human relationships with them better. As Brown and Switzer (1991) argued nearly two decades ago,

Women and men contribute to maintaining environmental, economic and social sustainability in distinctive ways. For women these contributions are made through:

- their public roles as the majority of the workforce in the health, education, welfare and service industries;

- their private roles as care-givers, farm managers, educators of children, and the principal purchasers of food and consumer goods; and
- the many public (paid) and private (unpaid) arenas where women have a major responsibility for the management of change and the transmission of social values.

(p. iv)

Such a perspective is neither intended to essentialize women as caretakers of the earth's household, obsessed with green cleaners, nor intended to cast women as symbols of nature (for further discussion of these aspects, see, e.g., MacGregor, 2006; Warren, 1997). Rather, the intention in developing a feminist perspective in environmental education research is to recognize the complexity of human roles and relationships with respect to environments and that there are multiple subjectivities and multiple ways of knowing and interacting with environments that cannot be encapsulated within the notion of universalized subjects.

Most importantly, by pursuing a feminist perspective in environmental education research, we are able to construct "less partial, less distorted" (Harding, 1991) accounts of environments. Such a pursuit is also consistent with feminist praxis – a term that recognizes "a continuing feminist commitment to a political position in which 'knowledge' is not simply defined as 'knowledge *what*' but also as 'knowledge *for*'" (Stanley, 1990, p. 15, emphasis in original). It indicates rejection of the theory/research divide, "seeing these as united manual and intellectual activities which are symbiotically related (for all theorising requires 'research' of some form or another)" (p. 15). And it centers interest on methodological/epistemological concerns: "'how' and 'what' are indissolubly interconnected and . . . the shape and nature of the 'what' will be a product of the 'how' of its investigation" (p. 15). A central concern of feminist praxis is, thus, the reconstituting of knowledge. As Spender (1985) argues, "at the core of feminist ideas is the crucial insight that there is no one truth, no one authority, no one objective method which leads to the production of pure knowledge" (p. 5). I, therefore, argue that, through feminist research in environmental education, we will reconstitute knowledge about environments, and for environments.

Evidence that men and women do think differently about environmental issues is apparent in Brown's (1995) comparison of Australian women's priority concerns about the environment with a survey of issues prioritized in the scientific literature. Similarly, Zelezny et al.'s (2000) review of 32 research studies on gender and environmental attitudes and behaviors between 1988 and 1998 found that, in most studies, females reported more proenvironmental behaviors than males and females expressed greater environmental concern.

That women have a different role with respect to the environment was recognized nearly two decades ago in *Agenda 21*, the report from the 1992 United Nations Conference on Environment and Development (UNCED), but there was a lack of reciprocity between the "Global Action for Women Towards Sustainable and Equitable Development" and the "Promoting

Education, Public Awareness, and Training" chapters in *Agenda 21* (UNCED, 1992, Chapters 24 and 36, respectively). The "Women" chapter has as its overall goal achieving active involvement of women in economic and political decision making, with emphasis on women's participation in national and international ecosystem management and control of environmental degradation. One of its objectives for national governments was:

> To assess, review, revise and implement, where appropriate, curricula and other educational material, with a view to promoting the dissemination to both men and women of gender-relevant knowledge and valuation of women's roles through formal and non-formal education.
> (UNCED, 1992, para. 24.2(e))

Other objectives addressed topics such as increasing the proportion of women decision makers, eliminating obstacles to women's full participation in sustainable development, achieving equality of access to opportunities for education, health, and so on for women, equal rights in family planning, and prohibiting violence against women. The activities for governments related to such objectives are broadly concerned with:

- achieving equality of opportunity for women, such as by eliminating illiteracy;
- increasing proportions of women as decision makers in implementing policies and programs for sustainable development; and
- recognizing women as equal members of households, both with respect to workloads and finance.

Consumer awareness is particularly mentioned, as are "programmes to eliminate persistent negative images, stereotypes, attitudes and prejudices against women through changes in socialization patterns, the media, advertising, and formal and nonformal education" (UNCED, 1992, para. 24.3(i)). Here, women's contributions to society are recognized and valued as something different, rather than assuming that women will achieve equality simply through equal opportunity, although there are some elements of a liberal feminist view.[1]

Unfortunately, attention to gender was not matched in the "education" chapter of *Agenda 21*. Here, women were generally included in all sectors of society, although specific mention is made of the high illiteracy levels among women that need to be addressed (UNCED, 1992, para. 36.4(a)) in the objectives. In the activities, women are mentioned in the following terms: "Governments and educational authorities should foster opportunities for women in nontraditional fields and eliminate gender stereotyping in curricula" (UNCED, 1992, para. 36.5(m)). No mention is made of recognizing and valuing women's roles in promoting and achieving sustainable development, and the perspective seems, once more, that of liberal feminism, although indigenous peoples' experiences with and understanding of sustainable development are affirmed as playing a part in education and training (UNCED, 1992, para. 36.5(n)).

Gender equity is more closely related to education in the International Implementation Scheme for the United Nations Decade of Education for Sustainable Development (ESD) 2005–2014:

> Environmental issues like water and waste affect every nation, as do social issues like employment, human rights, gender equity, peace and human security. . . . Such issues are highly complex and will require broad and sophisticated educational strategies for this and the next generation of leaders and citizens to find solutions.
>
> (United Nations Educational, Scientific, and Cultural Organization [UNESCO], 2005b, Annex. I, p. 3)

However, much of this higher profile comes because gender equity is integral to the World Declaration on Education for All and the Millennium Development Goals, which are part of the decade agenda: "Education for sustainable development is based on ideals and principles that underlie sustainability, such as intergenerational equity, gender equity, social tolerance, poverty alleviation, environmental preservation and restoration, natural resource conservation, and just and peaceable societies" (UNESCO, 2005b, Annex. II, p. 3). The earlier Draft International Implementation Scheme provided a more direct message: "In terms of ESD specifically, the full and equal engagement of women is crucial, first, to ensuring balanced and relevant ESD messages and, second, to give the best chance for changed behaviours for sustainable development in the next generation" (UNESCO, 2005a, p. 19). This statement definitely puts gender on the agenda – but how do we make it happen?

What is feminist research?

According to Lather (1991), "to do feminist research is to put the social construction of gender at the centre of one's inquiry . . . feminist researchers see gender as a basic organising principle which profoundly shapes and/or mediates the concrete conditions of our lives" (p. 17). Feminist research is, thus, openly ideological, aiming to correct both the invisibility of female experience and its distortion.

Feminist educational research methods comprise many traditional forms, including statistical, interview, ethnographic, survey, cross-cultural, oral history, content analysis, case studies, and action research, and some feminist researchers have invented or are using newer forms as well (see Hesse-Biber, 2007). As Stanley (1990) argued, "there is no one set of methods or techniques . . . which should be seen as distinctly feminist. Feminists should use any and every means available" (p. 12).

According to Reinharz (1992), in her still useful text on feminist methods in social research, "feminist methodology is the sum of feminist research methods" (p. 240). In making this statement, Reinharz epitomizes a problem identified by Harding (1987) "that social scientists tend to think about methodological issues primarily in terms of methods of inquiry" (p. 2). Indeed, "method" and "methodology" are terms frequently either intertwined, used interchangeably, or confused in feminist (and other genres of) research scholarship, and contestation abounds as to whether or not there is a feminist research method or methodology. For example, Harding (1987, 2007) argues against the idea of a distinctive feminist method of research, because it is a distraction from discussing the more interesting aspects of feminist research processes: the differences between method, methodology, and epistemology. As Harding (1987) argues,

> it is new methodologies and new epistemologies that are requiring these new uses of familiar research techniques. If what is meant by a "method of research" is just this most concrete sense of the term, it would undervalue the transformations feminist analyses require to characterize these in terms only of the discovery of distinctive methods of research.
>
> (p. 2)

The confusion between method (techniques for gathering evidence), methodology (a theory and analysis of how research should proceed), and epistemology (issues about an adequate theory or justificatory strategy) is not the sole province of feminist research. Such confusions abound in nonfeminist research as well. In both feminist and nonfeminist research, "method" is often used to refer to all aspects of research, thus, making discussions about distinctiveness particularly difficult in regard to feminist research, because it undervalues the transformations happening in feminist research methods, methodologies, and epistemologies. According to Harding (1987), what makes feminist research distinctive is that it opens up:

- new empirical and theoretical resources (women's experiences),
- new purposes of social science research (for women), and
- new subject matter of inquiry (locating the researcher in the same critical plane as the overt subject matter).

Although early feminist research was largely positivistic, recent methodologies have been more concerned with "generating and refining . . . more interactive, contextualised methods in the search for pattern and meaning rather than for prediction and control . . . Hence feminist empirical work is multiparadigmatic" (Lather, 1991, p. 18). Feminist research methodologies now include the whole range, from postpositivistic concerns with prediction through interpretive, constructivist, phenomenological, and ethnographic concerns to understand, to emancipatory methodologies, such as critical, participatory, and action research, and postmodern concerns, such as poststructuralism and deconstruction (Hesse-Biber, 2007). Reinharz (1992, p. 240) proposed ten characteristics of feminist research (which she calls "methodology," but I prefer to call "approach"):

1. Feminism is a perspective, not a research method.
2. Feminists use a multiplicity of research methods.
3. Feminist research involves an ongoing criticism of nonfeminist scholarship [*To this I would add criticism of feminist scholarship too!*].
4. Feminist research is guided by feminist theory.
5. Feminist research may be transdisciplinary.
6. Feminist research aims to create social change.
7. Feminist research strives to represent human diversity.
8. Feminist research frequently includes the researcher as a person.
9. Feminist research frequently attempts to develop special relations with the people studied (in interactive research).
10. Feminist research frequently defines a special relation with the reader.

These characteristics share much in common with some environmental education research, particularly the use of a multiplicity of methods, adopting a transdisciplinary focus, and aiming to create social change.

What feminist research has already been undertaken in environmental education?

The English-language literature on feminist research in environmental education comes from the past two decades, but it is sparse and generally Australian or Canadian. One of the earliest articles is by Di Chiro (1987), which was written and published in Australia, even though she is American. Another is by Salleh (1989), an Australian ecofeminist and social theorist. Other early research that drew attention to gender differences in environmental knowledge, concerns, and behaviors includes Kremer et al. (1991, American), Fawcett et al. (1991, Canadian), Hallam and Pepper (1991, British), Hampel et al. (1996, Australian), Pawlowski (1996, Polish), Russell and Bell (1996, Canadian), and Storey et al. (1998, UK/Brazil). Other relevant literature from the 1990s, which specifically relates gender to environmental education, is Australian (including Barron, 1995; Brown & Switzer, 1991; Department of Prime Minister and Cabinet/Office of the Status of Women, 1992; Gough, 1994, 1997, 1999a, 1999b, 1999c; Greenall Gough, 1993; National Women's Consultative Council [NWCC], 1992; Peck, 1992; Whitehouse & Taylor, 1996). During the past decade, Canadian writing about feminist environmental education research has continued with the work of Barrett (2005), Fawcett (2000, 2002), McKenzie (2004, 2005), and Russell (2003, 2005, 2006). Other relevant research has continued to be sparse (see, e.g., Carrier, 2007; Gough & Whitehouse, 2003; Hallen, 2000; Newbery, 2003; Wane & Chandler, 2002).

The paucity of feminist research in environmental education could be considered surprising when related fields, such as science education, have a history of feminist research (see, e.g., Parker et al., 1996). Feminist scholarship in science started about the same time as the ecofeminist movement[2] (in the early 1970s), but, perhaps because of the greater social status of science or because of some of the more extreme[3] writings of some ecofeminists, feminist research in science education has received a much higher profile and generated many more studies than in environmental education.

In her keystone paper, Di Chiro (1987) places a feminist perspective on environmental education within a socially critical framework. She grounds her ecofeminist perspective in radical and socialist feminism and asserts:

> A feminist perspective [on] environmental education offers a more complete analysis of environmental problems and therefore a better understanding of those problems and their potential solutions. Such an analysis is political, in that it examines how power relations (in, for example, gender, class, race) shape the world in which we live; it asserts that the "polity" (human social world) determines and controls how this social world is and has been historically constructed and organised, and hence refutes the myth that the past and present state of the world is a "natural" and therefore justifiable progression. Moreover, environmental education's analysis of ocio-environmental problems is political in that it believes that if human social relations create the problems they can also change and improve them.
>
> (p. 40)

In particular, she argues that the environmental problem is socially constructed and should be viewed as a social problem, that environmental education should engage in a feminist critique of environmental problems and that it should engage in self-criticism "in order to understand how it is responsible as an educational enterprise for maintaining certain 'un-environmental' values and ideologies" (p. 41). Although others also argued the first two points to varying degrees, Di Chiro seems to be alone in asserting the need for environmental education to be self-critical, as well as socially critical.

An example of the type of feminist research that informs the discussions around the need for a gender agenda in ESD is Salleh's study from two decades ago. Salleh (1989) describes a group of upper-working-class women coming together "to see what might be done about household waste recycling in their local community" (p. 27). Few of the group had completed high school (only one had a university degree), all were older than thirty, and either were or had been married with children. All were already involved in some kind of ecologically sound practice at home. Opening up the issue with the technique of consciousness-raising as a catalyst for moving from personal to political concerns, Salleh found that:

> it was the tensions growing out of the consciousnessraising process itself that undermined the possibility of their participation in an environmental program. . . . Their workforce and personal marginality were so

severe that they lacked the necessary human support and self-assurance to transform their critical stance into a collaborative praxis.

(p. 30)

Responses to the 1992 NWCC consultations on Australian women's priorities for environmental action support Salleh's findings. In Salleh's group, few women had completed high school and they continued to choose individual action, rather than social or political action, as their focus. In the NWCC (1992) study, with respect to women's priorities for action on environmental issues, there were noticeable differences in responses, depending on education level:

> The less education, the more likelihood that respondents would choose education or individual actions as the principal action for women, and the less likely they would suggest changing social frameworks or political action. More of the respondents who had not proceeded beyond school level gave individual or personal answers, than women from the other three levels of education.
>
> (p. 66)

Similar findings were reported by Wane and Chandler (2002) in their work with elderly rural Kenyan women who had no formal education but did have a deep understanding of their ecological situation. Wane and Chandler also raise questions about valuing indigenous women's knowledge: "What would transpire if indigenous women's ecological knowledges were included in environmental discourse and influenced curriculum, teaching, and learning?" (p. 92).

In posing this challenge, Wane and Chandler echo Hutcheon's (1991) question, "How do we construct a discourse, which displaces the effects of the colonizing gaze while we are still under its influence?" (p. 176). The task is to dismantle colonialism's system, expose how it has silenced and oppressed its subjects, and find ways for its subjects' voices to be heard, whether referring to the colonized or women anywhere. Many of these and related issues, such as the education levels of women and the importance of local knowledges, have been taken up in the rhetoric around the implementation of the Decade of ESD (see, e.g., UNESCO, 2005a, pp. 19–20); however, they have not had much influence on the environmental education research agenda to date.

The potential of feminist research in environmental education

Although my personal disposition is toward critical and poststructuralist feminist research, all feminist research methodologies (and methods) can be applied to environmental education research contexts. We can participate in many types of research that put the social construction of gender at the center of the inquiry, whether seeking to predict, understand, emancipate, or deconstruct, and we need more stories from women relating to environments that we can use in environmental education.

Like McKenzie (2004, 2005) and Barrett (2005), I argue for poststructuralist research as the most promising approach for achieving the potential of feminist research in environmental education (Gough, 1997, 1999a, 1999b, 1999c). The power of feminist poststructuralist research in environmental education is that it calls into play a deconstructionist impulse that provokes consideration of the gendered positions made available to students and of understandings of gender identity. This approach is consistent with Davies (1994), who suggests a pedagogy informed by poststructuralist theory might begin

> with turning its deconstructive gaze on the fundamental binarisms of pedagogy itself: teacher/student, adult/child, internal/external, society/individual, reality/fiction, knower/known, nature/culture, objective/subjective [because] each of these underpin or hold together both what we understand as pedagogy and the discourses through which pedagogy is done.
>
> (p. 78)

Such an examination of the binaries of environmental education practices could also form the basis for a research agenda.

Another aspect of the appeal of feminist poststructuralist research as an approach for environmental education is summarized by Agger (1992), who argues that

> [a]nother primary aim of feminist cultural criticism is to decenter men from their dominance of various official canons and genres. Equally as troubling as the omission of women from the canon and from criticism is the installation of men as *those who speak for women* – universal subjects of world history. A good deal of poststructural feminist criticism has focused on the issue of the voices in which culture is expressed, the standpoints from which knowledge is claimed.
>
> (p. 118, emphasis in original)

Decentering the male perspectives that dominate environmental education discourses is a challenge for the future and will not be easy. As Felicity Grace (1994) argues, "Men's interests, women's interests and the common interest are all difficult contingent alliances, and the ideological claim to them forms part of continuing negotiations of power" (p. 19). Nevertheless, there is a need to tell "less partial, less distorted" stories in our research. As I previously noted, by studying women's experiences, we open up new empirical and theoretical resources and provide a new purpose for social science (for women rather than men), and a new subject matter of inquiry; although studying women is not new, it is new to study them "from the perspective of their own experiences so that women can understand themselves and the world" (Harding, 1987, p. 8).

Empowerment, and the search for more empowering ways to know, is a central concern of both critical and poststructuralist theorizing; it is also a goal shared by many environmental educators. Emancipatory (empowerment) research involves "analyzing ideas about the causes of powerlessness, recognizing systemic oppressive forces, and acting both individually and collectively to change the conditions of our lives" (Lather, 1991, p. 4), and Ellsworth (1989), for example, argues that "critical pedagogies employing this strategy prescribe various theoretical and practical means for sharing, giving or redistributing power to students" (p. 306). However, empowerment is not something done *to* or *for* someone; it is a process one undertakes for oneself in the development of a new relationship within his or her own particular contexts. We need research in environmental education that focuses on how women have been empowered, and are empowering themselves, rather than looking at universalized subjects; traditional research has focused on men's experiences and "asked only the questions about social life that appear problematic from within the social experiences that are characteristic for men" (Harding, 1987, p. 6).

Because of its political nature and concerns with empowerment, it is important to look at power and knowledge in the context of environmental education. This can be a focus of both critical and poststructuralist research. Power and knowledge, as they are exercised through discourses, are central aspects of poststructuralist theory: "discourses . . . are ways of constituting knowledge, together with the social practices, forms of subjectivity and power relations which inhere in such knowledges and the relations between them" (Weedon, 1987, p. 108).

The texts, myths, and meanings of our culture and our relationships with nature need to be deconstructed so we know the stories of which we are a part. Such deconstruction and critical analysis helps practitioners and students recognize whose interests are served at particular moments in environmental issues. It helps them understand that

> it *does* make a difference who says what and when. When people speak from the opposite sides of power relations, the perspective from the lives of the less powerful can provide a more objective view than the perspective from the lives of the more powerful.
>
> (Harding, 1991, pp. 269–270, emphasis in original)

The challenge is to encourage the development of alternative stories (discourses) by drawing attention to whose knowledge is legitimized and valorized in the power/knowledge structures of the dominant discourses. In particular, women's knowledge needs to be recognized and valued.

The dominant discourses in environmental education treat the subject of knowledge as homogeneous and unitary. Thus, in the behaviorist/individualist model that dominates much of these discourses, there is an emphasis on individuals having "the right behavior" and the knowledge of how to "get it right." This implies a power relationship in which some take it as their role to set out what those "right behaviors" are. However, it is no longer possible to find a universal subject: as subjects/agents of knowledge, we are all part of multiple, heterogeneous, and contradictory or incoherent positionings of race, class, gender, sexuality, disability, and ethnicity, and there is no one right way of knowing or behaving. The recent turn to "intersectional analyses" (e.g., Kahn & Humes, 2009) is relevant here, but it does raise the tension of how to do intersectional analyses without losing sight of particular concerns, such as gender.

Such multiple subjectivities are constantly achieved through relations with others (both real and imagined), which are themselves made possible through discourse. Accepting that the subjects of knowledge are multiple, rather than homogeneous, unitary, and universal, has implications for curriculum, pedagogy, and research in environmental education. Exploring and developing such possibilities for environmental education is a challenge for the future, and it is one that must include the power and promise of feminist research.

Although poststructuralist research in environmental education is a challenge, I believe it offers much promise consistent with the stated goals of environmental education. The dominant discourses of environmental education recognize that the environment and environmental problems are complex, not simple. For example, the guiding principles of environmental education from Tbilisi (UNESCO, 1978, p. 27) refer to the multidisciplinary nature of the environment – "consider the environment in its totality – natural and built, technological and social (economic, political, technological, cultural-historical, moral, aesthetic)" – and "the complexity of environmental problems." Although not overt or necessarily intended and recognizing that they have also been interpreted as consistent with positivist and behaviorist perspectives, such a perspective could involve multiple readings or interpretations of the environment consistent with adopting poststructuralist pedagogy and research approaches. The multiple readings I suggest are not only of nature but also of the individuals and groups concerned with the particular environment or environmental issue: there is a need to develop local or situated knowledges that disrupt oppressions. We should be listening to multiple stories in the spirit of a partnership ethic (and its precepts), rather than following, and an egocentric or homocentric ethic (Merchant, 1996).

Whatever the issue or environment, there are multilevel meanings of narratives and texts and multiple stories that can be told. There is not "one true story" about the environment. The knowledges involved in dealing with environments are multiple, involving both humans (in which each human is a multiple subject) and nonhuman nature (which also has a multiple subjectivity), and must be considered as such. Thus, poststructuralist pedagogy and research is also consistent with a partnership ethic[4] (and feminist research) in that it is concerned with listening to the voices of the marginalized, as well as those of the dominant discourses.

Poststructuralist research is also concerned with deconstructing power/knowledge relationships, which is also a goal of a partnership ethic in environmental education. As in critical research, it is important to analyze who has the power and what can be done to dismantle or subvert that power through developing counter-hegemonic and oppositional discourses. We need to know the stories of which we are a part and develop local knowledges.

Poststructuralist research consistent with a partnership ethic is also concerned with the liberation of nature and people. The goal is "to work toward a socially-just, environmentally sustainable world" (Merchant, 1996, p. 222). It is time to stop trying "from the outside, to dictate to others, to tell them where their truth is and how to find it" (Foucault, 1990, p. 9). By engaging in feminist and poststructuralist research in environmental education, we can come closer to achieving this goal, because we will have less partial and less distorted stories.

Notes

1 The term "liberal feminism" is often used to characterize the dominant form of feminism up to the 1960s. Its current form, inspired by the works of Simone de Beauvoir (*The Second Sex*, 1949/1953/1972) and Friedan (*The Feminine Mystique*, 1963), "emanates from the classical liberal tradition that idealizes a society in which autonomous individuals are provided maximal freedom to pursue their own interests . . . [and] endorses a highly individualistic conception of human nature" (Warren, 1987, p. 8). This conception locates our uniqueness as humans in our capacity for rationality and/or the use of language (Jaggar, 1983), and "when reason is defined as the ability to comprehend the rational principles of morality, then the value of individual autonomy is stressed" (Tong, 1989, p. 11). For liberal feminists, the attainment of knowledge is an individual project, and their epistemological goal is "to formulate value-neutral, intersubjectively verifiable, and universalizable rules that enable any rational agent to attain knowledge 'under a veil of ignorance'" (Warren, 1987, p. 9).

  Historically, liberal feminists have argued that women do not differ from men as rational agents and that it is only their exclusion from educational and economic opportunities that has prevented women from realizing their potential (Jaggar, 1983). Critiques of liberal feminism focus on its alleged tendencies to accept male values as human values, to overemphasize the importance of individual freedom over that of the common good, to adhere to normative dualism, and to valorize a gender-neutral humanism over a gender-specific feminism (Jaggar, 1983; Tong, 1989).

2 The term "ecofeminism" was coined in 1974 by Françoise d'Eaubonne "who called upon women to lead an ecological revolution to save the planet. Such an ecological revolution would entail new gender relations between women and men and between humans and nature" (Merchant, 1996, p. 5).

3 Here, I particularly refer to radical ecofeminism, because it overlooks "the historical and material features of women's oppression (including the relevance of race, class, ethnic, and national background), [and] it insufficiently articulates the extent to which women's oppression is grounded in concrete and diverse social structures" (Warren, 1987, p. 15).

4 I argue that it is time to move beyond utilitarian homocenterism and toward an environmental ethic for society along the lines of what Merchant (1992) calls a "partnership ethic" that "treats humans (including male partners and female partners) as equals in personal, household, and political relations and humans as equal partners with (rather than controlled-by or dominant-over) nonhuman nature" (p. 188).

References

Agger, B. (1992). *Cultural studies as critical theory*. London, UK: Falmer Press.

Barrett, M. J. (2005). Making (some) sense of feminist poststructuralism in environmental education research and practice. *Canadian Journal of Environmental Education, 10*, 79–93.

Barron, D. (1995). Gendering environmental education reform: Identifying the constitutive power of environmental discourses. *Australian Journal of Environmental Education, 11*, 107–120.

Bodzin, A. M., Shiner Klein, B., & Weaver, S. (2010). *The inclusion of environmental education in science teacher education*. Dordrecht: Springer.

Brown, V. A. (1995, February 9–11). *Women who want the earth: Managing the environment is a gender issue*. Paper presented at the Conference "Towards Beijing – Women, Environment and Development in the Asian and Pacific Regions," Victoria University of Technology, Melbourne, Victoria.

Brown, V. A., & Switzer, M. A. (1991). *Engendering the debate: Women and ecologically sustainable development*. Canberra, ACT: Office of the Status of Women, Department of the Prime Minister and Cabinet.

Carrier, S. J. (2007). Gender differences in attitudes toward environmental science. *School Science and Mathematics, 107*(7), 271–278.

Cherryholmes, C. H. (1993). Reading research. *Journal of Curriculum Studies, 25*(1), 1–32.

Davies, B. (1994). *Poststructuralist theory and classroom practice*. Geelong, Victoria: Deakin University Press.

de Beauvoir, S. (1972). *The second sex*. Harmondsworth: Penguin. (Original work published in French 1949 and English 1953)

Department of Prime Minister and Cabinet/Office of the Status of Women. (1992). *Women and the environment*. Canberra, Australian Capital Territory: Australian Government Publishing Service.

Di Chiro, G. (1987). Environmental education and the question of gender: A feminist critique. In I. Robottom (Ed.), *Environmental education: Practice and possibility* (pp. 23–48). Geelong, Victoria: Deakin University Press.

Ellsworth, E. (1989). Why doesn't this feel empowering? Working through the repressive myths of critical pedagogy. *Harvard Educational Review, 59*(3), 297–324.

Fawcett, L. (2000). Ethical imagining: Ecofeminist possibilities and environmental learning. *Canadian Journal of Environmental Education, 5*, 134–149.

Fawcett, L. (2002). Children's wild animal stories: Questioning interspecies bonds. *Canadian Journal of Environmental Education, 7*(2), 125–139.

Fawcett, L., Marino, D., & Raglon, R. (1991). Playfully critical: Reframing environmental education. In J. H. Baldwin (Ed.), *Confronting environmental challenges in a changing world. Selected papers from the twentieth annual conference of the North American association for environmental education* (pp. 250–254). Troy, OH: NAAEE.

Fontes, P. J. (2002). The stories (woman) teachers tell: Seven years of community-action-oriented environmental education in the north of Portugal. *Canadian Journal of Environmental Education, 7*(2), 256–268.

Foucault, M. (1990). *The use of pleasure (Volume 2 of the history of sexuality)*. New York, NY: Vintage. (Original work published in French 1984 and English 1985)

Friedan, B. (1963). *The feminine mystique*. New York, NY: W.W. Norton.

Gough, A. (1994). *Fathoming the fathers in environmental education: A feminist post-structuralist analysis* (Unpublished doctoral dissertation). Deakin University, Geelong, Australia.

Gough, A. (1997). *Education and the environment: Policy, trends and the problems of marginalisation.* Melbourne, Victoria: Australian Council for Educational Research.

Gough, A. (1999a). Recognizing women in environmental education pedagogy and research: Toward an ecofeminist poststructuralist perspective. *Environmental Education Research, 5*(2), 143–161.

Gough, A. (1999b). Kids don't like wearing the same jeans as their mums and dads: So whose 'life' should be in significant life experiences research? *Environmental Education Research, 5*(4), 383–394.

Gough, A. (1999c). The power and the promise of feminist research in environmental education. *Southern African Journal of Environmental Education, 19*, 28–39.

Gough, A. (2004). Blurring boundaries: Embodying cyborg subjectivity and methodology. In H. Piper & I. Stronach (Eds.), *Educational research: Difference and diversity* (pp. 113–127). Farnham: Ashgate.

Gough, A., & Whitehouse, H. (2003). The "nature" of environmental education research from a feminist poststructuralist standpoint. *Canadian Journal of Environmental Education, 8*, 31–43.

Gough, N., & Gough, A. (2003). Tales from Camp Wilde: Queer(y)ing environmental education research. *Canadian Journal of Environmental Education, 8*, 44–66.

Grace, F. (1994). Do theories of the state need feminism? *Social Alternatives, 12*(4), 17–20.

Greenall Gough, A. (1993). *Founders in environmental education.* Geelong, Victoria: Deakin University Press.

Hallam, N., & Pepper, D. (1991). Feminism, anarchy, and ecology: Some connections. *Contemporary Issues in Geography and Education, 3*(2), 151–167.

Hallen, P. (2000). Ecofeminism goes bush. *Canadian Journal of Environmental Education, 5*, 150–166.

Hampel, B., Holdsworth, R., & Boldero, J. (1996). The impact of parental work experience and education on environmental knowledge, concern and behaviour among adolescents. *Environmental Education Research, 2*(3), 287–300.

Harding, S. (1987). *Feminism and methodology.* Bloomington, IN: Indiana University Press.

Harding, S. (1991). *Whose science? Whose knowledge? Thinking for women's lives.* Ithaca, NY: Cornell University Press.

Harding, S. (2007). Feminist standpoints. In S. N. Hesse-Biber (Ed.), *Handbook of feminist research: Theory and praxis* (pp. 45–69). Thousand Oaks, CA: Sage.

Heilbrun, C. G. (1999). *Women's lives: The view from the threshold.* Toronto, ON: University of Toronto Press.

Hesse-Biber, S. N. (2007). Feminist research: Exploring the interconnections of epistemology, methodology, and method. In S. N. HesseBiber (Ed.), *Handbook of feminist research: Theory and praxis* (pp. 1–26). Thousand Oaks, CA: Sage.

Hutcheon, L. (1991). Circling the downspout of Empire. In I. Adam & H. Triffin (Eds.), *Past the last post: Theorizing post-colonialism and post-modernism* (pp. 167–189). Hemel Hempstead: Harvester Wheatsheaf.

International Union for the Conservation of Nature and Natural Resources (IUCN). (1970). *International working meeting on environmental education in the school curriculum, Nevada, USA.* Final Report. https://portals.iucn.org/library/node/10447

90   *Annette Gough*

Jaggar, A. M. (1983). *Feminist politics and human nature*. Totowa, NJ: Rowman and Allanheld.

Kahn, R., & Humes, B. (2009). Marching out from Ultima Thule: Critical counterstories of emancipatory educators working at the intersection of human rights, animal rights, and planetary sustainability. *Canadian Journal of Environmental Education, 14,* 179–195.

Kremer, K. B., Mullins, G. W., & Roth, R. E. (1991). Women in science and environmental education: Need for an agenda. *Journal of Environmental Education, 22*(2), 4–6.

Lather, P. (1991). *Getting smart: Feminist research and pedagogy with/in the postmodern*. New York, NY/London, UK: Routledge.

Linke, R. D. (1980). *Environmental education in Australia*. Sydney, New South Wales: Allen & Unwin.

Lotz-Sisitka, H., & Burt, J. (2002). Writing environmental education research texts. *Canadian Journal of Environmental Education, 7*(1), 132–151.

Lousley, C. (1999). (De)politicizing the environment club: Environmental discourses and the culture of schooling. *Environmental Education Research, 5*(3), 293–304.

MacGregor, S. (2006). *Beyond mothering earth: Ecological citizenship and the politics of care*. Vancouver, BC: University of British Columbia Press.

Malone, K. (1999). Environmental education researchers as environmental activists. *Environmental Education Research, 5*(2), 163–177.

McKenzie, M. (2004). The 'willful contradiction' of poststructural socio-ecological education. *Canadian Journal of Environmental Education, 9,* 177–190.

McKenzie, M. (2005). The 'post-post period' and environmental education research. *Environmental Education Research, 11*(4), 401–412.

Merchant, C. (1992). *Radical ecology: The search for a livable world*. New York, NY: Routledge.

Merchant, C. (1996). *Earthcare: Women and the environment*. New York, NY: Routledge.

National Women's Consultative Council. (1992). *A question of balance: Australian women's priorities for environmental action*. Canberra, Australian Capital Territory: National Women's Consultative Council.

Newbery, L. (2003). Will any/body carry that canoe? A geography of the body, ability, and gender. *Canadian Journal of Environmental Education, 8,* 204–216.

Parker, L. H., Rennie, L. J., & Fraser, B. J. (1996). *Gender, science and mathematics: Shortening the shadow*. Dordrecht: Kluwer.

Pawlowski, A. (1996). Perception of environmental problems by young people in Poland. *Environmental Education Research, 2*(3), 287–300.

Peck, D. (1992). Environmental education: Time to branch out. *The GEN, 1*(4).

Reinharz, S. (1992). *Feminist methods in social research*. New York, NY: Oxford University Press.

Rich, A. (1972). When we dead awaken: Writing as re-vision. *College English, 34*(1), 18–30.

Russell, C. (2003). Minding the gap between methodological desires and practices. In D. Hodson (Ed.), *OISE papers in STSE education* (Vol. 4, pp. 485–504). Toronto, ON: University of Toronto Press.

Russell, C. (2005). 'Whoever does not write is written': The role of 'nature' in post-post approaches to environmental education research. *Environmental Education Research, 11*(4), 433–443.

Russell, C. (2006). Working across and with methodological difference in environmental education research. *Environmental Education Research, 12*(3–4), 403–412.

Russell, C., & Bell, A. (1996). A politicized ethic of care: Environmental education from an ecofeminist perspective. In K. Warren (Ed.), *Women's voices in experiential education* (pp. 172–181). Dubuque, IA: Kendall Hunt.

Russell, C., Sarick, T., & Kennelly, J. (2002). Queering environmental education. *Canadian Journal of Environmental Education, 7*(1), 54–66.

Salleh, A. K. (1989). Environmental consciousness and action: An Australian perspective. *Journal of Environmental Education, 20*(2), 26–31.

Sandilands, C. (1999). *The good-natured feminist: Ecofeminism and the quest for democracy.* Minneapolis, MN: University of Minnesota Press.

Spender, D. (1985). *For the record: The making and meaning of feminist knowledge.* London, UK: Women's Press.

Stanley, L. (1990). *Feminist praxis: Research, theory and epistemology in feminist sociology.* New York, NY: Routledge.

Storey, C., da Cruz, J. G., & Camargo, R. F. (1998). Women in action: A community development project in Amazonas. *Environmental Education Research, 4*(2), 187–199.

Tong, R. (1989). *Feminist thought: A comprehensive introduction.* Boulder, CO: Westview Press.

United Nations Conference on Environment and Development. (1992). *Agenda 21.* Rio de Janeiro: UNCED. Final, advanced version as adopted by the Plenary on June 14.

United Nations Educational, Scientific, and Cultural Organization. (1978). *Intergovernmental conference on environmental education: Tbilisi (USSR)* (pp. 14–26). Final Report. Paris: UNESCO. (Original work published October 1977)

United Nations Educational, Scientific, and Cultural Organization. (2005a). *United Nations decade of education for sustainable development 2005–2014. Draft implementation scheme.* Paris: UNESCO.

United Nations Educational, Scientific, and Cultural Organization. (2005b). *Report by the Director-General on the United Nations decade of education for sustainable development: International implementation scheme and UNESCO's contribution to the implementation of the decade.* Executive Board Meeting Paper 172 EX/11. Retrieved from portal.unesco.org/education/en/ev.php-URL_ID=36025&URL_DO=DO_TOPIC& URL_SECTION=201 html.

Wane, N., & Chandler, D. J. (2002). African women, cultural knowledge, and environmental education with a focus on Kenya's indigenous women. *Canadian Journal of Environmental Education, 7*(1), 86–98.

Warren, K. J. (1987). Feminism and ecology: Making connections. *Environmental Ethics, 9*(1), 3–20.

Warren, K. J. (1997). *Ecofeminism: Women, culture, nature.* Bloomington, IN: Indiana University Press.

Weedon, C. (1987). Feminist practice and poststructuralist theory. Oxford, UK: Blackwell.

Whitehouse, H., & Taylor, S. (1996). A gender inclusive curriculum model for environmental studies. *Australian Journal of Environmental Education, 12*, 77–83.

Zelezny, L., Chua, P., & Aldrich, C. (2000). Elaborating on gender differences in environmentalism. *Journal of Social Issues, 27*(3), 41–44.

6 Centering gender on the agenda for environmental education research

Annette Gough and Hilary Whitehouse

Gough, A., & Whitehouse, H. (2019). Centering gender on the agenda for environmental education research. *The Journal of Environmental Education*, 50(4–6), 332–347. doi:10.1080/0 0958964.2019.1703622

Abstract

The environmental education movement developed in the 1970s at the same time as the environment movement, the feminist movement and the ecofeminist movement. However, while environmental education feels a close affinity with the environment movement the relationship with the (eco)feminist movement in environmental education research has been less than robust. Although there was some feminist environmental education research in the 1990s and 2000s, until the two recent special issues of *The Journal of Environmental Education* there had been a prolonged, even a deafening, silence around gender, eco/feminism and environmental education research. This chapter traces the history of feminist environmental education research across this and other environmental education research journals to argue that it is time for gender to be much higher on the agenda of environmental education researchers and of journals if we are to better achieve gender equality and more fully address the climate emergency within the field.

Introduction

Gender was not on the agenda when the field of environmental education research and praxis became established (Gough, 2013). However, as women's status and gender equality became part of international discourses – through significant events such as the United Nations International Women's Year in 1975 – the universalized human, male, subject became differentiated at least into the binary of male and female. For example, the Declaration from the 1977 UNESCO-UNEP Intergovernmental conference on environmental education states, "the environment concerns all men and women in every country and . . . its preservation and improvement require the support and active participation of the population of those countries" (UNESCO, 1978, p. 28).

DOI: 10.4324/9781003390930-7

The ecofeminist movement became established during the 1970s (Gough & Whitehouse, 2018), arguing for the recognition of the complexity of human roles and relationships with respect to multiple environments. That women have to be recognized for their differing life roles became apparent internationally in *Agenda 21*, the outcomes document from the 1992 United Nations Conference on Environment and Development (United Nations, 1993), and this recognition has continued through successive United Nations conferences on environment and development. Gender equality is now one of the seventeen Sustainable Development Goals (SDGs) of the United Nations (2015) 2030 agenda for sustainable development. As we enter the third decade of the 21st century ecofeminist research is engaging with intersectionality and new materialisms, to better illuminate diverse material practices in rapidly changing environmental conditions.

Achieving gender equality is now seen as "the best chance we have in meeting some of the most pressing challenges of our time – from economic crisis and lack of health care to climate change, violence against women and escalating conflicts" (UN Women, 2018b, np). It would therefore seem wise for environmental education to respond to and reflect this imperative. However, as we demonstrate in this chapter, gender and feminism have never been high priorities in the environmental education literature, including *The Journal of Environmental Education*, and, although there have been some recent changes in this regard, more effort is needed for a truer equality to be achieved.

The eco/feminist turn in environmental education research 1970–2020

In the early 1970s there arose a number of so-called "new social movements" – environmental, women's, gay rights and the various multicultural and multiracial movements – which directed "new political claims on the democratizing of the distribution of economic, political, cultural and social resources" to the state (Yeatman, 1990, p. xi). Although the second wave of feminism (focusing on more than achieving political enfranchisement) arose at the same time as the environmental movement in the 1960s, with its associated environmental education movement, "the notion that the collective voices of women should be central to the greening of the Earth did not blossom until the mid to late 1970s" (Diamond & Orenstein, 1990, p. ix). Charlene Spretnak (1990, p. 5) wrote that "the first tendrils of ecofeminism appeared not in the exhuberant [*sic*] season of Earth Day 1970 – for feminists were quite preoccupied with the birthing of our own movement then – but in mid-decade".

Diamond and Orenstein (1990) point out it was not coincidental that a woman, Rachel Carson, "a passionate voice of conscience in protest against the pollution and degradation of nature. . . was the first to respond both emotionally and scientifically to the wanton human domination of the natural world" (p. ix) and, with *Silent Spring* (Carson, 1962), pre-figured a powerful environment movement. Yet her critics at the time "depicted her as hysterical, mystical and witchy" (Gershon, 2019) – these, of course, were typically gendered responses, and sadly remain familiar to women environmentalists today.

New Woman/New Earth was one of the first texts to focus upon the intimate connection between the women's movement and the ecology movement, without using the term "ecofeminism", which had been coined the year before (d'Eaubonne, 1974). In 1975, Rosemary Radford Ruether (1975, p. 204) argued that both movements required "transforming the worldview which underlies domination and replacing it with an alternative value system", meaning:

> Women must see that there can be no liberation for them and no solution to the ecological crisis within a society whose fundamental model of relationships continues to be one of domination. They must unite the demands of the women's movement with those of the ecological movement to envision a radical re-shaping of the basic socioeconomic relations and the underlying values of this society.

Ecofeminism is a system of thinking that explicitly draws attention to deep social, economic, and political structures resulting in the parallel domination of the human over the non-human with male domination over women. For Stephanie Lahar (1991, p. 28), ecofeminism was "the convergence of ecology and feminism into a new social theory and political movement [that] challenges gender relations, social institutions, economic systems, sciences, and views of our place as humans in the biosphere". Ecofeminism is decidedly "transformative rather than reformist in orientation, in that ecofeminists seek to radically restructure social and political institutions" (Lahar, 1991, p. 30). Connie Russell and Anne Bell (Russell & Bell, 1996, p. 172) explain

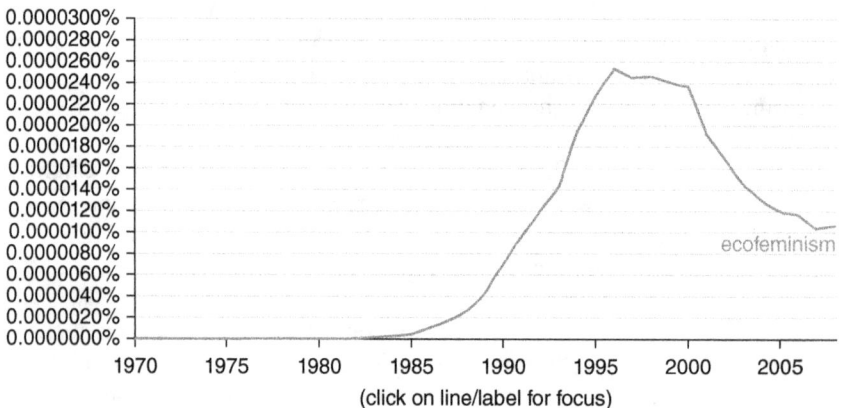

Figure 6.1 Ngram of use of "ecofeminism" in books 1970–2008.

that ecofeminism is "a theory and a movement that makes explicit the links between the oppression of women and the oppression of nature in patriarchal cultures. This means coming to terms with a number of cherished and widespread beliefs, most notably the anthropocentric assumption that humans are different from, and superior to, the rest of [a categorized] nature, and therefore ought to dominate" the socially othered beings that share this beautiful, living planet of ours.

While the literature around ecofeminism bloomed during the 1980s and 1990s (see, for example, Caldecott & Leland, 1983; Diamond & Orenstein, 1990; King, 1987, 1989, 1990; Plant, 1989; Plumwood, 1986, 1993; Spretnak, 1990; Warren, 1987) (see Figure 6.1 for the trend in usage of the term "ecofeminism" in books between 1970 and 2008), it was not until 1987 that the first English-language article relating feminism, ecofeminism and environmental education was published by Giovanna Di Chiro (1987) in the *Australian Journal of Environmental Education* (*AJEE*). Di Chiro (1987) argued the close connection between the socially and politically constructed nature of environmental problems and the importance of developing a feminist perspective in environmental education for a more complete analysis. She wrote that, as environmental education is problem-solving focused,

> the feminist perspective offers a more complete analysis of the environmental issue and thereby a better understanding of the problem and its potential solutions. Such an analysis is a political one, in that it looks at how power relations (in, for example, gender, class, race) shape the world in which we live. It is political in that it asserts that the "polity" (human social world) determines and controls how this social world is and has been historically constructed and organised and hence, refutes the myth that the past and present state of the world is a "natural" and therefore justifiable progression.
>
> (Di Chiro, 1987, p. 15)

By contrast with the *AJEE*, the first mention of ecofeminism in *The Journal of Environmental Education* (*JEE*) was made by Peter B. Corcoran and Eric Sievers in 1994, 25 years after the establishment of this journal, in their discussion of ways for "reconceptualising" environmental education. Noting that approaching the subject of ecofeminism was a problem because the writers were both men, and being aware of "no common agreement" (Corcoran & Sievers, 1994, p. 7) as to a single definition for ecofeminism, they saw ecofeminist thinking as having much to offer in challenging "traditional male thinking" and "bringing in a profound understanding of the deeper implications, for nature and society, of our male-centered culture" (p. 7).

Environmental education researchers did not readily engage with feminist theory and ecofeminism (Gough, 2013), and the sparseness of feminist research in environmental education is surprising when related fields such as science education have a much stronger record of feminist research (see, for example, Hildebrand, 1989; Kenway & Gough, 1998; Parker et al., 1996; Sjoberg & Imsen, 1988). Feminist scholarship in science emerged about the same time as the ecofeminist movement in the early 1970s, and feminist research in science education has generated many more studies than in environmental education. Perhaps this is because of the greater social status of science education. Or perhaps this is due to a historical reluctance to integrate the isms of feminism, ecofeminism and environmentalism into environmental education. As late as 1998 Paul Zeph stated without hesitation that environmentalism must be kept out of environmental education (see Jickling, 2003), the same year that Annie Booth (1998) – the first woman to write about ecofeminism in the *JEE* – argued for inclusions of ecofeminist perspectives in recognition that "feelings of caring, perhaps even love . . . must be recovered as legitimate sources of knowledge" (p. 5) in environmental education.

The inevitable pushback occurred with the marginalization and rejection of ecofeminism, firstly due to its transgressive and transformational stance, and secondly, due to accusations of essentialism (Gaard, 2011; Phillips & Rumens, 2016a). In inquiring into discourses of the (near) past, we may be encouraged to see the history of women's activism and scholarship as a leading along a path to progress, an idea that, when systematically examined, proves simply not the case. Charlotte Riley (2018) reminds us not to be seduced by the temptation "to tell a story about things getting better, about women's lives improving, and increasing female power in politics and society" (p. 294). Riley cautions that "for every battle won, there is a fight that still needs to be fought. And for every critic won over to the cause, there has been one whose resolve to resist progress has only been strengthened" (p. 294).

Ecofeminism and feminist writings in environmental education have resurfaced recently, particularly in the context of the emergence of new materialism and intersectionality (see, e.g., Gough & Whitehouse, 2018; Hart, 2017; Maina-Okori et al., 2018; Piersol & Timmerman, 2017). As discussed in this article and in Gough and Whitehouse (2018), there have been many discussions around interpretation. Ecofeminism is not seen as a theory and movement that belongs in the past, due, in part, to the "durability of sexism" (Bartos & Ives, 2019, p. 778); the prevailing large societal trends bent on unraveling the fabric of life; and our collective experience of heating into chaos a formerly, fairly stable, Holocenic, climate. In this century, many feminisms (i.e., no one, single fit for purpose, universal feminism) representing the differing interests of women from all over the world are making their way into the literature, but not so quickly into the environmental education research literature.

Of particular relevance to discussions here is the term "decolonial", which is applied to feminist movements and theory that explicitly recognize difference in locale, social settings, histories and interests across the humanized globe. Decolonial feminism does not only focus on gender division and oppression but is a more complex and broader political project "which struggles for the de-privatization of water and territories, against extractivism, and against violence" (Villanueva, 2019, n.p.). The "supply chain of violence" proceeds apace and has very real consequences in that environmental defenders are being murdered (a large proportion being Indigenous activists), "dying to protect their livelihoods, along with the forests, lands and ecosystems that are essential for all our futures" (Butt et al., 2019, np).

Decolonial eco/feminism recognizes a plurality of experiences and agendas in opposing oppression in accordance with the realities of different territories, societies, cultures and places. Villanueva (2019, n.p.) quotes Lolita Chávez, an Indigenous and communitarian feminist from Guatemala, who says,

> Estamos aquí alzando nuestra voz para que los feminismos sean plurales y diversos y no se limiten a ciertas agendas individuales o de un solo sector.[We are here to raise our [collective] voice up to make feminisms plural and diverse, so they [feminisms] aren't limited to only certain individual agendas or to specific sectors.]

While feminist environmental education scholarship has tended to be in English, we should be taking up the challenge to make feminisms plural and engage our scholarship across agendas and sectors. As with politics, activism and advocacy, lasting changes in advancing feminist scholarship in environmental education, education for sustainability and education for sustainable development will not be achieved without sustained and significant effort well into the 21st century.

In the next sections we discuss the silences around women and (eco)feminist theories in key environmental education journals as a first step in addressing this challenge, focusing first on the publishing history of this journal.

Gender matters in *The Journal of Environmental Education*

In its origin, environmental education was not conceptualized as having a place for women as a single group, nor for women with all their differing experiences and perspectives (Gough, 1997, 1999a, 1999b). These insights were to come later, due to the considerable efforts of feminist scholars. Quite literally, the texts that informed the establishment of the global environmental and environmental education movement from the middle of the 20th century were written in terms of a universal subject "Man" and "his" relationship with a singular "Environment", or alternatively, with a singular "Nature". Far-sightedness in general ecological predictions did not extend to a nuanced understanding of gendered relations. An article published in 1960 quotes a proverb, "to survive, all men must hold hands" while simultaneously arguing for a "new creed" for understanding and respecting the interrelationships between all living things (Royal Bank of Canada, 1960). And students of ecology were encouraged to read texts with titles such as *Man and the Environment* (Boughey, 1974) well into the 1980s.

The Journal of Environmental Education (*JEE*) was founded in 1969 with the aim to promote and professionalize research and development in the new field of environmental education and ecological communications. On the first page of the very first issue, the editor, Clay Schoenfeld, rhetorically asked, what's new about environmental education? "Is there anything really new about environmental education or is it simply conventional conservation education in a new bottle? Are we merely 'word-merchandising' or are we creating new dimensions in ecological communications?" (Schoenfeld, 1969, p. 1). Schoenfeld put these questions to his postgraduate students who were "unanimous in the feeling that environmental education does indeed represent a significant new scale in the interpretation of man-land relationships". One woman, Beverly Southern (carefully described as "Mrs"), published in the first issue – a one-page article on vitalizing natural resource education (Southern, 1969) – however, authorship by women remained very low (between 10% and 13%) for almost two decades.

The Journal of Environmental Education was no exception in its reflection of dominant norms of just who was constituted as human. In 1974, five years in, and also a year in which no women published in the journal, Matthew Brennan (1974, p. 16) reproduced this language of masculinity in order to argue the case for holistic pedagogy at the Pinchot Institute for Studies in Conservation, in Pennsylvania:

> Man has come face to face with the most serious problem he has ever encountered – the total destruction of his environment . . . man has subjugated the land; claimed its renewable resources; used its non-renewable resources at a reckless rate; even invaded the inexhaustible resources of air and space; and has begun to multiply beyond the capacity of the earth to support him. . . . If those changes continue without control or planning, man may soon be locked with man in mortal struggle for space in which to live, for fresh air to breathe for fuel to power his body and his machines, for water to drink, in short, for an environment fit for life and fit for living.
>
> (Brennan, 1974, p. 16)

In the same issue, David LeHart and C. Richard Tillis (LeHart & Tillis, 1974, p. 43) argued the case for teaching environmental values in these terms:

> Man is an animal yet he alone makes the moral and ethical decisions determining the fate of a myriad of other creatures; all fellow passengers on Spaceship Earth. Man must develop the intellectual humility to allow him – compel him – to save the finite resources of our planet. Many men are moved to action by the knowledge that some species of life . . . are now threatened by artificial conditions introduced by man. . . . [However] knowledge about these endangered species is not enough because most men are insulated from the realities of natural systems. These men are products of technological and societal developments. . . . These creations of man serve him in many ways yet seem to channel his

vision and his sensitivities away from the natural world that produced and sustained him.

A year later, William Stapp (1975, p. 7) wrote about UNESCO's environmental education program from the Stockholm Conference and managed to turn "people" into the universalized male subject, "man":

> All recent authoritative studies on present day environmental problems conclude that there is no hope of finding viable solutions unless the content of general education at all levels is suitably modified so that from childhood, people, particularly in the industrialized countries and in urbanized areas, grasp the fundamental interrelations between man and his environment.

UNESCO did stop using gendered language. As previously noted, the 1977 Tbilisi Declaration referred to "all men and women" (1978, p. 28). However, the terms "man" and "he" were also in the Declaration: the opening sentence stated, "In the last few decades, man has, through his power to transform his environment, wrought accelerated changes in the balance of nature" ((UNESCO, 1978, p. 24). Otherwise the Declaration did not use gendered language, and even "man-made" in the guiding principles for environmental education in the Belgrade Charter (UNESCO, 1975, p. 4) became "built" in the Tbilisi Declaration (1978, p. 27). While it can be argued that the use of "man" at this time was consistent with the species and encompassed both genders, as used in scientific discourses (and most of those who were influential in the field at this time had science backgrounds (Greenall Gough, 1993a)), the male domination at these meetings could also be relevant: at the 1977 Tbilisi Conference there were only 55 females out of a total of 340 participants.

Despite the changes in the international discourses, effective gender blindness was a feature throughout the 1970s. Robert Horvak and Alan Voekler's (Horvat & Voelker, 1976, p. 46) large study on the environmental orientations of responding preadolescent girls' and boys' concluded "that the immediate surroundings of a student should affect *his* environmental orientation is not surprising", especially given that "urban kids do live largely in a world of man, where nature impinges only peripherally". Girls became boys in the researcher's final analysis, and "nature" was excluded to the suburbs and beyond.

An article on gathering and preparing wild foods (Curry & Williams, 1975, p. 45) blithely stated of the educator, that "[h]e can involve the child with the 'wild' edible plant life that is in his environment all the time" (p. 45). Food gathering was described as "sharing in the free bounties of nature" (p. 47) and referred to as "a primitive lifestyle [and] a stage of human development which still exists in remote areas of the world". (You can see why, at some point, there had to be decolonial, ecological feminisms.)

What John C. Miles (1976) called "the problem of humanization" in the new field of environmental education was predicated on the assumption that all classes of person, teachers and learners, could be collectively named as men. Miles (1976, p. 9) wrote that "the environmentalist seeks a rather fundamental reordering of thought and action away from growth, control and mastery over nature and progress traditionally defined, and such humaneness "must be regarded as a part of the challenge of environmental education". The discourses drawn upon to make such an argument are entirely silent on the matter of gender.

Looking back, this may not seem materially important except, as Marilyn Waring (1988/1999) pointed out, at that time, globally, almost every woman was denied possession of legislatively determined equal rights, including financial rights. And while there have been many improvements in a generation, the World Bank (IBRD, 2019) reports that great legal inequalities still persist across the globe at the cost of diminishing women's lives and losing the world economy trillions of dollars.

Dale Spender (1990) stepped into this weird silence with her influential text *Man-Made Language*, where she properly pointed out that "the English language has been literally man made and that it is still primarily under male control" (1990, p. 12), and that men's control over language and meanings creates men's power and denies women's experiences and values. Spender (1990) argued that men control and define meanings from their vantage point of difference and dominance and through such privilege impose their view of the world, while women, without the ability to symbolize their experience in the male language, either internalize male reality and see things through male eyes (alienation) or reject existing words and find themselves unable to speak at all (silence).

With respect to dominance and difference, Spender argued men encode in language their worldview together with their conviction that they are superior to women:

A patriarchal society is based on the belief that the male is the superior sex and many of the social institutions and much social practice is then organized to reflect this belief: in one sense a patriarchal society is organized so that the belief in male supremacy "comes true".

(p. 1)

For women to represent themselves, the power structure of patriarchy must be disturbed, and: "If and when sufficient women agree that they no longer subscribe to the rules and patterns of patriarchy, then the rules and patterns are likely to be transformed . . . we have the power to obstruct patriarchy and we can use it" (Spender, 1990, pp. 3–4).

The history of scholarship on gender and ecofeminist thought in *JEE* is not a track of linear progress but one of brief eruptions against prolonged periods of uninterest. It took a goodly while for gender analysis to make its breakthrough. Even in the 1980s, Larry Nielsen and Marie Schroeder

(Nielsen & Schroeder, 1983) researched a "comparative analysis of textbooks for Man-Environment courses", stating that "with the environment movement have come college [university] courses covering the broad range of man-environment interactions" (p. 7) such as natural resource management, environmental conservation, environmental science and human population problems.

In this same issue Jim Swan (1983) wrote about the need to research with "other cultures and their attitudes and lifestyles regarding nature, especially those which appear to live in greater harmony with nature than we do now in modern Western society" (p. 32). In undertaking such an interrogation, Swan reported only on his conversations with male, American, "Indian" informants. Remaining unproblematized for years and years was the concept of American wilderness, and of wilderness education (Miles, 1991) about education in landscapes from which the original landowners had been forcibly removed.

By the 1990s, the universal subject ("man") did disappear from the journal, to be replaced by gender-neutral language ("humans", "people") as Dale Spender's and others' critique seeped into academic consciousness evidenced by our own early publications (see Greenall Gough, 1993a, 1993b; Gough, 1999a, 1999c; Whitehouse & Taylor, 1996). One of the earliest articles which took "a close look at how the typically "feminine" ideological set with its value on nurture and survival relates to a more generalized environmental sensibility, and what constraints, if any, cut across the translation of this consciousness into an organized praxis" (Salleh, 1989, p. 26) was published in *JEE*.

This did not mean that investigatory tropes questioning women's abilities disappeared. Research was conducted into whether gender (binary, based on sexual characteristics, not self-identified) had an effect on a phenomenon termed "environmentally responsible behavior" (Hines et al., 1987) – it did not. A 1986 paper showed that when American women forestry students were given equal learning opportunities, they achieved slightly higher test scores than their male counterparts, despite an absence of female role models – the authors pointed out "even Bambi is a male" (Burrus-Bammel & Bammel, 1986, p. 10). A study investigating a "potential" relationship between (binary) gender and the cognitive ability to "discriminate between facts and opinions" showed that "as in previous studies on sex differences in environmental competencies (Hines et al., 1987), gender did not affect the ability to differentiate between facts and opinions" (Corral-Verdugo, 1993, p. 27).

A 1998 *JEE* study showed female students had significantly higher perceptions of a number of environmental risks, and while the authors argued against gender stereotyping, they did express some mystification with the results (Riechard & Peterson, 1998). Similar surveys from the 1990s show women's general understanding of environmental concerns tended to be higher; however, they were no more likely to participate in activism as men (Blocker & Eckberg, 1997), a phenomenon explained by the social and cultural restraints on women at this time (Mohai, 1992). Studies of "the gender gap" (identified or otherwise) in environmental education proved a robust topic in *JEE* well into the 21st century (see, for example, Liefländer & Bogner, 2014; Olsson & Gericke, 2016, 2017) However, these studies rarely stray far into the terrains of ecofeminist analysis.

Such erratic attention to feminism as a dimension of environmental education research led Annette Gough, Connie Russell and Hilary Whitehouse, in mid-2015, to propose a special issue of *JEE* on gender and environmental education, which was accepted by the editorial team. The special issue editors, Gough et al. (2017, pp. 5–6),

> called for manuscripts that respond to the need for promotion of social equity and enhancing gender equality and women's empowerment within environmental education. We welcomed theoretical and empirical research papers and we explicitly aimed to gather together a collection of articles that reflects a broad scope of perspectives, methodological approaches, and research foci.

They received enough suitable submissions to argue for two special issues, and this was agreed to by the editor-in-chief and an executive editor; they were published as Volumes 48(1) and 49(4). The submissions reflected a resurgence of interest in ecofeminist analysis as well as bringing other theoretical perspectives – such as new materialisms, assemblages and intersectionality – into the discussions. Such research is needed now, more than ever, and we look forward to more research in this vein in *JEE* in the future.

Gender matters in other EE journals

Gender blindness and inattention to ecological feminism have not just been the province of *JEE*. Journals such as the *Australian Journal of Environmental Education* (started 1984), *Environmental Education Research* (started 1995) and the *Canadian Journal of Environmental Education* (started 1996) are also sparse in publishing research related to feminism, gender and women.

The *Australian Journal of Environmental Education* has published one article with feminist in the title – the previously mentioned Di Chiro (1987), which was reprinted in Volume 30 as part of a special 30th-anniversary issue with a response (Di Chiro, 2014a, 2014b). There has also been one article with feminism in the title (Stovall et al., 2015), two with gender (Barron, 1995; Whitehouse & Taylor, 1996) and one with women (a book review, Gough, 1996).

Environmental Education Research has published five articles with women in the title (with one reprinted) (Gough, 1999a; Kaufman et al., 2001/2006; Quigley et al., 2017; Skanavis & Sakellari, 2012; Storey et al., 1998), two articles with feminist in the title (Gough, 1999a; Lloro-Bidart, 2017) and one book review (Kennedy, 2018); and one article (Tuncer et al., 2005) and one book review (Bell, 2017) with gender in the title. There have also been a few articles which had a gender dimension to their research (Ceaser, 2015; Gough, 1999b; Hampel et al., 1996; Lousley, 1999; Malone, 1999; McKenzie, 2005; O'Donoghue, 2006; Pawlowski, 1996; Russell, 2005, 2006).

The *Canadian Journal of Environmental Education* has published articles with titles referring to ecofeminism (Hallen, 2000; Fawcett, 2000; Fry, 2000), feminism (Barrett, 2005; Gough & Whitehouse, 2003), gender (Newbery, 2003) and women (Fontes, 2002; Wane & Chandler, 2002). There have also been articles which have adopted other gender dimensions (Dunlop, 2002; Fawcett, 2002; Gough et al., 2003; Lotz-Sisitka & Burt, 2002; Martusewicz, 2014; McKenzie, 2004; Russell et al., 2002). While this journal has more articles than the other journals, with the exception of Martusewicz (2014), all were published between 2002 and 2005.

That environmental education journals have been so silent around gender and feminist issues – with only three articles and two book reviews as exceptions since 2006 – is consistent with the trend in usage of the term ecofeminism on Google (see Figure 6.2), and with the feminist backlash ecofeminism suffered in the late 1990s. As discussed by Gough and Whitehouse (2018) ecofeminism was criticized as essentialist, elitist, and ethnocentrist, "and effectively discarded" (Gaard, 2011, p. 26). However, recent studies at the intersection of feminism and environmentalism have seen a resurgence of the label of ecofeminism (see, for example, Adams & Gruen, 2014; Gaard, 2011, 2015; Phillips, 2016, 2019; Phillips & Rumens, 2016a, 2016b; Thompson, 2006). Indeed, Greta Gaard (2011), a prominent ecofeminist, argues that what is needed now to address the climate emergency and the biodiversity crisis is an intersectional ecological-feminist approach that "frames these issues [problems] in such a way that people can recognize common cause across the boundaries of race, class, gender, sexuality, species, age, ability, nation – and affords a basis for engaged theory, education, and activism" (p. 44). We are optimistic that such an approach will soon be reflected in environmental education journals.

Interset over time

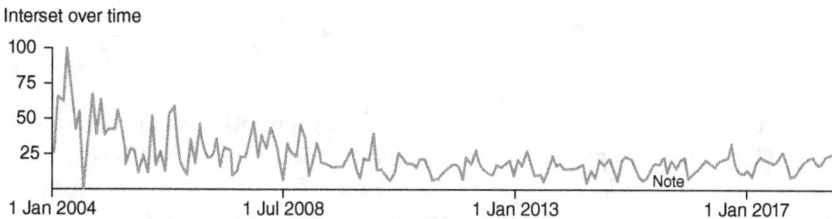

Figure 6.2 Google trend for interest in "ecofeminism" 1 January 2004–15 April 2019.

Ecofeminisms in the age of environmental breakdown

We have chosen to use the term "environmental breakdown" (Laybourn-Langton et al., 2019) as one which has increasing currency given recent experiences and observations from all over Earth's biosphere. The breakdown of Earth's living systems has many origins and is constituted by several intersecting global disasters: global atmospheric and oceanic heating now known as the climate emergency (Sato & Hansen, 2019); the great magnitude of losses associated with biodiversity collapse; ocean acidification and plastication; soil losses and rapid deforestation; changes in biogeochemical flows; increasingly high levels of air, land, fresh water and marine pollution; and the concomitant increase in area and volume of "dead zones" on a living planet.

The consequences of environmental breakdown tend to fall most heavily on the most vulnerable, those already denied power within the current structures of commerce and society. The vulnerable include women and their children, the poorer/poorest people of the world, and all animals and plants and their assemblages in rivers, lakes swamps, savannahs, forests, tundras and oceans who are presently denied life rights under dominant forms of human law. The British barrister Polly Higgins (2011, 2012) led the movement to recognize and codify ecocide in international law, where ecocide is defined as "the extensive damage to, destruction of or loss of ecosystem(s) of a given territory, whether by human agency or by other causes, to such an extent that peaceful enjoyment by the inhabitants of that territory has been or will be severely diminished" (Higgins, 2012, p. 1). Higgins is very clear that ecocide is essentially a justice issue, and matters of social and environmental justice are prominent in most contemporary feminist analyses of our current (and dangerous) situation.

Numerous studies have been published documenting the phenomenon of the unequal impact of the climate emergency on women and their children (see, for example, Bätthge, 2010; Global Gender and Climate Alliance (GGCA)), 2013; Mann, 2015; Mignaquy, 2016; Red Cross, 2014; UN Women, 2014; UN WomenWatch, 2009; Women & Gender Constituency, 2015; World Health Organization, 2014). It is a distressing situation that, "in many ways women, with their strength and love, [who] are the people who hold the world together yet [who] are also the people most neglected in climate crisis" (OMW, 2015). And this phenomenon, as Elizabeth Kolbert (2014) documented in her Pulitzer Prize–winning book *The Sixth Extinction*, is the cause of disturbing levels of existential and material distress among populations of humans and others (Albrecht, 2019).

Most concerning is the acceleration of destruction is "at unprecedented scale" (Laybourn-Langton et al., 2019) and has occurred in the 50 years since the founding of *The Journal of Environmental Education*. By implication, this raises many challenges for how we conceive of environmental education praxis in the third decade of the 21st century. As Higgins (2012, p. xiv) writes,

> This is a story with two possible endings; one is fertile and abundant with life, the other is arid and speaks of death. We have a choice: to make

the leap into the new and leave the old ways behind as distant memories, or follow the current route into the ecocide of the earth.

Analysis by Edgar Gonzalez-Gaudiano and Pablo Meira-Cartea (this special issue) is that over these last 50 years, conventional environmental education may have "contributed little" to slowing down dizzying and dangerous levels of environmental change at global scale, though, within some nations, environmental educators and researchers have held some ramparts against the sweeping tides of neoliberalism (see Henderson et al., 2017). At this point in human history, with the Arctic turning to slush and tundra on fire, we have less than a decade to materially transform our societies away from the 200-year "carbon party" (see Simmens, 2017). One way to counteract deathly hegemonies is to harness the energy of women and their children in mobilizing action and activism. Ecofeminist scholars, determinedly working from the margin of the margins in environmental/education, have been pointing out for years that deep structural change is required to alter the course of complete biosphere destruction.

David Wallace-Wells (2019a, 2019b) gives us all permission to panic by raising the specter of much increased human, animal and plant suffering in the coming years.

> The majority of the burning has come in the last 25 years – since the premiere of Seinfeld. Since the end of the second world war, the figure is about 85%. The story of the industrial world's kamikaze mission is the story of a single lifetime . . . all of the paths projected from the present will be defined by what we choose to do now. . . . Climate nihilism is, in fact, another of our delusions. What happens, from here, will be entirely our own doing.
>
> (Wallace-Wells, 2019b, np)

One of the most notable features is the energy of women and their children quickly emerging in the face of environmental collapse. Writers of the One Million Women blog (devoted to the climate emergency) urge that all is not lost, and there is

> a lot to be said for the power of one, and every woman on the planet has it in her hands to prevent climate change – That's a fact. We know this from the way women have shown their strength and resilience in the developing world's climate battle. Women have demonstrated not just physical power, but strong intellectual ability in creating solutions.
>
> (OMW, 2015, np)

A generation ago, in *JEE*, Abour Cherif (1992) was arguing for an environmental education that developed "the ability to engage in responsible environmental behavior in a productive manner [where] children, especially those in societies where economic decisions are based on anthropocentric, market considerations, should be prepared to take responsible action in order to bring about change" (p. 42). In 2019, a huge climate movement led by teenage girls swept Europe, the Americas and Australasia (Feder et al., 2019, np). Jamie Margolin, a founder of Zero Hour, a group largely led by young women of color, says, "If you're a victim of a system of oppression, you're more affected by the climate crisis – that goes for women . . . Nobody is going to hand us this. We have to step up and raise our voices" (Feder et al., 2019, np). However, in many countries, some white middle-aged men (the "stale, pale males". of common vernacular) poured scorn on young women activists and it's far from uplifting politics (Meade, 2019; Baggini, 2019). High levels of attack on young women's activism in the face of the climate emergency is dispiriting, but such *argentum ad feminam* attacks provide evidence that the climate emergency can be ignored no longer. For Mary Phillips (2019, p. 1151), meaningful responses to environmental breakdown cannot be sought within capitalist systems underpinned by a "logics of appropriation" promulgated through "sets of hierarchical and interrelated dualisms which define the human in opposition to the realm of "nature" leading to the resilience of ecosystems, social reciprocity and care being unvalued or undervalued". What is required is morally care-full activism, informed by the bracing, transgressive perspectives of ecofeminism (Phillips & Rumens, 2016a, 2016b).

Directions for ecofeminist research in environmental education – third decade 21st century

Gender justice and ecological justice were always at the heart of the ecofeminist project, and they are encompassed within the 2030 agenda for sustainable development (United Nations, 2015; UN Women, 2018a). As Isis Alvarez and Simone Lovera (Alvarez & Lovera, 2016) succinctly state, this is "New Times for Women and Gender Issues in Biodiversity Conservation and Climate Justice". It is increasingly understood that the climate emergency must be addressed through gender. However, it is clear that a different approach is needed (Koehler, 2016; O'Manique & Fourie, 2016; Salleh, 2016). For example, Colleen O'Manique and Pieter Fourie (2016, p. 121) argue:

> The Sustainable Development Goals (SDGs) are presented as the new global framework to rid the world of poverty and inequality. While emerging from widespread consultation, we argue that they perpetuate rather than challenge the systemic drivers of gender injustice, silencing feminist critiques which demand systemic transformation. . ... Unless new forms of agency emerge through truly transformative local strategies and global alliances, inequality and gender injustice will remain the norm.

Gabrielle Koehler (2016, p. 53) similarly argues that "by incorporating other, more progressive conventions and declarations, the SDGs can be used creatively and subversively, to move toward gender and climate justice". And Ariel Salleh (2016, p. 954), an early ecofeminist scholar (Salleh, 1989), argues that the continuation of "silo thinking – the hegemonic separation between economics, sociology, and ecology – means that the transdisciplinary problem-solving needed to remedy these contradictions is not available".

A situational analysis of SDG 4 with a gender lens in the Asia-Pacific region conducted by the United Nations Girls' Education Initiative (2019) shows gender equality is underrepresented in national curricula and that the region's female potential in the sciences is mostly left dormant, an unfortunate situation given nurturing global citizenship is one way to achieve gender equality and education for social transformation. Other key findings related to gender and education are that girls are still likely to be excluded from school (53% out of school primary aged children are girls), there is a need for gender-responsive learning environments, there is a gender imbalance in tertiary and technical and vocational education (TVET) enrollments, and there is dearth of research on gender in TVET and tertiary education. Women remain disadvantaged in educational attainment, they lack ICT skills, and poverty and disability are major barriers to engagement with formal education, and much greater gender responsiveness can be integrated into teacher education. Such problems will not be resolved without deploying transformative strategies.

This problematization of the SDGs and the calls for different approaches and new forms of agency that question both the feminist and economic theories underpinning current directions provide the challenge for ecofeminist environmental education in the 21st century. We need to be paying attention to decolonial feminism and intersectionality as well as the possibilities offered by new materialism for a resurgence of ecofeminist approaches.

Most obviously, we need to stop ignoring gender as a dimension in environmental education research, which is perhaps our biggest challenge. As David Wallace-Wells (in Meyer, 2019, np) recently said, we need to stop thinking that we can sit by and witness history, we need to act – and act now:

> I knew that there were political fights to be had there. But in general, I just intuited in a deep emotional way . . . that history did move forward and therefore my life was going to be an experience of witnessing progress. I feel very profoundly *not* that way anymore.

The Journal of Environmental Education (and other environmental education journals) needs to be foregrounding gender issues and problematizing the directions of society that systemically drive gender injustice and silence feminist critiques which demand systemic transformation. The climate emergency and the concurrent biodiversity emergency are material and need to be addressed through gender equality. Environmental education research can have a meaningful analytical, action and advocacy role toward achieving a just transformation. To return to Clay Schoenfeld's original question, "are we creating new dimensions?", the answer has to be yes; constantly, there is a climate emergency going on. Transformation includes transforming relationships that presently discriminate against all silenced members of the world.

Dedication

This article is dedicated to Polly Higgins (1968–2019), the British barrister who led a decade-long campaign for "ecocide" to be recognized as a crime against humanity and died much too young while we were writing this article. We will continue to work to make her dream become a reality: "What is required is an expansion of our collective duty of care to protect the natural living world and all life. International ecocide crime is a law to protect the Earth" (https://eradicatingecocide.com/).

References

Adams, C., & Gruen, L. (2014). Groundwork. In C. Adams & L. Gruen (Eds.), *Ecofeminism: Feminist intersections with other animals and the earth* (pp. 7–36). New York, NY: Bloomsbury.

Albrecht, G. A. (2019). *Earth emotions: New words for a new world.* Ithaca, NY: Cornell University Press.

Alvarez, I., & Lovera, S. (2016). New times for women and gender issues in biodiversity conservation and climate justice. *Development, 59*(3–4), 263–265. doi:10.1057/s41301-017-0111-z

Baggini, J. (2019, August 19). Greta Thunberg's attackers are morally bankrupt, but her deification isn't helpful. *The Guardian.* Retrieved from www.theguardian.com/commentisfree/2019/aug/19/greta-thunberg-attackers- climate-crisis-activist

Barrett, M. J. (2005). Making (some) sense of feminist poststructuralism in environmental education research and practice. *Canadian Journal of Environmental Education, 10,* 79–93.

Barron, D. (1995). Gendering environmental education reform: Identifying the constitutive power of environmental discourses. *Australian Journal of Environmental Education, 11,* 107–120. doi:10.1017/S0814062600003001

Bartos, A. E., & Ives, S. (2019). Learning the rules of the game': Emotional labor and the gendered academic subject in the United States. *Gender, Place and Culture, 26*(6), 778–794. doi:10.1080/0966369X.2018.1553860

Bätthge, S. (2010). *Climate change and gender: Economic empowerment of women through climate mitigation and adaptation?* Working Paper. Eschborn, Germany: Governance and Democracy Division, The Governance Cluster, Programme Promoting Gender Equality and Women's Rights. Retrieved from www.oecd.org/dac/gender-development/46975138.pdf

Bell, K. (2017). Book review: Gender and the environment by Nicole Detraz. *Environmental Education Research, 23*(10), 1510–1512. doi:10.1080/13504622.2017.1324619

Blocker, T. J., & Eckberg, D. L. (1997). Gender and environmentalism: Results from the 1993. *General Social Survey. Social Science Quarterly, 78*(4), 841–858.

Booth, A. L. (1998). Caring for nature 101, or alternative perspectives on educating natural resource managers and ecologically conscious citizens. *The Journal of Environmental Education, 29*(3), 4–9. doi:10.1080/00958969809599113

Boughey, A. S. (1974). *Man and the environment: An introduction to human ecology and evolution.* New York, NY: Macmillan.

Brennan, M. J. (1974). Total education for the total environment. *The Journal of Environmental Education, 6*(1), 16–19. doi:10.1080/00958964.1974.9941480

Burrus-Bammel, L. L., & Bammel, G. (1986). Gender test differences during an environmental camp. *The Journal of Environmental Education, 17*(3), 8–11. doi:10.1080/00958964.1986.9941412

Butt, N., Lambrick, F., Menton, M., & Renwick, A. (2019). The supply chain of violence. *Nature Sustainability, 2*(8), 742–747. doi:10.1038/s41893-019-0349-4

Caldecott, L., & Leland, S. (Eds.). (1983). *Reclaim the earth: Women speak out for life on earth*. London: Women's Press.

Carson, R. (1962). *Silent spring*. Greenwich, CT: Fawcett.

Ceaser, D. (2015). Significant life experiences and environmental justice: Positionality and the significance of negative social/environmental experiences. *Environmental Education Research, 21*(2), 205–220. doi:10.1080/13504622.2014.910496

Cherif, A. H. (1992). Barriers to ecology education in North American high schools: Another alternative perspective. *The Journal of Environmental Education, 23*(3), 36–46. doi:10.1080/00958964.1992.9942800

Corcoran, P. B., & Sievers, E. (1994). Reconceptualizing environmental education: Five possibilities. *The Journal of Environmental Education, 25*(4), 4–8. doi:10.1080/00958964.1994.9941958

Corral-Verdugo, V. (1993). The effect of examples and gender of third grader's ability to distinguish facts from opinions. *The Journal of Environmental Education, 24*(4), 26–28. doi:10.1080/00958964.1993.9943506

Curry, A. D., & Williams, R. A. (1975). Gathering and preparing wild foods. *The Journal of Environmental Education, 7*(2), 44–47. doi:10.1080/00958964.1975.9941526

d'Eaubonne, F. (1974). *Le feminisme ou la mort [Feminism or death]*. Paris, France: Pierre Horay.

Di Chiro, G. (1987). Applying a feminist critique to environmental education. *Australian Journal of Environmental Education, 3*(1), 10–17. doi:10.1017/S0814062600001270

Di Chiro, G. (2014a). Applying a feminist critique to environmental education. *Australian Journal of Environmental Education, 30*(1), 9–16. doi:10.1017/aee.2014.17

Di Chiro, G. (2014b). Response: Reengaging environmental education in the 'anthropocene'. *Australian Journal of Environmental Education, 30*(1), 17–17. doi:10.1017/aee.2014.18

Diamond, I., & Orenstein, G. F. (Eds.). (1990). *Reweaving the world: The emergence of ecofeminism*. San Francisco, CA: Sierra Club Books.

Dunlop, R. (2002). In search of tawny grammar: Poetics, landscape and embodied ways of knowing. *Canadian Journal of Environmental Education, 7*(2), 23–37.

Fawcett, L. (2000). Ethical imagining: Ecofeminist possibilities and environmental learning. *Canadian Journal of Environmental Education, 5*, 134–149.

Fawcett, L. (2002). Children's wild animal stories: Questioning inter-species bonds. *Canadian Journal of Environmental Education, 7*(2), 125–139.

Feder, J. L., Hirji, Z., & Mueller, P. (2019, February 11). A huge climate change movement led by teenage girls is sweeping Europe. And it's coming to the US next. *BuzzFeed.News*. Retrieved from www.buzzfeednews.com/article/lesterfeder/europe-climate-change-protests-teens.?

Fontes, P. J. (2002). The stories (woman) teachers tell: Seven years of community-action-oriented environmental education in the north of Portugal. *Canadian Journal of Environmental Education, 7*(2), 256–268.

Fry, K. (2000). Learning, magic, & politics: Integrating ecofeminist spirituality into environmental education. *Canadian Journal of Environmental Education, 5*(1), 2500–2212.

Gaard, G. (2011). Ecofeminism revisited: Rejecting essentialism and re-placing species in a material feminist environmentalism. *Feminist Formations, 23*(2), 26–53. doi:10.1353/ff.2011.0017

Gaard, G. (2015). Ecofeminism and climate change. *Women's Studies International Forum, 49*, 20–33. doi:10.1016/j.wsif.2015.02.004

Gershon, L. (2019, February 21). Rachel Carson's critics called her a witch. *JSTOR Daily.* Retrieved from https://daily.jstor. org/rachel-carsons-critics-called-her-a-witch/.

Global Gender and Climate Alliance (GGCA). (2013). *Overview of linkages between gender and climate change.* Retrieved from https://issuu.com/undp/docs/pb1_asiapacific_capacity_final.

Gough, A. (1996). Women and the environment. *Australian Journal of Environmental Education, 12*, 93–94. doi:10.1017/S0814062600004249

Gough, A. (1997). *Education and the environment: Policy, trends and the problems of marginalisation.* Melbourne: Australian Council for Educational Research.

Gough, A. (1999a). Recognising women in environmental education pedagogy and research: Toward an ecofeminist poststructuralist perspective. *Environmental Education Research, 5*(2), 143–161. doi:10.1080/1350462990050202

Gough, A. (1999b). Kids don't like wearing the same jeans as their mums and dads: So whose 'life' should be in significant life experiences research? *Environmental Education Research, 5*(4), 383–394. doi:10.1080/1350462990050404

Gough, A. (1999c). The power and the promise of feminist research in environmental education. *Southern African Journal of Environmental Education, 19*, 28–39.

Gough, A. (2013). Researching differently: Generating a gender agenda for research in environmental education. In R. B. Stevenson, M. Brody, J. Dillon, & A. Wals (Eds.), *International handbook of research on environmental education* (pp. 375–383). New York: Routledge for the American Educational Research Association. doi:10.4324/9780203813331.ch35

Gough, A., Russell, C., & Whitehouse, H. (2017). Moving gender from margin to center in environmental education. *The Journal of Environmental Education, 48*(1), 5–9. doi:10.1080/00958964.2016.1252306

Gough, A., & Whitehouse, H. (2003). The "nature" of environmental education research from a feminist poststructuralist standpoint. *Canadian Journal of Environmental Education, 8*, 31–43.

Gough, A., & Whitehouse, H. (2018). New vintages and new bottles: The "nature" of environmental education from new material feminist and ecofeminist viewpoints. *The Journal of Environmental Education, 49*(4), 336–349. doi:10.1080/0095896 4.2017.1409186

Gough, N., Gough, A., Appelbaum, P., Appelbaum, S., Doll, M. A., and Sellers, W. (2003). Tales from Camp Wilde: Queer(y)ing environmental education research. *Canadian Journal of Environmental Education, 8*, 44–66.

Greenall Gough, A. (1993a). *Founders in environmental education.* Geelong: Deakin University Press.

Greenall Gough, A. (1993b). Globalizing environmental education: What's language got to do with it? *Journal of Experiential Education, 16*(3), 32–39, 46. doi:10.1177/105382599301600306

Hallen, P. (2000). Ecofeminism goes bush. *Canadian Journal of Environmental Education, 5*, 150–166.

Hampel, B., Holdsworth, R., & Boldero, J. (1996). The impact of parental work experience and education on environmental knowledge, concern and behaviour among adolescents. *Environmental Education Research, 2*(3), 287–300. doi:10.1080/1350462960020303

Hart, C. (2017). En-gendering the material in environmental education research: Reassembling otherwise. *The Journal of Environmental Education, 48*(1), 46–55. doi:10.1080/00958964.2016.1249328

Henderson, J., Hursh, D., & Greenwood, D. (Eds.). (2017). *Neoliberalism and environmental education.* New York, NY: Routledge.

Higgins, P. (2011/2015). *Eradicating ecocide: Laws and governance to stop the destruction of the planet.* London, UK: Shepheard-Walwyn Publishers.

Higgins, P. (2012). *Earth is our business: Changing the rules of the game.* London, UK: Shepheard-Walwyn Publishers.

Hildebrand, G. (1989). Creating a gender-inclusive science education. *Australian Science Teachers' Journal, 35*(3), 7–16.

Hines, M., Hungerford, H. R., & Tomera, A. N. (1987). Analysis and synthesis of research on responsible environmental behavior: A meta-analysis. *The Journal of Environmental Education, 18*(2), 1–8. doi:10.1080/00958964.1987.9943482

Horvat, R. E., & Voelker, A. M. (1976). Using a Likert scale to measure environmental responsibility. *The Journal of Environmental Education, 8*(1), 36–47. doi:10.1080/00958964.1976.9941556

International Bank for Reconstruction and Development (IBRD). (2019). *Women, business and the law 2019: A decade of reform.* Washington, DC: World Bank. Retrieved from https://openknowledge.worldbank.org/bitstream/handle/10986/31327/WBL2019.pdf

Jickling, B. (2003). Environmental education and environmental advocacy: Revisited. *The Journal of Environmental Education, 34*(2), 20–27. doi:10.1080/00958960309603496

Kaufman, J. S., Ewing, M. S., Hyle, A. E., Montgomery, D., & Self, P. A. (2001). Women and nature: Using memory-work to rethink our relationship to the natural world. *Environmental Education Research, 7*(4), 359–377. Reprinted in *Environmental Education Research, 12*(3–4) (2006). doi:10.1080/13504620600942774

Kennedy, J. (2018). Book review: The Palgrave international handbook of women and outdoor learning. *Environmental Education Research, 24*(12), 1699–1701. Published online 23 December 2018. doi:10.1080/13504622.2018.1545157

Kenway, J., & Gough, A. (1998). Gender and science education in schools: A review 'with attitude'. *Studies in Science Education, 31*(1), 1–29. doi:10.1080/03057269808560110

King, Y. (1987). What is ecofeminism? *The Nation, 702*, 730–731.

King, Y. (1989). The ecology of feminism and the feminism of ecology. In J. Plant (Ed.), *Healing the wounds: The promise of ecofeminism* (pp. 18–28). Philadelphia, PA/Santa Cruz, CA: New Society Publishers.

King, Y. (1990). Healing the wounds: Feminism, ecology, and the nature/culture dualism. In I. Diamond & G. F. Orenstein (Eds.), *Reweaving the world: The emergence of ecofeminism* (pp. 106–121). San Francisco: Sierra Club Books.

Koehler, G. (2016). Tapping the sustainable development goals for progressive gender equity and equality policy? *Gender & Development*, *24*(1), 53–68. doi:10.1080/13552074.2016.1142217

Kolbert, E. (2014). *The sixth extinction. An unnatural history.* New York, NY: Bloomsbury.

Lahar, S. (1991). Ecofeminist theory and grassroots politics. *Hypatia*, *6*(1), 28–45. doi:10.1111/j.1527-2001.1991.tb00207.x

Laybourn-Langton, L., Rankin, L., & Baxter, D. (2019). *This is a crisis: Facing up to the age of environmental breakdown.* London, UK: Institute for Public Policy Research. Retrieved from www.ippr.org/research/publications/age-of-environmental-breakdown

LeHart, D., & Tillis, C. R. (1974). Using wildlife to teach environmental values. *The Journal of Environmental Education*, *6*(1), 43–48. doi:10.1080/00958964.1974.9941486

Liefländer, A., & Bogner, F. (2014). The effects of children's age and sex on acquiring pro-environmental attitudes through environmental education. *The Journal of Environmental Education*, *45*(2), 105–117.

Lloro-Bidart, T. (2017). A feminist posthumanist political ecology of education for theorizing human-animal relations/relationships. *Environmental Education Research*, *23*(1), 111–130. doi:10.1080/13504622.2015.1135419

Lotz-Sisitka, H., & Burt, J. (2002). Writing environmental education research texts. *Canadian Journal of Environmental Education*, *7*(1), 132–151.

Lousley, C. (1999). (De)politicizing the environment club: Environmental discourses and the culture of schooling. *Environmental Education Research*, *5*(3), 293–304. doi:10.1080/1350462990050304

Maina-Okori, N. M., Koushik, J. R., & Wilson, A. (2018). Reimagining intersectionality in environmental and sustainability education: A critical literature review. *The Journal of Environmental Education*, *49*(4), 286–296. doi:10.1080/00958964.2017.1364215

Malone, K. (1999). Environmental education researchers as environmental activists. *Environmental Education Research*, *5*(2), 163–177. doi:10.1080/1350462990050203

Mann, T. (2015, December 31). *COP21 insights: Climate wise women.* Retrieved from http://gender-climate.org/resource/cop21-insights-climate-wise-women/

Martusewicz, R. A. (2014). Letting our hearts break: On facing the "hidden wound" of human supremacy. *Canadian Journal of Environmental Education*, *9*, 31–46.

McKenzie, M. (2004). The 'willful contradiction' of poststructural socio-ecological education. *Canadian Journal of Environmental Education*, *19*, 177–190.

McKenzie, M. (2005). The 'post-post period' and environmental education research. *Environmental Education Research*, *11*(4), 401–412. doi:10.1080/13504620500169361

Meade, A. (2019, August 2). Greta Thunberg hits back at Andrew Bolt for 'deeply disturbing' column. *The Guardian.* Retrieved from www.theguardian.com/environment/2019/aug/02/greta-thunberg-hits-back-at-andrew-boltfor-deeply-disturbing-column

Meyer, R. (2019, February 22). The 3 big things that people misunderstand about climate change. *The Atlantic.* Retrieved from www.theatlantic.com/science/archive/2019/02/david-wallace-wells-climate-change-interview/583360/?

Mignaquy, J. (2016). *Gender perspectives on climate change.* Retrieved from www.sprc.unsw.edu.au/media/SPRCFile/1_Gender_perspectives_on_climate_change_Jazmin_Mignaquy.pdf

Miles, J. C. (1976). Humanism and environmental education. *The Journal of Environmental Education, 7*(3), 2–10. doi:10.1080/00958964.1976.9941530

Miles, J. C. (1991). Teaching in wilderness. *The Journal of Environmental Education, 22*(4), 5–9. doi:10.1080/00958964.1991.9943055

Mohai, P. (1992). Men, women and the environment: An examination of the gender gap in environmental concern and activism. *Society & Natural Resources, 5*(1), 1–19. doi:10.1080/08941929209380772

Newbery, L. (2003). Will any/body carry that canoe? A geography of the body, ability, and gender. *Canadian Journal of Environmental Education, 8,* 204–216.

Nielsen, L. A., & Schroeder, M. (1983). A comparative analysis of textbooks for man-environment courses. *The Journal of Environmental Education, 14*(4), 7–11. doi:10.1080/00958964.1983.9943474

O'Donoghue, R. (2006). Locating the environmental in environmental education research: A review of research on nature's nature, its inscription in language and recent memory work on relating to the natural world. *Environmental Education Research, 12*(3–4), 345–357. doi:10.1080/13504620600799117

Olsson, D., & Gericke, N. (2016). The adolescent dip in students' sustainability consciousness-Implications for education for sustainable development. *The Journal of Environmental Education, 47*(1), 35–51.

Olsson, D., & Gericke, N. (2017). The effect of gender on students' sustainability consciousness: A nationwide Swedish study. *The Journal of Environmental Education, 48*(5), 357–371. doi:10.1080/00958964.2017.1310083

O'Manique, C., & Fourie, P. (2016). Affirming our world: Gender justice, social reproduction, and the sustainable development goals. *Development, 59,* 121–126. doi:10.1057/s41301-017-0066-0

One Million Women (OMW). (2015, September 2). *Why climate change is a woman's fight.* Retrieved from https://1millionwomen.com.au/blog/why-climate-change-womans-fight/?

Parker, L. H., Rennie, L. J., & Fraser, B. J. (1996). *Gender, science and mathematics: Shortening the shadow.* Dordrecht: Kluwer.

Pawlowski, A. (1996). Perception of environmental problems by young people in Poland. *Environmental Education Research, 2*(3), 279–300. doi:10.1080/1350462960020302

Phillips, M., & Rumens, N. (2016a). Introducing contemporary ecofeminism. In M. Phillips & N. Rumens (Eds.), *Contemporary perspectives on ecofeminism* (pp. 1–16). London, England: Routledge. doi:10.4324/9781315778686-1

Phillips, M., & Rumens, N. (Eds.). (2016b). *Contemporary perspectives on ecofeminism.* London, England: Routledge.

Phillips, M. E. (2016). Developing ecofeminist corporeality: Writing the body as activist poetics. In M. E. Phillips & N. Rumens (Eds.), *Contemporary perspectives on ecofeminism* (pp. 57–75). London, UK: Routledge. doi:10.4324/9781315778686-4

Phillips, M. E. (2019). "Daring to care": Challenging corporate environmentalism. *Journal of Business Ethics, 156*(4), 1151–1164.

Piersol, L., & Timmerman, N. (2017). Reimagining environmental education within academia: Storytelling and dialogue as lived ecofeminist politics. *The Journal of Environmental Education, 48*(1), 10–17. doi:10.1080/00958964.2016.1249329

Plant, J. (Ed.). (1989). *Healing the wounds: The promise of ecofeminism.* Philadelphia, PA, Santa Cruz, CA: New Society Publishers.

Plumwood, V. (1986). Critical review – ecofeminism: An overview and discussion of positions and arguments. *Australasian Journal of Philosophy, 64*(Suppl 1), 120–138. Supplement to Volume. doi:10.1080/00048402.1986.9755430

Plumwood, V. (1993). *Feminism and the mastery of nature.* London, New York: Routledge.

Quigley, C. F., Che, S. M., Achieng, S., & Liaram, S. (2017). Women and the environmental are together": Using participatory rural appraisal to examine gendered tensions about the environment. *Environmental Education Research, 23*(6), 773–796. doi:10.1080/13504622.2016.1169511

Red Cross. (2014). Gender and climate change. *Guidance Note.* Retrieved from www.redcross.org.au/files/2014_Gender_and_Climate_Change.pdf

Riechard, D. E., & Peterson, S. B. (1998). Perception of environmental risk related to gender, community, socio-economic setting and locus of control. *The Journal of Environmental Education, 30*(1), 11–19. doi:10.1080/00958969809601858

Riley, C. (2018). The history of feminism: A look to the past? *IPPR Progressive Review, 24*(4), 292–298. doi:10.1111/newe.12068

Royal Bank of Canada (RBC). (1960). The relationship of man and nature. *The Royal Bank of Canada Monthly Letter, 41*(4), 1–4.

Ruether, R. R. (1975). *New woman/new earth: Sexist ideologies and human liberation.* New York, NY: Seabury Press. doi:10.1017/s0360966900015802

Russell, C. (2005). Whoever does not write is written': The role of 'nature' in post-post approaches to environmental education research. *Environmental Education Research, 11*(4), 433–443. doi:10.1080/13504620500169569

Russell, C. (2006). Working across and with methodological difference in environmental education research. *Environmental Education Research, 12*(3–4), 403–412. doi:10.1080/13504620600799141

Russell, C., & Bell, A. (1996). A politicized ethic of care: Environmental education from an ecofeminist perspective. In K. Warren (Ed.), *Women's voices in experiential education* (pp. 172–181). Dubuque, IA: Kendall/Hunt Publishing.

Russell, C., Sarick, T., & Kennelly, J. (2002). Queering environmental education. *Canadian Journal of Environmental Education, 7*(1), 54–66.

Salleh, A. (2016). Climate, water, and livelihood skills: A post-development reading of the SDGs. *Globalizations, 13*(6), 952–959. doi:10.1080/14747731.2016.1173375

Salleh, A. K. (1989). Environmental consciousness and action: An Australian perspective. *The Journal of Environmental Education, 20*(2), 26–31. doi:10.1080/009589 64.1989.9943028

Sato, M., & Hansen, J. (2019). *Updating the climate science: What path is the real world following?* New York, NY: Columbia University. Retrieved from www.columbia.edu/~mhs119/

Schoenfeld, C. (1969). What's new about environmental education? *Environmental Education, 1*(1), 1–4.

Simmens, H. (2017). *A climate vocabulary of the future.* Tucson, AZ: Wheatmark.

Sjoberg, S., & Imsen, G. (1988). Gender and science education: I. In P. J. Fensham (Ed.), *Development and dilemmas in science education* (pp. 218–248). London: Falmer Press.

Skanavis, C., & Sakellari, M. (2012). Free-choice learning suited to women's participation needs in environmental decision-making processes. *Environmental Education Research, 18*(1), 1–17. doi:10.1080/13504622.2011.572158

Southern, B. H. (1969). Revitalizing natural resources education. *Environmental Education, 1*(1), 29–29. doi:10.1080/0013 9254.1969.10801478

Spender, D. (1990). *Man made language.* London: Pandora.

Spretnak, C. (1990). Ecofeminism: Our roots and flowering. In I. Diamond & G. F. Orenstein (Eds.), *Reweaving the world: The emergence of ecofeminism* (pp. 3–14). San Francisco, CA: Sierra Club Books.

Stapp, W. B. (1975). UNESCO's environmental education program. *The Journal of Environmental Education, 6*(4), 6–8. doi:10.1080/00958964.1975.9941997

Storey, C., Da Cruz, J. G., & Camargo, R. F. (1998). Women in action: A community development project in Amazonas. *Environmental Education Research, 4*(2), 187–199. doi:10.1080/1350462980040206

Stovall, H. A., Baker-Sperry, L., & Dallinger, J. M. (2015). A new discourse on the kitchen: Feminism and environmental education. *Australian Journal of Environmental Education, 31*(1), 110–131. doi:10.1017/aee.2015.11

Swan, J. (1983). Sacred places in nature: A unitive theme for a transpersonal approach to environmental education. *The Journal of Environmental Education, 14*(4), 32–37. doi:10.1080/00958964.1983.9943479

Thompson, C. (2006). Back to nature? Resurrecting ecofeminism after poststructuralist and third-wave feminisms. *Isis, 97*(3), 505–512. doi:10.1086/508080

Tuncer, G., Ertepinar, H., Tekkaya, C., & Sungur, S. (2005). Environmental attitudes of young people in Turkey: Effects of school type and gender. *Environmental Education Research, 11*(2), 215–233. doi:10.1080/1350462042000338379

UN Women. (2014). World survey on the role of women in development 2014: Gender equality and sustainable development. *United Nations.* Retrieved from https://sustainabledevelopment.un.org/content/documents/1900unwomen_surveyreport_advance_16oct.pdf

UN Women. (2018a). *Turning promises into action: Gender equality in the 2030 agenda for sustainable development.* Retrieved from www.unwomen.org/sdg-report

UN Women. (2018b). *Women and the sustainable development goals (SDGs).* Retrieved from www.unwomen. org/en/news/in-focus/women-and-the-sdgs

UN WomenWatch. (2009). Women, gender equality and climate change. *Fact Sheet.* Retrieved from www.un.org/womenwatch/feature/climate_change/downloads/Women_and_Climate_Change_Factsheet.pdf

UNESCO. (1975). *The Belgrade charter: A global framework for environmental education.* Retrieved from https://unesdoc.unesco.org/ark:/48223/pf0000017772

UNESCO. (1978). *Intergovernmental conference on environmental education: Tbilisi (USSR), 14–26 October 1977.* Final Report. Paris: UNESCO.

United Nations. (1993). *Agenda 21: Earth summit: The United Nations programme of action from Rio.* New York: United Nations. doi:10.18356/a9d63da6-en

United Nations. (2015). *Transforming our world: The 2030 agenda for sustainable development.* Retrieved from https://sustainabledevelopment.un.org/post2015/transformingourworld

United Nations Girls' Education Initiative. (2019). *Situation analysis of SDG 4 with a gender lens.* Bangkok: UNESCO. Retrieved from https://bangkok.unesco.org/content/situation-analysis-sdg4-gender-lens

Villanueva, P. (2019, February 9). Why decolonial feminism: New possibilities from Abya Yala (trans. Earth Alive). *Toward Freedom.* Retrieved from https://towardfreedom. org/archives/women/why-decolonial-feminism-new-possibilities- from-abya-yala/

Wallace-Wells, D. (2019a). *The uninhabitable earth: A story of the future.* London, UK: Allen Lane.

Wallace-Wells, D. (2019b, February 2). The devastation of human life is in view: What a burning world tells us about climate change. *The Guardian.* Retrieved from www.

theguardian.com/environment/2019/feb/02/the-devastation-   of-human-life-is-in-view-what-a-burning-world-tells-us-about-climate-change-global-warming

Wane, N., & Chandler, D. J. (2002). African women, cultural knowledge, and environmental education with a focus on Kenya's indigenous women. *Canadian Journal of Environmental Education, 7*(1), 86–98.

Waring, M. (1988/1999). *Counting for nothing: What men value and what women are worth*. Toronto, Canada: University of Toronto Press. doi:10.7810/97808 68615714

Warren, K. J. (1987). Feminism and ecology: Making connections. *Environmental Ethics, 9*(1), 3–20. doi:10.5840/enviroethics19879113

Whitehouse, H., & Taylor, S. (1996). A gender inclusive curriculum model for environmental studies. *Australian Journal of Environmental Education, 12*, 77–83. doi:10.1017/S0814062600001609

World Health Organization. (2014). *Gender, climate change and health*. Retrieved from www.who.int/globalchange/publications/reports/gender_climate_change/en/

Yeatman, A. (1990). *Bureaucrats, technocrats, femocrats: Essays on the contemporary Australian state*. Sydney: Allen & Unwin. doi:10.1177/072551369103000111

Section II

Feminisms and nature in environmental education

This second thematic grouping (with colleague Hilary Whitehouse and on my own) focuses on the construction of and relationships with "nature" and the "nature" of environmental education research from a feminist perspective, particularly how poststructuralist, new materialist and ecofeminist perspectives can enrich how we research in environmental education.

As discussed in Chapter 1, since the 1960s environmental education discourses have been constructed around persistent Cartesian dualisms of modernist thought that divide an "othered" category of being from that of a constituted homogeneous human identity. This is evident in the Stockholm Declaration from the 1972 United Nations Conference on the Human Environment, which states:

> Man is both creature and moulder of his environment, which gives him physical sustenance and affords him the opportunity for intellectual, moral, social and spiritual growth. . . . Both aspects of man's environment, the natural and the man-made, are essential to his well-being and to the enjoyment of basic human rights – even the right to life itself.
>
> (United Nations 1973, 3)

The sexist language of these sentences is significant, and it is important to note the separation of "man" from "his" environment because this separation continued through to the definitions of environmental education that were framed around the same time. For example, the draft United States Environmental Education Act of 1970 (cited, for example, in Linke 1980, 26): "Environmental education is an integrated process which deals with man's interrelationship with his natural and man-made surroundings". In contrast, in Chapter 7, Hilary Whitehouse and I argue that adopting a research approach which moves away from representations of universalised subjects, such as the mythic "Man" and "His Environment," and towards a distinct recognition of multiple subjectivities will create research in environmental education that more accurately represents the diversities of lived experiences. Such a feminist poststructuralist approach is seen as one way to reveal certain dimensionalities

DOI: 10.4324/9781003390930-8

that may otherwise be ignored or silenced within the field of environmental education. Following Patti Lather (1991), what we saw as particularly attractive about feminist poststructuralism was that within it

> language moves from representational to constitutive; binary logic implodes, and debates about "the real" shift from radical constructivism to a discursively reflexive position is mediated by the concepts and categories of our understanding.
>
> (39)

By focusing on the language and meanings through which we constitute our ontological and epistemological understandings, we can disrupt the binaries man/woman and woman/nature (among others) and challenge the "nature" of environmental education.

Chapter 8, which is an article Hilary and I wrote 15 years later, explores the "nature" of environmental education as informed by new material feminist, ecofeminist and other viewpoints, interrogating their similarities and differences, their relationship to feminist poststructuralism and their implications for environmental education research and practice. We acknowledged that, in the intervening years between the articles, feminist new materialism has emerged, ecofeminism has had an enlivening resurgence, and feminist scholarship in environmental education has expanded. We also noted that "nature" seems to have declined within dominant discourses, supplanted by a more anthropocentric agenda that continues to create a binary between humans and their environments. We argued that it is important for environmental education researchers to draw on all the available ideas about new materialism – such as in the work of Karen Barad and Donna Haraway – more conclusively, comprehensively and coherently to advance thinking in environmental education research. In particular, environmental education researchers should draw more inclusively on international sociological and cultural thinking across ecofeminism and material discursive analyses. We also argued that there is a need for more research that explores the possibilities of an intersectional ecological-feminist approach as well as exploring the implications of postnature and post-humannature, particularly if we accept that a nature untouched by humanity no longer exists anywhere on the planet, that we can no longer look to any ice floe or mountain top or tropical forest and not find evidence of our impacts.

Chapter 9 is another article I wrote with Hilary Whitehouse for a special issue of the journal *Environmental Education Research* on new materialisms and environmental education, around the same time as we were writing Chapter 8 for the special issues of *The Journal of Environmental Education* on gender and environmental education. Both articles take up the challenge from Richard Twine (2010, 402) that "the emergence of a feminist new materialism ought to usher in a renewed conversation between feminism and ecofeminism due to shared interests" and added environmental education research to the conversation. This article discusses the apparent amnesia with regard to insights manifested in ecofeminist thought and applies a re-collective analysis to thinking on the implications of an ecofeminist new materialism for contemporary environmental education research and curriculum practice. We engage in a conversation between feminist new materialism and the tropes of ecofeminism at this very unusual time in human history, making visible such interactions. Drawing attention to this and other apparent amnesias and, arguing from a genealogical perspective, we argue the scholarly and conceptual disruption caused by rapidly changing environmental (hence social and cultural) conditions can be fruitfully understood and analysed through a reconceived new materialist ecofeminism.

Some of these ideas are pursued in Chapter 10, which was an invited essay on Stefan Helmreich's (2009) book *Alien Ocean*, in a special issue of the journal *Cultural Studies* <=> *Critical Methodologies*. However, rather than just focusing on Helmreich's book, I put it into conversation with Elspeth Probyn's (2016) *Eating the Ocean* and other works, such as Carolyn Merchant's (2016) *Autonomous Nature* and Steve Mentz's (2012) article "After sustainability", as background to rethinking possibilities for inquiries towards a more-than-human scientific inquiry curriculum. Each of these works challenges readers to think differently about how we conceive of "nature" and more-than-human. I draw on the other three works to discuss the silences in Helmreich's arguments related to sustainability, gender and new materialisms, and then I consider how we can conceptualise a more-than-human scientific inquiry curriculum, which is very different from the current image of scientific inquiry common in Western curricula.

References

Helmreich, Stefan. 2009. *Alien Ocean: Anthropological Voyages in Microbial Seas.* Berkeley: University of California Press.

Lather, Patti. 1991. *Getting Smart: Feminist Research and Pedagogy with/in the Postmodern.* New York/London: Routledge.

Linke, Russell. 1980. *Environmental Education in Australia.* Sydney: Allen & Unwin.

Mentz, Steve. 2012. "After sustainability." *PMLA 127*(3): 586–592.

Merchant, Carolyn. 2016. *Autonomous Nature: Problems of Prediction and Control from Ancient Times to the Scientific Revolution.* New York, NY: Routledge.

Probyn, Elspeth. 2016. *Eating the Ocean.* Durham: Duke University Press.

Twine, Richard. 2010. "Intersectional disgust? Animals and (eco)feminism." *Feminism & Psychology 20*(3): 397–406. doi:10.1177/0959353510368284.

United Nations. 1973. *Report of the United Nations Conference on the Human Environment, Stockholm, 5–16 June 1972.* New York: United Nations.

7 The "nature" of environmental education research from a feminist poststructuralist standpoint

Annette Gough and Hilary Whitehouse

Gough, A. & Whitehouse, H. (2003). The "nature" of environmental education research from a feminist poststructuralist standpoint. *Canadian Journal of Environmental Education, 8*, 31–43.

Abstract

For a generation or more, environmental education discourses have been constructed around persistent Cartesian dualisms of modernist thought that divide an "othered" category of being from that of a constituted homogeneous human identity. During the same period, both feminist and poststructuralist theorizing have acted to destabilize the constitution of identities, revealing knowledge, including environmental knowledge, to be multiple, subjective, contingent, and intimately tied in with embodied experiences of place. We explore some of the contingencies of environmental knowledge as revealed through a poststructuralist feminist research methodology and the place for such understandings within an early twenty-first-century vision for environmental education research and practice.

Résumé

Pendant au moins une génération, les discours de l'ERE ont été construits en fonction de dualismes cartésiens persistants issus de la pensée moderniste et opérant une séparation entre une catégorie "autre " de l'être et un ensemble constitué et homogène de l'identité humaine. Pendant cette même période, les théories féministes et post-structuralistes se sont employées à déstabiliser la constitution d'identités en montrant que le savoir, y compris le savoir environnemental, était multiple, subjectif, contingent et intimement lié à une expérience infuse des lieux. Nous explorons certaines des contingences propres au savoir environnemental, telles que révélées par une méthodologie de recherche féministe post-structuraliste. Nous sondons également quelle place peut être faite à de telles réflexions au sein de l'optique envisagée en ce début de XXIe siècle pour le domaine de la recherche et de la pratique de l'ERE.

(The rise of the global environmental movement in the 1960s coincided with the innovations of feminist theory and the contributions of European cultural

DOI: 10.4324/9781003390930-9

theorists whose insights have come to be collected under the umbrella term of poststructuralism. As part of these contributions, "the self-fulfilling autonomous subject-acting-on-an-object . . . associated with the concurrent domination of nature, women and non-European cultures" (Conley, 1997, p. 1) came under intense scrutiny as theorists found new ways for thinking through the ecological and socio-cultural complexities of twentieth-century life.)

The resultant shifts in thinking emerging in the last four decades have opened up new ways for doing research in educational fields. Poststructuralism, as a movement, owes many of its original ideas to an understanding of ecological awareness, although in the "tidal ebb and flow" of these ideas, the ecological connection has not always been apparent (Conley, 1997, p. 1). In this chapter we discuss the contributions to contemporary environmental education made by both poststructuralist and feminist ideas. We argue that, at the intersection of these three major intellectual movements of the twentieth century, we can find productive methodologies for undertaking environmental education research in the twenty-first century.

Harding (1987) defined methodologies as theories and analyses of how research should proceed. We have used feminist poststructuralist research methodology for a number of years to research alternatives for thinking through constitutions of environmental knowledge (Davies & Whitehouse, 1997; Whitehouse, 2000, 2001, 2002) and the subjects of international environmental education teaching and research (Gough, 1994, 1997a, 1999a, 1999b). An important contribution of poststructuralist thinking is that it brings into focus the subject of subjectivity to consider the ways in which we experience ourselves within space, place, and time (Probyn, 2003). The contribution of feminism has been to reveal gender as central to conceptualisations of the agentic subject (Davies, 1993). In this chapter, we use a blended narrative of personal research stories and analysis to expose how we have been thinking on the combined contribution a feminist *and* poststructuralist analysis can make to future environmental education research.

Annette's story

Having worked in the field of environmental education for nearly 20 years, I came to my doctoral research study (Gough, 1994) with an intention of looking at environmental education as a "man-made subject," drawing attention to the absence of female voices in environmental education discourses and proposing some strategies for their inclusion: an approach that could, perhaps, be caricatured as essentialist and liberal feminist. Through researching, reading and writing I increasingly came to recognize that women are one of many marginalized groups, absent and/or silenced, in the foundational discourses of environmental education, and that multiple subjectivities abound. The project I started was not where I ended, but I learned much about identity, subjectivity, and myself along the way.

In undertaking an analysis of the "foundations" of the field, I was inspired by the words of A.S. Byatt (1990) to create a reading of the texts of the field which had

> a sense that the text has appeared to be wholly new, never before seen, . . . followed, almost immediately, by the sense that it was always there, that we the readers, knew it was always there, and have always known it was as it was, though we have now for the first time recognized, become fully cognisant of, our knowledge.
>
> (p. 472)

And although I did not have these words at the time, Carolyn Heilbrun's (1999) notion of the state of liminality was also where I felt I was situated:

> The word "limen" means "threshold" and to be in a state of liminality is to be poised upon uncertain ground, on the brink of leaving one condition or country or self to enter upon another. But the most salient sign of liminality is its unsteadiness, its lack of clarity about exactly where one belongs and what one should be doing, or wants to be doing.
>
> (p. 3)

I embarked upon a study to explore the foundations of environmental education in terms of its grounding in modern science as well as the gender relationships in society. The discourses I related to environmental education were:

- its grounding in modern science;
- its relationship with behaviorist and critical research in education;
- the political and economic worldviews that are both explicit and implicit in its rhetoric;
- its colonialism; and,
- its relationship with developments in philosophy, particularly ecofeminism and feminist epistemology.

In analyzing and drawing attention to the relationships between the discourses of environmental education and other significant discourses I felt that I was providing a "traitorous" (Harding, 1991, p. 288) reading of the field by reading against the grain of my dominant experiences in the field, and against the founding fathers' stories with "a focal interest in signification, in power/ knowledge relationships, in the harm done by master-narratives, and in the way institutional structures are controlled" (Greene, 1992, p. ix).

The "founding tongues" of environmental education were males from scientific backgrounds, and I analyzed the work and words of these founders by adopting feminist research methodologies. In many ways my study was attempting, on a smaller scale, to apply to environmental education what Carolyn Merchant (1980) did in her study of "women, ecology and the Scientific Revolution." The founders and foundations of environmental education, particularly those in Australia, were therefore the focus of my study. There were also cross-cultural references to the situation in the United States of America because of the ongoing links between the two countries with respect to environmental education, and the influence that the earlier developments in the United States had on developments in Australia and on international discourses in the field (Gough, 1997b).

The study had as its goal an analysis of the foundational discourses of environmental education, the outcome of which was "neither unitary wholeness nor dialectical resolution" (Lather, 1991, p. 13) but rather the suggestion of some different research principles for environmental education which rejected traditional binaries of "Man" and "Environment" and incorporated perspectives from ecofeminisms, feminist epistemologies, and feminist research methodologies. In so doing, and in the spirit of the quotation from Byatt (1990) mentioned earlier, I provided another reading of the same text of environmental education, but one which, I hoped, was toward being "wholly new" but also recognized as always having been there.

A feminist standpoint

The major contributions of feminist research, in all its many forms, have been to raise the question of epistemological claims such as who can be an agent of knowledge, what counts as knowledge, what constitutes and validates knowledge, and what the relationship should be between knowing and being. Feminist questions put the social construction of gender at the center of research (Lather, 1991), and "what 'grounds' feminist standpoint theory is not women's experiences but the view from women's lives" (Harding, 1991, p. 269).

Feminism enables people to re-vision their world – "to know it differently than we have ever known it; not to pass on a tradition, but to break its hold on us" (Rich, 1990 in Crotty, 1998, p. 182). To quote Heilbrun (1999) again,

> Women began to portray the new possibilities that, as a result of feminism, they found themselves confronting. They began to question . . . all strictures about women and about the institutions in which women now, in even greater numbers, and in a state of awakening, found themselves.
>
> (p. 8)

Ecological feminists have embraced personal and political action to "fully engage in the interweaving of humour, irony, grace, resistance, struggle and transformation" (Sandilands, 1999, p. 210) to envision a more democratic future for all. And, increasingly, environmental education researchers whose work is informed by feminism are publishing their work (see, for example, Fawcett, 2000, 2002; Lotz-Sisitka & Burt, 2002; Lousley, 1999; Malone, 1999; Russell, 2003), after a rather dry spell (Gough, 2001). But despite these inroads, the subject of gender remains marginal to much environmental education research.

As we see it, many researchers still consider "human" subjectivity to be homogeneous, ungendered, and unproblematic when, in fact, a vast edifice of sociological research reveals the opposite to be the case. Environmental education research remains bound up with traditional epistemological frameworks of scientific research, which have, in Sandra Harding's (1987) words, "whether intentionally or unintentionally, systematically exclude(d) the possibility that women could be 'knowers' or *agents of knowledge*" (p. 3, emphasis in original). For example, recent writings on significant life experience research in *Environmental Education Research* can be critiqued as remaining blind to gendered subjectivities (Gough, 1999c).

Addressing the balance is simply not a matter of "adding women" to traditional analyses. What is needed is a transformative process where new empirical and theoretical resources are opened up to reveal new purposes and subjects for inquiry. It is our argument that what needs to come under scrutiny is the implicit constitution of the assured, homogeneous, and universalized human subject of much environmental research. "Human" identity as constituted through positivist research regimes is not inclusive of all the different ways of being in the world.

Much past environmental education research has analyzed only male experiences or has constructed universalized subjects, which are not distinguished as male or female. Yet, there is no universal "Man" who acts as a powerful agent on an equally symbolic "Environment" – except perhaps in the imaginations of writers who reproduce these discourses. "Man" is not a term that is logically inclusive of women. Early formulations of environmental education, such as the IUCN (1970) definition (as cited in Linke, 1980), referred to "the interrelatedness among man, his culture, and his biophysical surroundings" (pp. 26–27). Although more recent environmental education literature is gender neutral in its language, this too is a problem as the neutral voice is still interpreted as a male by readers of both genders. As Cherryholmes (1993) argues, "texts that deny gender present themselves as generic. They pretend to speak the truth and truth is gender-neutral. Authoritative texts are distanced, objective, have a single voice (otherwise they would not be authoritative), are value-neutral, dispassionate and controlling" (p. 10). Perhaps the shackles of the past are proving difficult to shrug off, but the practice of creating gender-blind binaries is exclusive of lived experience.

In reality, we have culturally, racially, socio-economically, sexually (and so on) different people with fragmented identities whose experiences and understandings can only be constituted through the lenses of subjectivity. However, to date, environmental education research has rarely addressed areas of different women's experiences and knowledges, which means that many useful insights have not been adequately pursued. We acknowledge that environmental education research has generally ignored other aspects of human identities too, but these are beyond the scope of this chapter. Our emphasis here is on women's experiences and knowledge.

The problem of binary thinking

Many ecofeminist researchers have discussed the "Man" and "Environment" binary and associated the destruction of nonhuman nature with the oppression of women (see, for example, Eckersley, 1992; Merchant, 1996; Plumwood, 1993; Salleh, 1997). As Eckersley (1992) notes, ecofeminists have embraced the association of women and nature "as a source of empowerment for women and the basis of a critique of the male domination of women *and* nonhuman nature" (p. 64, emphasis in original). However, ecofeminist writing to date has tended to be critical of postmodern and poststructuralist approaches. For example, Salleh (1997) argues "the tenets of deconstructive practice have been catechised and used as political rhetoric, resulting in an impractical nihilism when applied to everyday life" (p. 9). Thus, while our work is informed by ecofeminist writings, we reject such criticisms of the postmodern to argue that feminist poststructuralist methodologies can be productive for environmental education research.

From our perspective, the problem of the "nature" of environmental education research is further compounded if we look at the other side of the binary to the constitution of the "Environment," produced as an object of study, rather than as a subject for research. In support of our stance we draw upon Harding (1987), who argues that the best feminist analysis "insists that the inquirer her/himself be placed in the same critical plane as the overt subject matter" and that the researcher "must be placed within the frame of the picture that she/he attempts to paint" (p. 9). Research on an imagined "Environment" distanced and objectified and empirically impossible to determine, does not fall within this feminist research rubric. In addition we ground our work in the arguments of Taylor (1991), who makes the case for constituting "many women, many environments," in order to expand research thinking, and, extending this notion, Conley (1997) suggests the constitution of environmental, or ecological subjectivities as worthy of theorizing and study.

Jagtenberg and McKie (1997) argue that "the vocabularies of social theory are limited when it comes to characterising the relationships between humanity and other species" (p. 8). Analysis of recent environmental education research shows a minimal approach toward tackling the limiting vocabularies through which the world may become known. Most of this research remains bounded within the modernist/positivist constitution that "Man" (now transmogrified into "Human" through equally gender-blind language) is indeed a fixed and separate actor from a distinct and singular "Environment." Yet, if the world is indeed divided into these two categories of being, why are the boundaries between them impossible to locate?

Similar problems arise when making a close examination of the categories "Human" and "Nature." Soper (1995) argues that "nature" is impossible to define, because it is not a thing existent in itself, but a category of human identity. "Nature" exists to define what is or is not "us" in traditional Western thought. And membership of the category "us/human" has shifted constantly throughout Western history. Women, on the basis of sex alone, used to be confined to the category of "other" in the not-so-distant past.

Binary thinking traps research processes into a persistent stasis, erasing a complexity, which might otherwise be meaningful in the pursuit of elegant solutions to contemporary socio-environmental problems. Noel Gough (1991) asserts that to uncritically accept the positivist meta-narratives of Western knowledge represents a failure of responsibility for creatively "singing the world into existence," which may be one of the functions of innovative environmental research. While it was difficult for both of us to abandon binary thinking (being very well schooled in binary practices), we did manage (over time and in conversation with each other) to take up the challenge of doing our own singing – and we are willing to risk the consequences. In our view, being able to think more complexly opens up exciting possibilities for research. Complexity is an exciting invitation.

Hilary's story

I spent at least 30 years of my life talking through the familiar binary discourses of "man" and "nature," remaining quite ignorant of the power of language to shape the world. I hadn't realized how "naturalized" these discourses were, nor was I aware of how differently the world can be spoken into existence. Even when I shifted my own terminology from "man" to "human" in line with general feminist understandings and education policy, the nature of "nature" remained enigmatic. The term is so deeply embedded within Western culture that it becomes almost impossible for us (this indeterminately authored, whitey humanized "us") to think outside the binary categories of "human" and "other." Our fellow earth travelers, those multitudinous fleshy bodies who become constituted as the other, as the "non-human," and as "other species," get lumped together in a category of identity called "nature" and its twin "the environment," and these terms litter all the environmental literature as if they are indeed unproblematic and universally understood.

We (I am using this term most advisedly) all think "we" know what nature looks like and smells like. "Nature" is green and blue and gray and red. It smells green and fresh and salty and damp. "Nature" is composed of all those beings who do not fall into the (shifting and mobile) category of "human." "The environment" is a similar category of being, though perhaps a little less explicitly "natural." "Nature" and "the environment" are conceived in common socio-educational and politico-economic discourses as knowable and understandable terms. Hey, "we" all know what nature is, right?

No, "we" don't.

It took my meeting with Nora, a university lecturer from Papua New Guinea, to fully accept how blind I had been in initiating research into tropical environmental meanings without fully comprehending the discourses I was negotiating. Nora took it upon herself to explain to me, in her most graceful and perceptive way, how it was not only possible to think differently but, indeed, that many people do, as part of their own dominant socio-cultural practices.

Nora was born in the Western Province of Papua New Guinea and lived as a child on an island in the Fly River. She moved to Daru and then Port Moresby to attend school and university.

Nora explained to me that in her village language (as opposed to English, which she had learned for the purpose of getting herself an education) there is "no term for nature." A phrase such as "human relationships with Nature" (so common in environmental education curriculum discourses) does not carry any relevant meaning in her first language. As Nora told me: "Nature in my language, it's not there. I mean there is no such word as nature. It's not part of the language. The only way you can relate to the word is by the individual names of things, your experiences in living and interacting with the seas, the forests."

Nora explained that in the Fly River everybody has their own name. Personhood is not imagined as having special status over differently embodied forms. The crocodile owns the river so people must be careful. Trees are not classified together with birds on the basis of possessing a characteristic called naturalness. Trees are in and of themselves part of the collective imagination that binds space, place, and subjectivity together in ways that I (being so well schooled in binary thinking) can only hazily imagine.

Nora learned to talk nature as part of her high school experience of an Australian-designed curriculum. She figures that somewhere between Grade 9 and Grade 12 she had learned to "speak nature." Nora's story is not an isolated one. Many educators from the Pacific Island region have learned to speak nature at school (Whitehouse, 2000).

What I really learned from my research was how to question. If powerful binary discourses holding a "humanised identity" firmly in place remain the subjects of environmental education, how much are we really changing things? If current research and curriculum practices deliberately ignore different ways of speaking the world into existence, how democratic is international environmental education? If environmental education unproblematically reproduces discourses of "the natural" without explicitly recognizing that there are multiple ways in which to think about and comprehend the world, can democracy in practice truly be claimed?

These days the terms "human" and "nature" fail to convince me with their meaning. I read the environmental literature as being in a liminal state, as a mosaic of understandings. It was an intense intellectual struggle to come to terms with the binary habits of language through which I had been so well schooled. But it is to the power of language to shape our understandings that we now need to turn our attention.

Contributions from poststructuralism

A key Nietzschean-Foucauldian insight (to identify the genealogy) is that truth cannot be separated from the procedures of its production (Tamboukou, 1999). Any research methodology will reveal its own set of truths. In spite of a long history of such claims, there is no empirically discovered set of universal "large T" truths concerning the differential production of social and environmental knowledge. What can be discovered is that which can be revealed through the investigative methodology and the conditions of investigation. Modernist and positivist understandings will therefore differ from postmodernist and poststructuralist understandings as to the creation and applications of knowledge. This rich diversity of viewpoints needs to be recognized and celebrated for what it reveals about social and environmental meanings and actions.

We do not live in a mono-dimensional universe. According to recent research, a concept of 11 dimensions is the minimum needed to attempt an explanation of the evolution of the multiverse (Barrow, 2002). It therefore seems unrealistic to expect that any one research methodology, as a set of "small t" truth claims, will suffice to give a complete understanding of an identified field of human endeavor such as environmental education. No one methodology or way of constructing the world can either encompass or reveal all possibilities for knowing or for effective environmental action, and no favorite chosen research methodology can be the only way to (a singular) truth. When set alongside all other methodologies in education research, a feminist poststructuralist approach can be very informative and revealing of certain dimensionalities that may otherwise be ignored or silenced within the field.

There are many attractions to a poststructuralist approach. Quigley (1995) made the point that the ecological project "would benefit from a thorough reconsideration in light of poststructuralist philosophy" (p. 592) because the traditional postures of ecological thought shared too many features with the traditional power structures the ecological movement wished to oppose. Our own attraction to poststructuralism was the freedom and creativity to explore the texts, myths, stories, and meanings of which we are a part.

One of the key features of a poststructuralist research approach is the focus on the language and meanings through which we constitute our ontological and epistemological understandings. A poststructuralist analysis looks at "the work that language does to limit, shape, and make possible one kind of world or another" (Davies, 1993, p. xviii).

Weedon (1987, p. 167) provides a number of frames in which to conduct a feminist poststructuralist analysis that we have found useful:

- Literature is one specific site among many where the ideological constructions of gender along with other forms of subjectification takes place.
- The central focus of interest becomes the way in which texts construct meanings and subject positions for the reader, the contradictions inherent in this process, and its political implications, both in its historical context and in the present.
- The central humanist assumption that women or men have essential natures is denied.
- The social construction of gender in discourse is central.
- Feminist poststructural analysis refuses to fall back on general theories of the feminine psyche or biologically based definitions of femininity, which locate its essence in processes such as motherhood or female sexuality.

St. Pierre (2000) states that "feminists in education increasingly use post-structuralism to trouble both discursive and material structures that limit the ways we think about our work" (p. 477). Poststructuralist theorizing looks at the constitutive force of social structures and language within a historical context, to make more apparent how it is that (human) subjectivities have come into being. One of the major insights such social theorizing offers to environmental education is to make explicit the webs of power, agency, and desire in which we are caught and to illuminate which social forces are at work to either enhance or limit an individual's ability to act. As such, poststructuralist analysis presents an opportunity to challenge the privileged certainties of meta-narratives and the configurations of power carried within them. It also provides opportunity for exploration, deconstruction, and re-invention. And such analyses can have practical advantages.

As Doyle (2001) points out, "one of the strengths of postmodern analysis is that it has opened up possibilities for understanding the world in more ways than just simple dualisms" (p. 221). Politically this can have the effect of articulating a communal politics, which, to quote Wheeler (1995), "is not essentialist, fixed, separatist, divisive, defensive or exclusive" (p. 105). Within environmental education, feminist poststructuralist analysis allows the description of socio-cultural discursive practices that would otherwise be absent from the environmental literature, and as such, informs more completely our collective understandings of contemporary complexity (see, for example, Whitehouse, 2002).

Changing language practices can have powerful effects. One of the key learning areas identified in Australian curriculum documents is called "Studies of Society and Environment." One of Hilary's colleagues changed the name of the third-year university curriculum course designed to cover this learning area to "Social and Environmental Education." The difference is subtle and yet profound. Playing with language can shape curriculum possibilities. In this example, "Environment" is re-shaped from a logo-centric, disembodied concept to an actual practice of education. Add the word "tropical" and a location emerges along with ideas for re-writing curriculum through a pedagogy of place. Similarly, curriculum documents that construct a "human" and "environment" dualism can be re-thought and re-framed to reflect what Howitt and Suchet-Pearson (2003) call the "transformative energy" (p. 564) of ontological pluralism.

Awareness of the power of language to shape gendered subjectivities and the meanings of curriculum and pedagogy enables us to act by (and through) changing language practices. The outcomes of deconstruction are not nihilism, as many have argued, but the reconstruction of acute understandings of agency and power. With imagination these analyses provide recognition of different singing worlds.

Conclusion

In this chapter we argue that adopting a research approach which moves away from representations of universalized subjects, such as the mythic "Man" and

"His Environment," and toward a distinct recognition of multiple subjectivities will create research in environmental education that more accurately represents the diversities of lived experiences. We have provided our real-life narratives which trouble the discursive structures that limit our work as researchers. This viewpoint from "down under" seeks to broaden conceptions of the "nature" of environmental education research and demonstrates the potential of feminist poststructuralist research as a methodology for productive research in environmental education. By turning our gaze on some of the blind spots in environmental education research, we hope that we have opened up a space to move the field into a more democratic future.

Notes on contributors

Annette Gough is Associate Professor in the Faculty of Education, Deakin University, Australia. Her research interests are in environmental education, science education, and research methodology, with a particular emphasis on women and poststructuralist perspectives. She is the author of *Education and the Environment: Policy, Trends and the Problems of Marginalisation* (1997).

Hilary Whitehouse is a lecturer in the School of Education, James Cook University, Cairns, Australia. Her research and teaching interests are in tropical, social, and environmental education and science education.

References

Barrow, J. (2002). Enigma variations: Nature's constants may be changing, but nobody knows why. *New Scientist, 175*(2359), 30–33.

Byatt, A.S. (1990). *Possession: A romance*. London: Chatto & Windus.

Cherryholmes, C.H. (1993). Reading research. *Journal of Curriculum Studies, 25*(1), 1–32.

Conley, V.A. (1997). *Ecopolitics: The environment in poststructuralist thought*. London: Routledge.

Crotty, M. (1998). *The foundations of social research: Meaning and perspective in the research process*. St. Leonards, NSW: Allen & Unwin.

Davies, B. (1993). *Shards of glass: Children reading and writing beyond gendered identities*. St. Leonards, NSW: Allen & Unwin.

Davies, B. & Whitehouse, H. (1997). Men on the boundaries: Landscapes and seascapes. *Journal of Gender Studies, 6*(3), 237–254.

Doyle, T. (2001). *Green power: The environmental movement in Australia*. Sydney: UNSW Press.

Eckersley, R. (1992). *Environmentalism and political theory: Toward an ecocentric approach*. Albany, NY: State University of New York Press.

Fawcett, L. (2000). Ethical imagining: Ecofeminist possibilities and environmental learning. *Canadian Journal of Environmental Education, 5*, 134–149.

Fawcett, L. (2002). Children's wild animal stories: Questioning inter-species bonds. *Canadian Journal of Environmental Education, 7*(2), 125–139.

Gough, A. (1994). *Fathoming the fathers in environmental education: A feminist post-structuralist analysis.* Unpublished PhD dissertation, Deakin University, Geelong, Australia.

Gough, A. (1997a). *Education and the environment: Policy, trends and the problems of marginalisation.* Melbourne: Australian Council for Educational Research.

Gough, A. (1997b). Founders of environmental education: Narratives of the Australian environmental education movement. *Environmental Education Research, 3*(1), 43–57.

Gough, A. (1999a). Recognising women in environmental education pedagogy and research: Toward an ecofeminist poststructuralist perspective. *Environmental Education Research, 5*(2), 143–161.

Gough, A. (1999b). The power and promise of feminist research in environmental education. *Southern African Journal of Environmental Education, 19,* 28–39.

Gough, A. (1999c). Kids don't like wearing the same jeans as their mums and dads: So whose "life" should be in significant life experiences research? *Environmental Education Research, 5*(4), 383–394.

Gough, A. (2001). *Making sense of research in environmental education from a feminist standpoint: Moving beyond adding women.* Paper presented at the American Educational Research Association Annual Meeting, Seattle, USA, April.

Gough, N. (1991). Narrative and nature: Unsustainable fictions in environmental education. *Australian Journal of Environmental Education, 7,* 31–42.

Greene, M. (1992). Foreword. In C. Luke & J. Gore (Eds.), *Feminisms and critical pedagogy* (pp. ix–xi). New York, London: Routledge.

Harding, S. (1987). Introduction: Is there a feminist method? In S. Harding (Ed.), *Feminism and methodology: Social science issues* (pp. 1–14). Bloomington, Indianapolis: Indiana University Press.

Harding, S. (1991). *Whose science? Whose knowledge? Thinking from women's lives.* Ithaca: Cornell University Press.

Heilbrun, C.G. (1999). *Women's lives: The view from the threshold.* Toronto: University of Toronto Press.

Howitt, R. & Suchet-Pearson, S. (2003). Ontological pluralism in contested cultural landscapes. In K. Anderson, M. Domosh, S. Pile, & N. Thrift (Eds.), *Handbook of cultural geography* (pp. 557–569). London: Sage.

International Union for the Conservation of Nature and Natural Resources (IUCN). (1970). *International working meeting on environmental education in the school curriculum, Nevada, USA.* Final Report. https://portals.iucn.org/library/node/10447

Jagtenberg, T. & McKie, D. (1997). *Eco-impacts and the greening of postmodernity: New maps for communication studies, cultural studies and sociology.* Thousand Oaks, CA: Sage.

Lather, P. (1991). *Getting smart: Feminist research and pedagogy with/in the postmodern.* New York, London: Routledge.

Linke, R.D. (1980). *Environmental education in Australia.* Sydney: Allen and Unwin.

Lotz-Sisitka, H. & Burt, J. (2002). Writing environmental education research texts. *Canadian Journal of Environmental Education, 7*(1), 132–151.

Lousley, C. (1999). (De)politicizing the environment club: Environmental discourses and the culture of schooling. *Environmental Education Researcher, 5*(3), 293–304.

Malone, K. (1999). Environmental education researchers as environmental activists. *Environmental Education Research, 5*(2), 163–177.

Merchant, C. (1980). *The death of nature: Women, ecology and the scientific revolution.* New York: Harper and Row.

Merchant, C. (1996). *Earthcare: Women and the environment.* New York: Routledge.

Plumwood, V. (1993). *Feminism and the mastery of nature.* London, New York: Routledge.

Probyn, E. (2003). The spatial imperative of subjectivity. In K. Anderson, M. Domosh, S. Pile, & N. Thrift (Eds.), *Handbook of cultural geography* (pp. 290–299). London: Sage.

Quigley, P. (1995). Rethinking resistance: Environmentalism, literature and poststructuralist theory. In M. Oelschlaeger (Ed.), *Postmodern environmental ethics* (pp. 173–191). New York: State University of New York Press.

Rich, A. (1990). When we dead awaken: Writing as re-vision. In D. Bartholomae & A Petrosky (Eds.), *Ways of reading: An anthology for writers* (pp. 482–496). Boston: Bedford Books of St Martin's Press.

Russell, C.L. (2003). Minding the gap between methodological desires and practices. In D. Hodson (Ed.), *OISE papers in SMTE education, Volume 4.* Toronto: University of Toronto Press.

Salleh, A. (1997). *Ecofeminism as politics: Nature, Marx and the postmodern.* London: Zed Books.

Sandilands, C. (1999). *The good-natured feminist: Ecofeminism and the quest for democracy.* Minneapolis: University of Minnesota Press.

Soper, K. (1995). *What is nature?* Cambridge: Blackwell.

St. Pierre, E.A. (2000). Poststructural feminism in education: An overview. *Qualitative Studies in Education, 13*(5), 477–515.

Tamboukou, M. (1999). Writing genealogies: An exploration of Foucault's strategies for doing research. *Discourse: Studies in the Cultural Politics of Education, 20*(2), 201–217.

Taylor, S.G. (1991). Teaching environmental feminism: The potential for integrating women's studies and environmental studies. In K. Dyer & J. Young (Eds.), *Ecopolitics IV: Changing directions* (pp. 603–605). Adelaide: Mawson Graduate Centre for Environmental Studies, University of Adelaide.

Weedon, C. (1987). *Feminist practice and poststructuralist theory.* Oxford: Blackwell.

Wheeler, W. (1995). Nostalgia isn't nasty: The postmodernization of parliamentary democracy. In M. Perryman (Ed.), *Altered states: Postmodernism, politics, culture* (pp. 94–112). London: Lawrence & Wishart.

Whitehouse, H. (2001). Not greenies at school: Discourses of environmental activism in Australia. *Australian Journal of Environmental Education, 17,* 71–76.

Whitehouse, H. (2002). Landshaping: A concept for exploring the construction of environmental meanings within tropical Australia. *Australian Journal of Environmental Education, 18,* 57–62.

Whitehouse, H.L.W. (2000). *Talking up country: A study of the discursive production of environmental meanings in tropical Australia.* Unpublished PhD dissertation, James Cook University, Townsville, Australia.

8 The "nature" of environmental education from new material feminist and ecofeminist viewpoints

Annette Gough and Hilary Whitehouse

Gough, A., & Whitehouse, H. (2018). New vintages and new bottles: The "nature" of environmental education from new material feminist and ecofeminist viewpoints. *The Journal of Environmental Education*, 49(4), 336–349. doi:10.1080/00958964.2017.1409186

Abstract

Fifteen years ago we explored the implications of adopting a poststructuralist feminist research methodology in environmental education research and practice. We argued that speaking the world into existence provides multiple ways of thinking about and comprehending environmental knowledge and the way we experience ourselves in space, time, and place. In the intervening years, feminist new materialism has emerged, ecofeminism has had an enlivening resurgence, feminist scholarship in environmental education has expanded, and "nature" seems to have declined within dominant discourses, supplanted by a more anthropocentric agenda. The positioning of body/nature in this scholarship has become a point of contention, difference, and convergence. Here, we explore the "nature" of environmental education as informed by new material feminist, ecofeminist, and other viewpoints, interrogating their similarities and differences, their relationship to feminist poststructuralism, and their implications for environmental education research and practice.

Introduction

This chapter is situated in a space informed by feminist environmentalism, ecofeminism, feminist poststructuralism, new material feminism, and materialist feminism to explore what these mean for environmental education research and practice in the 21st century. When we last wrote on this topic (Gough & Whitehouse, 2003), our perspective was largely framed by feminist poststructuralism and the opportunities it provides for challenging privileged certainties of metanarratives and configurations of power. Our argument related to establishing a feminist environmentalist position within environmental education discourses because women were seen as a marginalized group (among others) within these discourses (Gough, 1999a, 1999b). At that time, ecofeminism

DOI: 10.4324/9781003390930-10

was being marginalized by accusations of essentialism (Gaard, 2011; Phillips & Rumens, 2016a), and feminist standpoint theory was being foregrounded for asserting questions such as who can be an agent of knowledge, what counts as knowledge, what constitutes and validates knowledge, and the relationship between knowledge and being. As Sandra Harding (1991) wrote, "what 'grounds' feminist standpoint theory is not women's experiences but the view from women's lives" (p. 269). Our relationship with ecofeminism was not straightforward: our work was informed by ecofeminist writings but we were concerned with the problems of binary thinking apparent in some versions of ecofeminism and we argued that feminist poststructuralist methodologies can be productive for environmental education research.

We begin this chapter by considering the history and evolution of ecofeminism in relation to current discussions and exploring how binary thinking continues to be a focus for re-thinking ecofeminism, as do material feminisms and other ways of (re)conceptualizing nature-culture relationships in research; we also consider the implications of these discussions for environmental education research. We reexamine our earlier focus on discourse in light of recent critiques of poststructuralist approaches by material feminists, and re-interpretations by ecofeminists, which has led us to consider the significance of (re)positioning both nature and the body in our work. We also discuss opportunities provided by new ways of studying human-nature relationships within the context of the resurgence of autonomous nature (Merchant, 2016), material feminist theory (Alaimo, 2008, 2010; Alaimo & Hekman, 2008; Barad, 2007), partnership ethics (Merchant, 1992, 2003, 2016), natures-cultures (Latour, 1993) or natureculture (Haraway, 2003, 2016), material-discursive practices (Haraway, 2008), and "intra-action" of matter and discourse (Barad, 1998), and provide examples of recent environmental education writings in these spaces. We are taking up the challenge from Richard Twine (2010) that, "the emergence of a feminist new materialism ought to usher in a renewed conversation between feminism and ecofeminism due to shared interests" (p. 402). Our hope, through this discussion, is to expand on these apparent and/or possible potentialities and tensions and provoke deliberations on whether we are looking at new wines, new bottles, neither, or both, and the implications of these for environmental education research.

Working with/for ecofeminism, past and present

"Ecofeminism" emerged in the 1970s, with the term first being used by Francoise D'Eaubonne (1974). Its roots are in activist social movements, particularly the anti-nuclear and peace movements of the time, and in the growing perception of gender blindness and sexism in other environmental groups (Gaard, 2011; Phillips & Rumens, 2016a). Ecofeminism developed as both an activist and academic/philosophical movement as "the convergence of ecology and feminism into a new social theory and political movement challenges gender relations, social institutions, economic systems, sciences, and views of our place as humans in the biosphere" (Lahar, 1991, p. 28). Anita Anand (1983) argued similarly that the "contribution of feminism to the environmental and development movements is the insistence that all forms of domination must be eradicated and that the living relations of subordination be a central problem for study, reflection and action for change" (p. 187).

Ecofeminism blossomed at the same time as the feminist movement. As Leonie Caldecott and Stephanie Leland (1983) commented then, "in many countries all over the world women are taking an increasingly prominent role in political struggles; in the peace, anti-nuclear, health and ecology movements. . ... The time has come for women to take a leading role to rectify the balance" (p. 5) between technology and the threats to health and life on the planet. There was a recognition and reclaiming of women's relationships with the environment, but there was also more than this: "ecofeminists have realized that we must question the entire civilization that mankind [*sic*] has contrived – all of its values, its goals, its achievements. It is not merely antifeminine, it is antihuman, antilife" (Russell, 1990, p. 225).

A conceived relationship between women and environment was enshrined in The Rio Declaration on Environment and Development (United Nations, 1992), one of the outcomes of the United Nations Conference on Environment and Development (UNCED), and in Agenda 21 (United Nations, 1993), the global action plan for achieving sustainable development. The actual wording referring to women given in the Rio Declaration – "Women have a vital role in environmental management and development. Their full participation is therefore essential to achieve sustainable development" (Principle 21, United Nations, 1992) – is problematic and can be read as condescending. More recent international statements shifted the discourse to focus more on gender equality. At the most recent UN conference on environment and development, the 2012 Rio+20 Conference (United Nations, 2012), the references to women in its declaration were linked with gender equality and human rights, "including the right to development and the right to an adequate standard of living, including the right to food, the rule of law, gender equality, women's empowerment and the overall commitment to just and democratic societies" (p. 2). Over time the focus has shifted from being about "women" and "the environment," particularly "women in the global South, whose real material needs for food security and productive agricultural land, forest resources, clean water and sanitation trumped more structural discussions about gendered environmental discourses," that constructed them as individuals and "victims of environmental degradation in need of rescue" (Gaard, 2015, p. 21, emphasis in original), to a new focus on "gender as a system structuring power relations" (Gaard, 2015, p. 22, emphasis in original). This shift has been an important development in terms of feminist responses to human-induced climate change and to the reconceptualizing of environmental feminism.

Ecofeminism suffered from a feminist backlash in the late 1990s, being then criticized as essentialist, elitist, and ethnocentrist, "and effectively discarded" (Gaard, 2011, p. 26). Recent studies at the intersection of feminism and environmentalism have seen a resurgence of the label of ecofeminism (see, for example, Adams & Gruen, 2014; Gaard, 2011, 2015; Phillips & Rumens, 2016a, 2016b; Thompson, 2006), as distinct from other labels such as "ecological feminism" (Warren, 1994; Warren & Cheney, 1991), "feminist environmentalism" (Seager, 1993), or "global feminist environmental justice" (Sturgeon, 2009). Greta Gaard (2011) argues that what is needed now, to address the climate emergency and the many problems of the Anthropocene, is an intersectional ecological-feminist approach that "frames these issues [problems] in such a way that people can recognize common cause across the boundaries of race, class, gender, sexuality, species, age, ability, nation – and affords a basis for engaged theory, education, and activism" (p. 44). This position is similar to what Carolyn Merchant (1992, 1995, 2016) describes as a partnership ethic. Merchant (2016) argues that, in order to live within the present-day chaos and complexity paradigm, a new, partnership ethics is needed to "understand nature as a complex system that includes humanity within it allows for the possibility that both the earth as we know it today and humanity can survive and thrive together in the coming decades" (p. 153). We will discuss this partnership ethic in greater detail in the next section when we compare it with new material feminist approaches.

Annette Gough (1997) conceived of using this partnership ethic for future developments of environmental education using a poststructuralist pedagogy, which she saw as offering ideas about new approaches that take into account the disorderly ways in which meanings are written, read, and rewritten. She encouraged a much greater and nuanced understanding of the multiple positionings of race, class, gender, ethnicity, and nonhuman communities, as well as the posthuman, to be taken account of in research. Other researchers have also overtly drawn on ecofeminism in and for their environmental education writings – see, for example, Fawcett (2000, 2013); Fawcett et al. (2002); Fry (2000); Gough (1999a, 2004b); Hallen (2000); Harvest and Blenkinsop (2011); Li (2007); Lloro-Bidart (2017); Lloro-Bidart and Semenko (2017); Martusewicz (2013); Piersol and Timmerman (2017); Russell and Bell (1996); Russell and Semenko (2016); Schwartz (1999); and Spencer and Nichols (2010). The grouping of the dates of these publications – up to 2000 and then recent (with the few exceptions that were published in the gap years drawing on ecofeminist writings from the 1990s) – is indicative of the fallow period for ecofeminism, and also of its resurgence, repositioning, and renewed potential for future research.

Dualisms, discourse, and the body

In common with much feminist theory, the focus of ecofeminism is on how what Marti Kheel (2008) calls "a culturally exalted hegemonic ideal" (p. 3) of masculinity "is promoted through a set of interrelated dualisms such as mind/body, reason/nature, masculine/feminine or human/nature" (Phillips, 2016, p. 59). Ecofeminism challenges this dualistic thinking "in order to undermine the alienation and disconnection from nature that are its outcome" (Phillips, 2016, p. 58) as well as the "interwoven and cross-cutting systems of domination that justify colonialism, racism, sexism and the subordination of nature" (Phillips, 2016, p. 59). In Gough and Whitehouse (2003), we distanced ourselves from the association of women and nature "as a source of empowerment for women and the basis of a critique of the male domination of women *and* nonhuman nature" (Eckersley, 1992, p. 64, emphasis in original), and argued that binary thinking "traps research processes into a persistent stasis, erasing a complexity, which might otherwise be meaningful in the pursuit of elegant solutions to contemporary socioenvironmental problems" (Gough & Whitehouse, 2003, p. 36). Although recognizing that no single research methodology will generate a complete understanding of any socioecological phenomenon, we saw a feminist poststructuralist approach as attractive. We liked the freedom and creativity this methodological inquiry provides for exploring the texts, myths, stories, and meanings of which we are a part. An exciting feature of feminist poststructuralism is the inquiry into the language and meanings by and through which we constitute our ontological and epistemological understandings. Elizabeth St. Pierre (2000) wrote that "feminists in education increasingly use poststructuralism to trouble both discursive and material structures that limit the ways we think about our work" (p. 477), and we asserted that "poststructuralist analysis allows the description of sociocultural discursive practices that would otherwise be absent from the environmental literature, and as such, informs more completely our collective understandings of contemporary complexity" (Gough & Whitehouse, 2003, p. 40).

The (female) body became a site of difference between ecofeminist and feminist poststructuralist theorists. Karen Warren (1987) and Val Plumwood (1993) argued that in the dominant dualistic system that privileges masculinity and reason, the body is devalued in opposition to the mind as it is the body that is more aligned with women and the expansive category of otherness named as "Nature." Such positioning led to ecofeminism being accused of essentialism, in the sense that these categories were immutable and unalterable, and led to academic ecofeminists tending to avoid engaging with the potential contributions to ecofeminist thought provided by poststructuralist approaches to the body (Phillips, 2016). However, although some like Mary Phillips (2016) see the positives of poststructuralist approaches to the body, others such as Stacy Alaimo (2008) argue that, although poststructuralist feminists have not denied the material existence of the body, they do tend to focus on how various bodies have been discursively produced and rendered abiological, severed from ongoing interconnections with the material world. This argument is incongruent with Hilary's story in Gough and Whitehouse (2003), where Nora helped her to understand that not everyone in every culture has a word for nature, some cultures "just live it," and it is also incongruent with Hilary's exploration of becoming a "wormbag" in her doctoral thesis (Whitehouse, 2000), an analysis formed after she fell ill with a hookworm infection. Similarly, and inspired by Denise Spencer's research with the Yolngu people and their experiences with the bauxite mine in their country, Annette wrote of breast cancer experiences in terms of constructing her body as a mine site (Gough, 2003, 2004a, 2005).

Some environmental educators working with poststructuralist feminism have focused more on the discursive (Barrett, 2005; Gough, 1999a, 1999b; Gough et al., 2003; Hart, 2017), but others have actively engaged the material world (Bell & Russell, 2000). In so doing these researchers are being consistent with Merchant's (2003) argument:

> The material world is mentally constructed and interpreted in terms of language, but it exists as biophysical entities. Only biophysical beings, such as humans, have the tongues and brains to create and perform speech acts – to invent the logos, or word. The real physical world and the constructed mental world thus exist in dialectical relation to each other. Reality and narrative, I argue, interact with each other.
>
> (p. 201)

There have been recent calls for ecofeminism to more directly embrace the body (rather than reject it in response to essentialist accusations), because "the body is an important reminder that humans are organically embedded beings whose future and wellbeing are inherently entwined with the future and wellbeing of the planet" (Phillips, 2016, p. 61). Phillips (2016) writes, "A productive dialogue between an ecofeminist approach to addressing looming ecological crises and Cixous' focus on writing that is multiplicitous and liberatory could encourage recognition that nature is intertwined with our corporeal existence" (p. 64). Drawing on the work of Helene Cixous, Phillips develops an argument for writing the body and for writing "nature" that will provide a means to shift understandings and encourage new perspectives that do not fall into the problems of poststructuralists ignoring the body.

In a similar but different vein, Alaimo (2008) argues that "human corporeality, in all its material fleshiness, is inseparable from 'nature' or environment'" and she proposes "trans-corporeality" as "a place where corporeal theories and environmental theories meet and mingle in productive ways" (p. 238) and a place where the body/nature are intertwined with "culture." By refusing to separate "culture" from "nature" we are able to destabilize the culture/nature binary and develop further ethical and political positions to contend with the ecological realities we are now facing. Alaimo's position is part of what has been called the "materialist turn" that avoids the gendered and essentialist culture/nature dualism. For example, Alaimo and Susan Hekman (2008) argue for the agency of nature and for a material feminism that reconceptualizes nature in ways that account for "'intra-actions' (in Karen Barad's terms) between phenomena that are material, discursive, human, more-than-human, corporeal, and technological" (p. 5).

Barad (2007) is the author of the much-cited book *Meeting the Universe Halfway: Quantum Physics and the Entanglement of Matter and Meaning*, which has strongly influenced the development of new materialism thinking. Interestingly, Barad does not reference ecofeminism or even environmental feminism, or any of the associated key writers such as Merchant (1992, 1995, 2003), Plumwood (1993), Warren (1987, 1994), or Maria Mies and Vandana Shiva (1993) in her text. Barad's argument is that

> [b]odies do not simply take their places in the world. They are not simply situated in, or located in, particular environments. Rather "environments" and "bodies" are intra-actively co-constituted. Bodies ("human," "environmental," or otherwise) are integral "parts" of, or dynamic reconfigurings of, what is.
>
> (2007, p. 170)

This entanglement of the world and everything in it leads Barad (2007) to argue that there are no separate "objects" with boundaries in nature, but there are identifiable "phenomena," which are the "*ontological* inseparability of agentially intra-acting components" and "basic units of reality" where "'intra-action' *signifies the mutual constitution of entangled agencies*" (p. 33, emphasis in original). Thus she writes that "[o]utside of particular agential intra-actions, 'words' and 'things' are indeterminate. Matter is therefore not to be understood as a property of things but, like discursive practices, must be understood in more dynamic and productive terms – in terms of intra-activity" (p. 150). When matter and meaning are conceived as mutually co-constituted at the moment of coming together, this is a different way of thinking about the body from the ways originally conceived by feminists and poststructuralists in the 20th century, and these ideas of co-entanglement have greatly informed the recent material turns in research thinking.

Material turns

In the 1990s, there was a divergence in the conceptualization of feminist environmentalism between ecofeminism and material/socialist feminism at the time Donna Haraway (1991) was articulating such in her writings around cyborgs. Alaimo (1994) described the difference this way:

> Whereas ecofeminism seeks to strengthen the bonds between women and nature by critiquing their parallel oppressions and encouraging an ethic of caring and a politics of solidarity, Haraway seeks to destabilize nature/culture dualism that grounds the oppression of both women and nature.
>
> (p. 133)

Drawing on Haraway (1988, 1989, 1991) and Merchant (1980), Alaimo (1994) argued for an environmental feminism that stressed a political alliance between women and nature and one that would not slide into essentialism. She argued that

> focusing on the agency of women and nature can help keep environmentalism in the political arena and can oppose the appropriation of nature as resource by stressing nature as an actor and by breaking down the nature/culture divide, thus undermining the systems of domination.
>
> (p. 150)

From such beginnings, material feminism has taken off in recent times with intersecting discussions of new materialisms, which, at times, can be confusing. Alaimo and Hekman (2008) distinguish what they call "material feminism" ("which is emerging primarily from corporeal feminism, environmental feminism, and science studies") from "new materialism" ("which emerges from, or is synonymous with[,] Marxist feminism"), while noting that many material feminist theorists have been influenced by Marxist theory, "their definition of materiality is not, or is not exclusively, Marxist" (p. 18). Alaimo and Hekman (2008) see "materialist feminism" as focusing on race, sexuality, imperialism and colonialism, and anthropocentrism, but not necessarily encompassing "the materiality of human corporeality or, certainly, of nonhuman nature" (p. 18) that material feminism addresses. However, as new materialism writings have developed over the past decade, these fine distinctions may have become blurred. Recently, Sonu and Snaza (2015) described new materialism as "a subset of the posthumanist drift in the fields of philosophy, biology, and the human sciences" that "attempts to rethink human subjectivity so that it accounts for its relationship with non-human affect and force" (p. 259).

Carol Taylor (in press) distinguishes new material feminism from Marxist materialism on the basis that signaling a hierarchy of bodies and beings (traditional philosophy has the (male) human at the top of that hierarchy) is becoming a much less productive way of thinking about our ways of being in the 21st century. Taylor argues that new material feminisms "shift the emphasis away from the human as the 'centre,' 'source,' and hierarchical 'top of the tree' in terms of knowledge-making" and "propose that all manner of bodies, objects and things have agency within a confederation of meaning-making" (p. 2). This is certainly a more nuanced and sophisticated approach to the vast problems of living in an age where the acidifying oceans are quite literally filling with plastic, and biodiversity is plummeting all over the globe (Kolbert (2014). Pulitzer Prize work *The Sixth Extinction* is a truly sobering investigation of this sad phenomenon). But, we ask, is this so new?

For Taylor, drawing on Barad (2007), it is the entanglement of human and non-human bodies as equal partners in knowledge-making that makes new materialism so potentially important. Gaard (2011) notes that materialist accounts of the woman-nature connection – such as those of Merchant (1992) for whom "nature is an active subject, not a passive object to be dominated, and humans must develop sustainable relationships with it" (p. 196) – "described a socially constructed association among women (sex), femininity (gender), and nature that was contextual and fluid, not ahistorical and static" (p. 31), which seems most similar to such descriptions of new material feminism.

The affinities between some ecofeminist theory, particularly Merchant's (1992, 1992, 1995, 2003, 2016) partnership ethic and Gaard's (2011) recon-ceptualizing of ecofeminism, and new material feminism are apparent in descriptions of new material feminism. For example, Taylor (in press) writes:

> New material feminism shares a social justice imperative with other modes of feminism. Like them, it is committed to finding ways to com-bat gender inequality, discrimination, and violence in education. More broadly, it shares with post-structuralism, post-colonialism, and intersec-tional studies a suspicion that the Enlightenment ideals of rationality, objectivity, and scientific progress have only delivered partial benefits for particular groups of people (mostly males, White, Western, able-bodied people) and that the narrative of "progress" it offers is also a partial affair designed to maintain the hegemony of those who benefit most from it. New material feminism, therefore, offers a radical set of tools for gen-erating new understandings of subjectivity, relationality, and ethics, and suggests that these tools offer ways of fundamentally rethinking what we mean by – and how we do – social justice in a more-than-human world.
>
> (p. 2)

As a comparison, Merchant (2016) describes her partnership ethics in these terms:

> By merging anthropocentric with ecocentric ethics – the ethic that includes all of nature within it – we can develop an integrated, interac-tive ethic based on partnership between the human and the nonhuman worlds. . . . *A partnership ethic holds that the greatest good for the human and nonhuman communities is in their mutual living interdependence.*
>
> (p. 162, italics in original)

Merchant's ethic contains five precepts for a human community in a sustain-able partnership with a nonhuman community – which is in a particular place, a place in which connections to the larger world are recognized through eco-nomic and ecological exchanges:

- equity between the human and nonhuman communities;
- moral consideration for both humans and other species;
- respect for both cultural diversity and biodiversity;
- inclusion of women, minorities, and nonhuman nature in the code of ethi-cal accountability; and
- an ecologically sound management that is consistent with the continued health of both the human and the nonhuman communities.

The present inequity in humans dealings with a recognized non-human (but highly agentic) realm is recognized by Kolbert (2014) and many others, but as Karen Emmerman (2014) points out, focusing research questions on "how we got into the mess we are in . . . is unhelpful for practical guidance in moving forward" (p. 160). Gaard (2011), drawing on a range of literature that she broadly calls "ecofeminists in the new millennium" (p. 41), notes that new fields such as posthumanism, postcolonial criticism, and animal studies are "moving forward with ideas initially developed in feminist and ecofeminist contexts often without acknowledging those contexts as foundations for their work" (p. 42). From this, she argues for an intersectional ecological-feminist approach, mentioned earlier, which is perhaps a "New Eco-feminism," but, she concludes, "it will need to be more cognizant of its rich and prescient history" (p. 44). (More recently, Gaard (2014) has made the argument for exploring "New Eco-masculinities" and "Eco-Genders" to take this thinking further.) The erasure of, or amnesia about, the foundations of the new fields also raises questions about whether intersectionality caught on because it was only about humans, or whether critical animal studies erased ecofeminism because of sexism (Russell & Semenko, 2016).

(De/Re)constructing "nature"

A significant point of tension between ecofeminism, poststructuralism, and new material feminism is around the construction of "nature." As noted previously, some ecofeminists and poststructuralists distanced themselves from "nature" in an attempt to avoid accusations of essentialism. As Alaimo (2008) argues,

> from an environmentalist-feminist standpoint, one of the most unfortunate legacies of poststructuralist and postmodern feminism has been the accelerated 'flight from nature' fueled by rigid commitments to social constructionism and the determination to rout out all vestiges of essentialism.
>
> (p. 237)

In our previous article (Gough & Whitehouse, 2003, p. 36) we wrote of our concern of the "environment" being "produced as an object of study rather than a subject for research" and drew on Kate Soper's (1995) argument that "nature" is impossible to define, because it is not an empirically defined thing, existent in itself, but a category of human identity. We asked the question, if the world is indeed divided into two categories of being ("human" and "nature" or "environment"), why are the boundaries between them impossible to locate? We concluded by arguing for broadening conceptions of the "nature" of environmental education research toward a distinct recognition of multiple subjectivities that more accurately reflect the diversities of lived experiences. In so doing we were probably guilty of an anthropocentric argument, as our focus was on making "explicit the webs of power, agency, and desire in which we are caught" (p. 40).

Phillips (2016) suggests that "in the West we have become disconnected from our bodies, and our bodies are disconnected from nature" (p. 57). There is certainly evidence for this in United Nations documents focused on the environment and sustainable development written over the past 45 years. The Declaration from the 1972 United Nations Declaration on the Human Environment, held in Stockholm, provides a vision and set of common principles focused on preserving and enhancing the human environment. The second paragraph of the Declaration highlights the importance of protecting and improving the human environment: "The protection and improvement of the human environment is a major issue which affects the well-being of peoples and economic development throughout the world. It is the urgent desire of the peoples of the whole world and the duty of all governments" (United Nations, 1972). *The Common Vision in The Future We Want* (United Nations, 2012) is grounded in a different orientation opening with a commitment to ensuring "the promotion of an economically, socially and environmentally sustainable future for our planet and for present and future generations" (p. 1, para 1), and promulgates sustainable development as "integrating economic, social and environmental aspects and recognizing their interlinkages, so as to achieve sustainable development in all its dimensions" (p. 2, para 3). The second paragraph states, "Eradicating poverty is the greatest global challenge facing the world today and an indispensible requirement for sustainable development" (p. 1, para 2), which is a very different focus from the Stockholm Declaration's concern for the protection and improvement of the human environment. More recently, the preamble to *Transforming our World: The 2030 Agenda for Sustainable Development* (United Nations, 2015) has reinforced this orientation, reiterating the focus on eradicating poverty and asserting:

> The 17 Sustainable Development Goals and 169 targets which we are announcing today demonstrate the scale and ambition of this new universal Agenda. They seek to build on the Millennium Development Goals and complete what they did not achieve. They seek to realize the human rights of all and to achieve gender equality and the empowerment of all women and girls. They are integrated and indivisible and balance the three dimensions of sustainable development: the economic, social and environmental.
>
> (p. 1)

Adequately recognizing that humans are part of this necessary other "nature" rather than separate from this category and understanding that nature is autonomous rather than predictable have been themes in recent research literature. These acts of rediscovery became necessary as the revelations of older and still thriving cultures have been ignored or obliterated instead of being actively integrated into common environmental education research thinking (see Lowan-Trudeau, 2013; Whitehouse, 2011; Whitehouse et al., 2014 as well as the special issues of the *Canadian Journal of Environmental Education* (Korteweg & Russell, 2012) and *Environmental Education Research* (Tuck et al., 2014) for articles related to the recent interest in decolonizing and Indigenizing environmental education).

Merchant (2016) notes that by the end of the 17th century, experimental knowledge and mathematics had become the pillars of an emerging mechanistic worldview, which together "would provide both power and knowledge leading to the possibility of prediction and control over a nature that could seemingly be wilful, recalcitrant, and unpredictable" (p. 97). She also discusses how, by the late 19th century, the mechanistic view of nature led to scientific and technological advances that gave rise to optimism over the control of nature. However, in the early 20th century, Newtonian mechanics was challenged by Einstein's Theory of Relativity and Heisenberg's Principle of Uncertainty, and later by chaos and complexity theories that challenged such views on predictability at the level of everyday life. This leads Merchant (2016) to argue that the 21st century has seen the return of autonomous nature:

> Here *natura naturans* (nature's creative force) and *natura naturata* (the created world) interact in complex, dynamic processes, many of which are beginning to have potentially irreversible effects on life on earth. . . .. The way in which nature as an autonomous system behaves depends on how humans behave in relationship to it.
>
> (pp. 149–150)

Merchant (2016) then argues that, in order to live within this new chaos and complexity paradigm, a new, partnership ethics is needed. Research must come to understand that "nature as a complex system that includes humanity within it allows for the possibility that both the earth as we know it today and humanity can survive and thrive together in the coming decades" (p. 153). Merchant also discusses new conceptualizations of nature, citing the work of Bruno Latour (1993) on natures-cultures, Soper (1995) on nature and the nonhuman, Hayles (1999) on nature and the posthuman, and Lynn Worsham and Gary Olson (2008) on the technological boundaries between humans and nature in a posthuman age. She sees each of these concepts as contributing to her partnership ethics, with each having positive aspects, but each also lacking in some aspect. From this discussion she concludes:

> Nature becomes postnature in ways that so thoroughly blur any human/ nature differences as to make a single interactive, mutually influential, and mutually interdependent post-human-nature . . . a new relationship between humanity and nature based on the idea of autonomous nature.
>
> (p. 161)

The dissipation of conceived binary categories in a posthuman-nature world is a characteristic of the Anthropocene. Like Barad (2007) and other new materialists, Merchant (2016) also refers to a posthuman world, and a different way of conceptualizing nature, where "the creating force that brought about the world civilized by humans – becomes the chaotic, uncontrollable force that dismantles it" (p. 156). From this position, she argues, "New concepts of nature are called for that are consistent with the new paradigm" (p. 157). Working from material feminism, Alaimo (2010, p. 4) writes that "'the environment' is not located somewhere out there, but is always the very substance of ourselves" and Alaimo and Hekman (2008) point out that nature "is an active, signifying force; an agent in its own terms; a realm of multiple, inter- and intra-active cultures" (p. 12). These new perceptions of nature, "a nature that is, expressly not the mirror image of culture – is emerging from the overlapping fields of material feminism, environmental feminism, environmental philosophy, and green cultural studies" (p. 12). Productive futures for research will include analyzing and critiquing the formation and circulation of potent discourses and also, as Alaimo (2010) suggests, "many of us would like to find ways to complement and complicate that sort of analysis with investigations that account for the ways in which nature, the environment, and the material world itself signify, act upon, or otherwise affect human bodies, knowledges and practices" (pp. 7–8).

Latour (1993) put forward the concept of natures-cultures as an interactive human/nature system because "the very notion of culture is an artefact created by bracketing Nature off. Cultures – different or universal – do not exist any more than Nature does. There are only natures-cultures, and these offer the only basis for comparison" (p. 104). Whitehouse (2011) has written that the concept of natureculture "is an enticing (and intellectually exciting) venture to consider in environmental education research" (p. 62) for people always live within both "nature" and "culture" and not on one side or the other of this discursive divide. She also suggested that pluralizing the concept, as natures-cultures, opens up interesting spaces for alternative forms of scholarship to be realized. Building on Haraway's work, Fawcett (2013) also discusses naturecultures and the rupturing of human relationships with various "natures." Clearly, many of the proposals discussed in this chapter merge and converge, so what does this mean for gender inquiry in environmental education research?

Implications for environmental education research

Developing a gender agenda has been discussed elsewhere (Gough, 2013), but here we are focusing on the particular import of ecofeminist and new material feminist theories. Ecofeminist theory has been taken up by environmental education researchers, but not extensively. New material feminism has gained some traction with environmental feminists and environmental education researchers, but, again, it is not widespread. Although some environmental feminist writing has drawn on Barad's work (Alaimo, 2010; Gaard, 2015), others have no references to her (Gaard, 2011; all of the authors in Phillips & Rumens, 2016b; Merchant, 2016; Thompson, 2006), but we argue that in a new material feminist context, she cannot be ignored. Similarly, Haraway's (2008) thinking on material-discursive practices is ignored by Merchant (2016) and Gaard (2011), but she is referenced by three authors in Phillips and Rumens (2016b) as well as by Alaimo (2008, 2010) and Gaard (2015). The lack of crossreferencing is bemusing when these researchers are writing in such converging spaces.

In environmental education research that references new material feminism, Barad's writings are overlooked in some writing (Fawcett, 2013; Hart, 2017; Lloro-Bidart, 2017; Lloro-Bidart & Semenko, 2017; Piersol & Timmerman, 2017) but referenced by others (Adsit-Morris & Gough, 2017; Blaise et al., 2017; Fawcett, 2014; Knox, 2015; Malone, 2016; Russell, 2005; Taylor et al., 2013). Similarly, some have ignored Haraway's work on material-discursive practices (Hart, 2017; Piersol & Timmerman, 2017), although her work is cited by others (see Adsit-Morris & Gough, 2017; Blaise et al., 2017; Fawcett, 2009; Lloro-Bidart, 2017; Lloro-Bidart & Semenko, 2017; Malone, 2016; Nxumalo & Pacini-Ketchabaw, 2017; Pacini-Ketchabaw & Nxumalo, 2015; Russell, 2005; Taylor et al., 2013; Taylor & Pacini-Ketchabaw, 2015; Whitehouse, 2011). The reason we mention this discrepancy is because what both Barad and Haraway are arguing warrants scrutiny and comparison with what ecofeminists and environmental education researchers are proposing in this decade and have written previously. It is important to draw on all the available ideas more conclusively, comprehensively, and coherently to advance thinking in environmental education research. We suggest environmental education researchers should draw more inclusively on international sociological and cultural thinking across ecofeminism and material discursive analyses.

Although some feminist environmental education writing has already explicitly named intersectionality in their work (see, for example, Lloro-Bidart, 2017; Lloro-Bidart & Semenko, 2017; Maina-Okori et al., 2017; Nxumalo & Cedillo, 2017; Russell, 2009; Russell & Semenko, 2016), there is a need for more research that explores the possibilities of an intersectional ecological-feminist approach (Gaard, 2011). Merchant's (2016) argument for a partnership ethic that "brings humans and nonhuman nature into a dynamically balanced more nearly equal relationship with each other" (p. 162) has useful application to practice. Such an ethic could be seen as consistent with the "General Capabilities of the Australian Curriculum" (ACARA, 2017a), which include Critical and Creative Thinking, Personal and Social Capability, Ethical Understanding, and Intercultural Understanding, and may be consistent with the Sustainability Cross-Curriculum Priority that promotes a "renewed and balanced approach to the way humans interact with each other and the environment" (ACARA, 2017b, n.p.). This suggests directions for insightful curriculum and praxis analyses, as education researchers continue to grapple with the emerging realities of teaching and learning in the Anthropocene.

Most importantly, though, is the need to explore the implications of post-nature and post-humannature. The whole notion of connecting children with urban and non-urban "nature" can be conceptualized differently if we accept that a nature untouched by humanity no longer exists anywhere on the planet, that we cannot no longer look to any ice floe or mountain top or tropical forest and not find evidence of our impacts (see Klein, 2014). There is some writing that draws on Haraway as well as postcolonial/decolonizing work that offers a useful extension in this space (see, for example, Nxumalo, 2015; Nxumalo & Cedillo, 2017; Nxumalo & Pacini-Ketchabaw, 2017; Pacini-Ketchabaw & Taylor, 2015), but more is needed.

Finally, it is important to remember Alaimo's (2010) caution that "there are no guarantees that emerging models of materiality will cultivate environmentalisms" (p. 9). There is little evidence of concern about the environment in Barad (2007), and although Alaimo does marry the two, not all new materialist or new material feminist scholars are doing so. A danger exists that environmental matters could well be forgotten again in the rush to argue that matter matters. So whether it is new vintages or new bottles may not make a difference if ecofeminism remains on the margins, and new materialism ignores present environmental realities, and, in particular, the current dangers – moral, ethical, and practical – posed by the biodiversity crisis (see Kolbert, 2014). In the introduction to this chapter, we took up the challenge from Twine (2010) that "the emergence of a feminist new materialism ought to usher in a renewed conversation between feminism and ecofeminism due to shared interests" (p. 402). We think that the ushering in of renewed conversations between ecofeminism and new material feminism is a challenge for all environmental education researchers, and one that we shall continue to explore.

References

Adams, C., & Gruen, L. (2014). Groundwork. In C. Adams & L. Gruen (Eds.), *Ecofeminism: Feminist intersections with other animals and the earth* (pp. 7–36). New York, NY: Bloomsbury.

Adsit-Morris, C., & Gough, N. (2017). It takes more than two to (multispecies) tango: Queering gender texts in environmental education. *The Journal of Environmental Education, 48*(1), 67–78. doi:10.1080/00958964.2016.1249330

Alaimo, S. (1994). Cyborg and ecofeminist interventions: Challenges for an environmental feminism. *Feminist Studies, 20*(1), 133–152.

Alaimo, S. (2008). Trans-corporeal feminisms and the ethical space of nature. In S. Alaimo & S. J. Hekman (Eds.), *Material feminisms* (pp. 237–264). Bloomington, IN: Indiana University Press.

Alaimo, S. (2010). *Bodily natures: Science, environment, and the material self.* Bloomington, IN: Indiana University Press.

Alaimo, S., & Hekman, S. (2008). Introduction: Emerging models of materiality in feminist theory. In S. Alaimo & S. J. Hekman (Eds.), *Material feminisms* (pp. 1–19). Bloomington, IN: Indiana University Press.

Anand, A. (1983). Saving trees, saving lives: Third world women and the issue of survival. In L. Caldecott & S. Leland (Eds.), *Reclaim the earth: Women speak out for life on Earth* (pp. 182–188). London, England: Women's Press.

Australian Curriculum and Assessment and Reporting Authority (ACARA). (2017a). The Australian curriculum. F-10 curriculum. *General Capabilities.* Retrieved from www.australiancurriculum.edu.au/f-10-curriculum/generalcapabilities/

Australian Curriculum and Assessment and Reporting Authority (ACARA). (2017b). The Australian curriculum. F-10 curriculum. Cross curriculum priorities. *Sustainability.* Retrieved from www.australiancurriculum.edu.au/f-10curriculum/cross-curriculum-priorities/sustainability/

Barad, K. (1998). Getting real: Technoscientific practices and the materialization of reality. *Differences, 10*(2), 87–128.

Barad, K. (2007). *Meeting the universe halfway: Quantum physics and the entanglement of matter and meaning.* London, England: Duke University Press.

Barrett, M. J. (2005). Making (some) sense of feminist poststructuralism in environmental education research and practice. *Canadian Journal of Environmental Education, 10,* 79–93.

Bell, A., & Russell, C. (2000). Beyond human, beyond words: Anthropocentrism, critical pedagogy, and the poststructuralist turn. *Canadian Journal of Education, 25*(3), 188–203. doi:10.2307/1585953

Blaise, M., Hamm, C., & Iorio, J. M. (2017). Modest witness(ing) and lively stories: Paying attention to matters of concern in early childhood. *Pedagogy, Culture & Society, 25*(1), 31–42. doi:10.1080/14681366.2016.1208265

Caldecott, L., & Leland, S. (Eds.). (1983). *Reclaim the earth: Women speak out for life on earth.* London, England: Women's Press.

d'Eaubonne, F. (1974). *Le Feminisme ou la mort [Feminisim or death].* Paris, France: Pierre Horay.

Eckersley, R. (1992). *Environmentalism and political theory: Toward an ecocentric approach.* Albany, NY: State University of New York Press.

Emmerman, K. (2014). Inter-animal moral conflicts and moral repair: A contexualised ecofeminist approach in action. In C. J. Adams & L. Gruen (Eds.), *Ecofeminism: Feminist interactions with other animals and the earth* (pp. 225–239). New York, NY: Bloomsbury.

Fawcett, L. (2000). Ethical imagining: Ecofeminist possibilities and environmental learning. *Canadian Journal of Environmental Education, 5,* 134–149.

Fawcett, L. (2009). Feral sociality and (un)natural histories: On nomadic ethics and embodied learning. In M. McKenzie, P. Hart, H. Bai, & B. Jickling (Eds.), *Fields of green: Re-storying culture, environment and education* (pp. 227–237). Cresskill, NJ: Hampton Press.

Fawcett, L. (2013). Three degrees of separation: Accounting for naturecultures in environmental education research. In R. Stevenson, M. Brody, J. Dillon, & A. Wals (Eds.), *International handbook of research on environmental education* (pp. 409–417). New York, NY: Routledge. doi:10.4324/9780203813331.ch39

Fawcett, L. (2014). Kinship imaginaries: Children's stories of wild friendships, fear, and freedom. In G. Marvin & S. McHugh (Eds.), *Routledge handbook of human-animal studies* (pp. 259–274). New York, NY: Routledge.

Fawcett, L., Bell, A. C., & Russell, C. L. (2002). Guiding our environmental praxis: Teaching for social and environmental justice. In W. Filho (Ed.), *Teaching sustainability at universities: Towards curriculum greening* (pp. 223–238). New York, NY: Peter Lang.

Fry, K. (2000). Learning, magic, and politics: Integrating ecofeminist spirituality into environmental education. *Canadian Journal of Environmental Education, 5*(1), 200–212.

Gaard, G. (2011). Ecofeminism revisited: Rejecting essentialism and re-placing species in a material feminist environmentalism. *Feminist Formations, 23*(2), 26–53. doi:10.1353/ff.2011.0017

Gaard, G. (2014). Toward new eco-masculinities, eco-genders and eco-sexualities. In C. J. Adams & L. Gruen (Eds.), *Ecofeminism: Feminist interactions with other animals and the earth* (pp. 225–239). New York, NY: Bloomsbury.

Gaard, G. (2015). Ecofeminism and climate change. *Women's Studies International Forum, 49*, 20–33. doi:10.1016/j. wsif.2015.02.004

Gough, A. (1997). *Education and the environment: Policy, trends and the problems of marginalisation.* Australian Education Review Series No.39. Melbourne, Victoria: Australian Council for Educational Research.

Gough, A. (1999a). Recognising women in environmental education pedagogy and research: Toward an ecofeminist poststructuralist perspective. *Environmental Education Research, 5*(2), 143–161. doi:10.1080/1350462990050202

Gough, A. (1999b). The power and promise of feminist research in environmental education. *Southern African Journal of Environmental Education, 19*, 28–39.

Gough, A. (2003). Embodying a mine site: Enacting cyborg curriculum. *Journal of Curriculum Theorizing, 19*, 33–47.

Gough, A. (2004a). Blurring boundaries: Embodying cyborg subjectivity and methodology. In H. Piper & I. Stronach (Eds.), *Educational research: Difference and diversity* (pp. 113–127). Burlington, VT: Ashgate.

Gough, A. (2004b). The contribution of ecofeminist perspectives to sustainability in higher education. In P. Corcoran & A. Wals (Eds.), *Higher education and the challenge of sustainability* (pp. 149–161). Dordrecht, Netherlands: Springer. doi:10.1007/0-306-48515-X_12

Gough, A. (2005). Body/mine: A chaos narrative of cyborg subjectivities and liminal experiences. *Women's Studies, 34*(3–4), 249–264. doi:10.1080/00497870590964147

Gough, A. (2013). Researching differently: Generating a gender agenda for research in environmental education. In R. B. Stevenson, M. Brody, J. Dillon, & A. Wals (Eds.), *International handbook of research on environmental education* (pp. 375–383). New York, NY: Routledge. doi:10.4324/9780203813331.ch35

Gough, A., & Whitehouse, H. (2003). The "nature" of environmental education from a feminist poststructuralist viewpoint. *Canadian Journal of Environmental Education, 8*, 31–43.

Gough, N., Gough, A., Appelbaum, P., Appelbaum, S., Aswell Doll, M., & Sellers, W. (2003). Tales from Camp Wilde: Queer(y)ing environmental education research. *Canadian Journal of Environmental Education, 8*, 44–66.

Hallen, P. (2000). Ecofeminism goes bush. *Canadian Journal of Environmental Education, 5*, 150–166.

Haraway, D. (2008). Otherwordly conversation, terrain topics, local terms. In S. Alaimo & S. J. Hekman (Eds.), *Material feminisms* (pp. 157–187). Bloomington, IN: Indiana University Press.

Haraway, D. (2016). *Staying with the trouble: Making kin in the Chtulucene.* Durham, NC: Duke University Press. doi:10.1215/9780822373780

Haraway, D. J. (1988). Situated knowledges: The science question in feminism and the privilege of partial perspective. *Feminist Studies, 14*(3), 575–600. doi:10.2307/3178066

Haraway, D. J. (1989). *Primate visions: Gender, race and nature in the world of modern science.* New York, NY: Routledge.

Haraway, D. J. (1991). *Simians, cyborgs, and women: The reinvention of nature.* London, UK: Free Association Books.

Haraway, D. J. (2003). *The companion species manifesto: Dogs, people, and significant otherness.* Chicago, IL: Prickly Paradigm Press.

Harding, S. (1991). *Whose science? Whose knowledge? Thinking from women's lives.* Ithaca, NY: Cornell University Press. doi:10.1080/00958964.2016.1249328

Hart, C. (2017). En-gendering the material in environmental education research: Reassembling otherwise. *The Journal of Environmental Education, 48*(1), 46–55. doi:10.1080/00958964.2016.1249328

Harvester, L., & Blenkinsop, S. (2011). Environmental education and ecofeminist pedagogy: Bridging the environmental and the social. *Canadian Journal of Environmental Education, 15*, 120–134.

Hayles, N. K. (1999). *How we became posthuman: Virtual bodies in cybernetics, literature, and informatics.* Chicago, IL: University of Chicago Press.

Kheel, M. (2008). *Nature ethics: An ecofeminist perspective.* Lanham, MD: Rowman & Littlefield.

Klein, N. (2014). *This changes everything: Capitalism versus the climate.* New York, NY: Simon and Schuster.

Knox, H. (2015). Thinking like a climate. *Distinktion: Journal of Social Theory, 16*(1), 91–109. doi:10.1080/1600910X.2015.1022565

Kolbert, E. (2014). *The sixth extinction. An unnatural history.* New York, NY: Bloomsbury.

Korteweg, L., & Russell, C. (2012). Decolonizing C indigenizing D moving environmental education towards reconciliation. *Canadian Journal of Environmental Education, 17*, 5–14.

Lahar, S. (1991). Ecofeminist theory and grassroots politics. *Hypatia, 6*(1), 28–45. doi:10.1111/j.1527-2001.1991.tb00207.x

Latour, B. (1993). *We have never been modern.* Cambridge, MA: Harvard University Press.

Li, H. L. (2007). Ecofeminism as a pedagogical project: Women, nature, and education. *Educational Theory, 57*(3), 351–368. doi:10.1111/j.1741-5446.2007.00262.x

Lloro-Bidart, T. (2017). A feminist posthumanist political ecology of education for theorizing human-animal relations/relationships. *Environmental Education Research, 23*(1), 111–130. doi:10.1080/13504622.2015.1135419

Lloro-Bidart, T., & Semenko, K. (2017). Toward a feminist ethic of self-care for environmental educators. *The Journal of Environmental Education, 48*(1), 18–25. doi:10.1080/00958964.2016.1249324

Lowan-Trudeau, G. (2013). Indigenous environmental education research in North America: A brief review. In R. B. Stevenson, M. Brody, J. Dillon, & A. Wals (Eds.), *International handbook of research on environmental education* (pp. 404–408). New York, NY: Routledge. doi:10.4324/9780203813331.ch38

Maina-Okori, N. M., Koushik, J. R., & Wilson, A. (2017, online). Reimagining intersectionality in environmental and sustainability education: A critical literature review. *The Journal of Environmental Education.* doi:10.1080/00958964.2017.1364215

Malone, K. (2016). Reconsidering children's encounters with nature and place using posthumanism. *Australian Journal of Environmental Education, 32*(1), 42–56.

Martusewicz, R. (2013). Toward an anti-centric ecological culture: Bringing a critical eco-feminist analysis to ecojustice education. In A. Kulnieks, D. Longboat, & K. Young (Eds.), *Contemporary studies in environmental and Indigenous pedagogies: A curricula of stories and place* (pp. 259–272). Rotterdam, Netherlands: Sense Publishers. doi:10.1007/978-94-6209-293-8_14

Merchant, C. (1980). *The death of nature: Women, ecology and the scientific revolution.* New York, NY: Harper and Row.

Merchant, C. (1992). *Radical ecology: The search for a livable world.* New York, NY: Routledge.

Merchant, C. (1995). *Earthcare: Women and the environment.* New York, NY: Routledge.

Merchant, C. (2003). *Reinventing Eden: The fate of nature in Western culture.* New York, NY: Routledge.

Merchant, C. (2016). *Autonomous nature: Problems of prediction and control from ancient times to the scientific revolution.* New York, NY: Routledge.

Mies, M., & Shiva, V. (1993). *Ecofeminism.* London, UK: ZEd Books.

Nxumalo, F. (2015). Forest stories: Restorying encounters with "natural" places in early childhood education. In V. Pacini-Ketchabaw & A. Taylor (Eds.), *Unsettling the colonial places and spaces of early childhood education* (pp. 21–42). New York, NY: Routledge.

Nxumalo, F., & Cedillo, S. (2017). Decolonizing "place" in early childhood studies: Thinking with Indigenous onto-epistemologies and black feminist geographies. *Global Studies of Childhood, 7*(2), 99–112. doi:10.1177/2043610617703831

Nxumalo, F., & Pacini-Ketchabaw, V. (2017, online). "Staying with the trouble" in child-insect-educator common worlds. *Environmental Education Research.* doi:10.1 080/13504622.2017.1325447

Pacini-Ketchabaw, V., & Nxumalo, F. (2015). Unruly raccoons and troubled educators: Nature/culture divides in a childcare centre. *Environmental Humanities, 7,* 151–168. doi:10.1215/22011919-3616380

Pacini-Ketchabaw, V. & Taylor, A. (Eds.). (2015). *Unsettling the colonial places and spaces of early childhood education.* New York, NY: Routledge.

Phillips, M. (2016). Developing ecofeminist corporeality: Writing the body as activist poetics. In M. Phillips & N. Rumens (Eds.), *Contemporary perspectives on ecofeminism* (pp. 57–75). London, England: Routledge.

Phillips, M., & Rumens, N. (2016a). Introducing contemporary ecofeminism. In M. Phillips & N. Rumens (Eds.), *Contemporary perspectives on ecofeminism* (pp. 1–16). London, England: Routledge.

Phillips, M., & Rumens, N. (Eds.). (2016b). *Contemporary perspectives on ecofeminism.* London, England: Routledge.

Piersol, L., & Timmerman, N. (2017). Reimagining environmental education within academia: Storytelling and dialogue as lived ecofeminist politics. *The Journal of Environmental Education, 48*(1), 10–17. doi:10.1080/00958964.2016.1249329

Plumwood, V. (1993). *Feminism and the mastery of nature.* London, England: Routledge.

Russell, C. (2005). "Whoever does not write is written": The role of "nature" in post-post approaches to environmental education research. *Environmental Education Research, 11*(5), 433–443. doi:10.1080/13504620500169569

Russell, C. (2009). Living and learning on the edge: Class, race, gender, animals and the environment in a university community outreach program. *Our Schools/Our Selves, 19*(1), 109–112.

Russell, C., & Bell, A. (1996). A politicized ethic of care: Environmental education from an ecofeminist perspective. In K. Warren (Ed.), *Women's voices in experiential education* (pp. 172–181). Dubuque, IA: Kendall Hunt.

Russell, C., & Semenko, K. (2016). We take "cow" as a compliment: Fattening humane, environmental, and social justice education. In E. Cameron & C. Russell (Eds.), *The fat pedagogy reader: Challenging weight-based oppression through critical education* (pp. 211–220). New York, NY: Peter Lang.

Russell, J. S. (1990). The evolution of an ecofeminist. In I. Diamond & G. F. Orenstein (Eds.), *Reweaving the world: The emergence of ecofeminism* (pp. 223–230). San Francisco, CA: Sierra Club Books.

Schwartz, E. G. (1999). Exploring children's picture books through ecofeminist literacy. In G. Smith & D. Williams (Eds.), *Ecological education in action: On weaving education and the environment* (pp. 103–116). Albany, NY: SUNY Press.

Seager, J. (1993). *Earth follies: Coming to feminist terms with the global environmental crisis.* New York, NY: Routledge.

Sonu, D., & Snaza, N. (2015). The fragility of ecological pedagogy: Elementary social studies standards and possibilities of new materialism. *Journal of Curriculum and Pedagogy, 12*(3), 258–277. doi:10.1080/15505170.2015.1103671

Soper, K. (1995). *What is nature?* Cambridge, UK: Blackwell.

Spencer, M. E., & Nichols, S. E. (2010). Exploring environmental education through ecofeminism: Narratives of embodiment of science. In A. Bodzin, B. Shine Klein, & S. Weaver (Eds.), *The inclusion of environmental education in science teacher education* (pp. 255–266). Dordrecht, Netherlands: Springer. doi:10.1007/978-90-481-9222-9_17

St. Pierre, E. A. (2000). Poststructural feminism in education: An overview. *Qualitative Studies in Education, 13*(5), 477–515.

Sturgeon, N. (2009). *Environmentalism in popular culture: Gender, race, sexuality, and the politics of the natural.* Tucson, AZ: University of Arizona Press.

Taylor, A., Blaise, M., & Giugni, M. (2013). Haraway's "bag lady story-telling": Relocating childhood and learning within a "post-human landscape." *Discourse: Studies in the Cultural Politics of Education, 34*(1), 48–62. doi:10.1080/01596306.2012.698863

Taylor, A., & Pacini-Ketchabaw, V. (2015). Learning with children, ants, and worms in the Anthropocene: Toward a common world pedagogy of multispecies vulnerability. *Pedagogy, Culture, and Society, 23*(4), 507–529. doi:10.1080/14681366.2015.1039050

Taylor, C. A. (in press). Diffracting the curriculum: Putting "new" material feminist theory to work to reconfigure knowledge-making practices in undergraduate higher education. In K. Scantlebury, C. A. Taylor, & A. Lund (Eds.), *Turning feminist theory into practice: Enacting material change in education.* Dordrecht, Netherlands: Sense Publishers. Retrieved from http://sites.udel.edu/kscantle/books/

Thompson, C. (2006). Back to nature? Resurrecting ecofeminism after poststructuralist and third-wave feminisms. *Isis, 97*(3), 505–512. doi:10.1086/508080

Tuck, E., McKenzie, M., & McCoy, K. (2014). Land education: Indigenous, postcolonial, and decolonizing perspectives on place and environmental education research. *Environmental Education Research, 20*(1), 1–23. doi:10.1080/1350462 2.2013.877708

Twine, R. (2010). Intersectional disgust? Animals and (eco)feminism. *Feminism & Psychology, 20*(3), 397–406. doi:10.1177/0959353510368284

United Nations. (1972). *Declaration of the United Nations conference on the human environment.* Retrieved from www. unep.org/Documents.Multilingual/Default.asp ?documentidD97&articleidD1503

United Nations. (1992). *Rio declaration on environment and development.* Annex 1 in report of the United Nations conference on environment and development. Rio de Janeiro, 3–14 June. Retrieved from www.un.org/documents/ga/conf151/aconf15126-1annex1.htm

United Nations. (1993). *Agenda 21: Earth summit: The United Nations programme of action from Rio.* New York, NY: United Nations. Retrieved from sustainabledevelopment.un.org/content/documents/Agenda21.pdf

United Nations. (2012). *The future we want: Outcomes document adopted at Rio C 20.* Rio de Janeiro, Brazil: United Nations. Retrieved from www.uncsd2012.org/content/documents/727The%20Future%20We%20Want%2019% 20June%201230pm.pdf

United Nations. (2015). *Transforming our world: The 2030 agenda for sustainable development.* Resolution adopted by the General Assembly on 25 September 2015. Retrieved from www.un.org/ga/search/view_doc.asp?symbolDA/RES/70/1&Lang D E

Warren, K. J. (1987). Feminism and ecology: Making connections. *Environmental Ethics, 9*(1), 3–20. doi:10.5840/enviroethics19879113

Warren, K. J. (Ed.). (1994). *Ecological feminism.* New York, NY: Routledge.

Warren, K. J., & Cheney, J. (1991). Ecological feminism and ecosystem ecology. *Hypatia, 6*(1), 179–197. doi:10.1111/j.1527-2001.1991.tb00216.x

Whitehouse, H. (2000). Talking up country: A study of the discursive production of environmental meanings in tropical Australia (Doctoral dissertation). James Cook University, Cairns, Australia.

Whitehouse, H. (2011). Talking up country: Language, nature culture and interculture in Australian environmental education research. *Australian Journal of Environmental Education, 27*(1), 56–67. doi:10.1017/S0814062600000070

Whitehouse, H., Watkin-Lui, F., Sellwood, J., Barrett, M. J., & Chigeza, P. (2014). Sea country: Navigating Indigenous and colonial ontologies in Australian environmental education. *Environmental Education Research, 20*(1), 56–69. doi:10.1080/13504 622.2013.852655

Worsham, L., & Olson, G. (2008). *Plugged in: Technology, rhetoric, and culture in a posthuman age.* Cresskill, NJ: Hampton Press.

9 Challenging amnesias

Feminist new materialism/ ecofeminism/women/climate/ education

Annette Gough and Hilary Whitehouse

Gough, A. and H. Whitehouse. 2020. "Challenging Amnesias: Feminist New Materialism/ Ecofeminism/Women/Climate/Education." *Environmental Education Research* 26(9–10): 1420–1434. doi:10.1080/13504622.2020.1727858

Abstract

This chapter discusses the apparent amnesia with regard to insights manifested in ecofeminist thought and applies a re-collective analysis to thinking on the implications of an ecofeminist new materialism for contemporary environmental education research, and curriculum practice. We engage with a conversation between feminist new materialism and the tropes of ecofeminism at this very unusual time in human history, making visible such interactions. Drawing attention to this and other apparent amnesias and, arguing from a genealogical perspective, we argue the scholarly and conceptual disruption caused by rapidly changing environmental (hence social and cultural) conditions can be fruitfully understood and analysed through a reconceived new materialist ecofeminism.

This is especially important given the unequal impact of the climate emergency on women and the continued absence of a truly coherent focus on women's interests – another amnesia – at this moment when the climate dominates all human and other-than-human life. In exploring the relationships between feminist new materialism, ecofeminism and the more-than-human, we theoretically and materially consider the conceptual challenges of confronting the climate emergency as viewed through the lens of articulating feminisms, and we promote possibilities for further conceptual and practical environmental education research.

Introduction

"Amnesia" is a term with its etymological origins in *amnēsía*, a variant of *amnēstía* meaning oblivion and obliviousness. The term is related to the Greek terms *mnemnon* (mindful), *mneme* (memory) and *mnasthai* (to remember). In social analysis, amnesia is a forgetfulness, a failure to remember, or recall;

DOI: 10.4324/9781003390930-11

or more seriously, the loss of memory. We use the term cautiously in order to indicate that there are conceptual and practical intersections in environmental education research and practice where forgetting to make the connections is not, or will not prove, useful for the third decade of the twenty-first century. In arguing for an unforgetting (an undoing of forgetfulness), we draw on the act of recollection, or re-collection, an act of regathering from the past to manifest useful theoretical + material concepts for understanding the deep challenges of the Anthropocene, particularly the climate emergency. The noun recollection has its etymological roots in the Latin *recollectus*, past participle of *recolligere*, meaning to take up again, to regain, to collect once more, from *re* (again) + *colligere* (gather). Recollection is defined as the ability to remember things, to possess a memory of past events (and concepts). The verb, recollect, refers to the act of remembering, to recover or recall knowledge and to bring back to the mind or memory. As Karen Barad (2012, 48) argues, "Matter feels, converses, suffers, desires, yearns and remembers".

Ecofeminism + feminist new materialism

The first amnesia we draw attention to concerns the intersections between ecofeminism and feminist new materialism (Gaard 2015, 2017; Gough and Whitehouse 2018; Merrick 2017; Sandilands 2017; Stevens et al. 2017a, 2017b). Richard Twine (2010, 402) has argued that "the emergence of a feminist new materialism ought to usher in a renewed conversation between feminism and ecofeminism due to shared interests" and such a conversation would be a very productive space for environmental education research. It is our analysis that historically ecofeminism and the new materialist feminisms have much to offer each other, and that "a more robust and inclusive feminism" (Gaard 2017, 117) is possible,

> one that replaces a culture of domination with a world of participatory economics grounded in communalism and social democracy, a world without discrimination based on race or gender, a world where recognition of mutuality and interdependency would be the dominant ethos, a global ecological vision of how the planet can survive and how everyone on it can have access to peace and well-being.
>
> (hooks 2000, 10)

However, as discussed later in this article, this analysis needs to acknowledge the many erasures of ecofeminist arguments through the genealogical shift from feminist philosophy to Euro-patriarchal thought that underpins new materialism (Gaard 2011, 2015, 2017; Sandilands 2017).

Historically, the term *ecoféminisme* was coined by Francoise d'Eaubonne in her book, *Le féminisme ou la mort* (1974). Ecofeminism was intended "to represent women's potential for bringing about an ecological revolution to ensure human survival on the planet. Such an ecological revolution would entail new gender relations between women and men and between humans and nature" (Merchant 1990, 100). According to Greta Gaard (2011, 28), a significant expansion of ecofeminist analysis "emerged from the intersections of feminist research and the various movements for social justice and environmental health, explorations that uncovered the linked oppressions of gender, ecology, race, species, and nation". The importance of ecofeminist thinking was its direct focus on 'nature' (and 'Nature') (see Mcphie and Clarke, 2018), how women were historically positioned, and explicitly embracing the world beyond the human. Rosemary Ruether (1975) argued that the liberation of women and the project of addressing the (then) emerging ecological crisis were intricately linked through societal practices of domination. Her theoretical solution was to envision a "radical re-shaping" (204) of socio-economic and environmental relations. Although, as Karen Warren (1993) pointed out, there are many feminisms and many ecofeminisms, a general conception of ecofeminism suffered from a backlash in the late 1990s, being criticised as essentialist, elitist and ethnocentrist, and ecofeminism was "effectively discarded" (Gaard 2011, 26). However, studies at the intersection of feminism and environmentalism have seen a resurgence of the label ecofeminism during this century (see for example, Gaard 2011, 2015; MacGregor 2017; Phillips and Rumens 2016a, 2016b; Stevens et al. 2017a; Thompson 2006), as distinct from other labels such as "ecological feminism" (Warren 1994; Warren and Cheney 1991), "feminist environmentalism" (Seager 1993), "feminist ecologies" (Stevens et al. 2017b) or "global feminist environmental justice" (Sturgeon 2009).

While clarifying labels, we also note that there is possible confusion between material feminism and feminist new materialism. Stacy Alaimo and Susan Hekman (2008, 18) distinguish what they call "material feminism" emerging "primarily from corporeal feminism, environmental feminism, and science studies" and "new materialism" which is "from, or is synonymous with[,] Marxist feminism". While many material feminist theorists have been influenced by Marxist theory, "their definition of materiality is not, or is not exclusively, Marxist". Alaimo and Hekman see "materialist feminism" as focusing on race, sexuality, imperialism and colonialism, and anthropocentrism, but not necessarily encompassing "the materiality of human corporeality or, certainly, of nonhuman nature" (18). As new materialism developed over the past decade these fine distinctions have become blurred. We do not see this as necessarily problematic for two reasons.

The first is that strictly delineated 'schools of thought' are possibly more relatable to the modernist project rather than to the post, postmodernist turn in social thinking. The privilege of arguing over categories may come to belong to a time before we properly understood we were going to be overwhelmed by the chemical, geo-physical and eco-biological realities of climate change (which we call the climate emergency in this article). This leads us to the second reason, that, in practice, scholars are running out of the luxury of time in which to debate the fixities of categories. The climate emergency is forcing a realisation there is no one conceptual model that is going to see us through. In a globalised situation where it is becoming evident that we need an 'all hands on deck' approach, scholars can draw on the strengths of relevant theoretical perspectives in order to generate means for more fully comprehending our present moment. This is why we argue against certain amnesias concerning past active theories and argue for a recollection of the strengths of historical perspectives, including those insights generated through ecofeminism.

From philosophical and theoretical perspectives, feminist new materialism attempts to 'denature' bodies (human, other, organic and inorganic) as a means to, "establish a radical break with both universalism and dualism as the co-constitutive-ness of cultural discourse and materiality" (Lenz Taguchi 2013, 707). Importantly, new materialism throws up provocative challenges for interdisciplinary feminist scholarship in that matter (including the body) is conceived as being in possession of "a distinctive kind of agency, one that is neither a direct nor an incidental outgrowth of human intentionality but rather one with its own impetus and trajectory" (Frost 2011, 70). New materialist feminism challenges some of the politically useful foundations of ecofeminism, the purposes of which were to undo the normalised power of gendered essentialisms that kept women and 'Nature' oppressed. In considering the relationship between new material feminism and ecofeminism, the challenge is to denaturalise nature *and* deculturalise culture, as "matter is not inert, neither does it form an empty stage for, or background space to, human activity. Instead, matter is conceptualised as agentic and all sorts of bodies, not just human bodies, are recognised as having agency" (Taylor and Ivinson, 2013, 666).

Another amnesia that is relevant to this discussion is what was being articulated by ecofeminists in the 1990s is not fully acknowledged (Gaard 2017, 118), and that

> feminist engagement with theories of posthumanism (e.g., Barad 2003) and the emergence of "new materialist feminists" (e.g., Hird 2004) do not address the relationship between feminism and ecofeminism: many new materialists do not acknowledge ecofeminist scholarship, despite its foundational contributions to new materialist feminisms and the continuing intersections of these two theoretical perspectives.

New materialist feminists are embracing posthumanism (Gaard 2017, 124) but not acknowledging its associations with ecofeminism, and also "disavow[ing] the history of biopolitics in feminism" and forgetting, ignoring, or turning away "from questions of matter, biology, and nature" (Sandilands 2017, 232). Yet posthumanism is relevant, as Karen Barad (2007, 32) argues, for "posthumanism marks a refusal to take the distinction between 'human' and 'nonhuman' for granted, and to found analysis on this presumably fixed and inherent set of categories". However, rather than use the term 'posthumanism', we follow Elspeth Probyn (2016) and adopt the term 'more-than-human'.

Probyn argues that, while the terms "posthuman" and "more-than-human" "are generative in that they seek to shake up any assumptions that we might have had about what conjoins and what separates us, not to mention what that profoundly confusing 'us' might be". Probyn (2016) prefers "more-than-human" because it is "ontologically and materially relational, and opens up new epistemologies as it narrows the diverse and shifting relations between and among humans, and the many different aspects of that are so much more-than-human" (p. 110).

The "renewed conversation" (Twine 2010, 402) between feminism and ecofeminism is not yet happening to any great extent, but it is important because without it there is an amnesia in new material feminism towards what had been discussed and achieved by ecofeminist theorising. Both of these forms of feminism share a concern with challenging what Bronwyn Davies (2018, 114) calls "the taken-for-granted ascendance of all things human . . . and the co-implication of humans with non-human matter" and "the apparent indifference . . . to the ways human thinking-doing affects what will matter in the world". Before Barad (2007) offered her new concepts and new ways of thinking in new materialism, Carolyn Merchant (2003, 201) recognised that epistemology and ontology are always relational and interacting:

The material world is mentally constructed and interpreted in terms of language, but it exists as biophysical entities. Only biophysical beings, such as humans, have the tongues and brains to create and perform speech acts – to invent the logos, or word. The real physical world and the constructed mental world thus exist in dialectical relation to each other. Reality and narrative, I argue, interact with each other.

So why is it that new materialist feminists have tended to ignore the genealogical connections with ecofeminism? Historical amnesia seems to be a growing issue with new generation academics who perhaps do not have the time to immerse themselves in historical literature from theirs and related fields; however, we believe it is important to look at where we have come from, where we are now and where we can go from here in seeking understanding and changes: as Alasdair MacIntyre (1984, 216) wrote, "I can only answer the question 'What am I to do?' if I can answer the prior question 'Of what story or stories do I find myself a part?'". In an environmental education context, the stories of which we are a part include ecofeminism, and these stories must not be forgotten. More than this, ecofeminism describes "a socially constructed association among women (sex), femininity (gender), and nature that was contextual and fluid, not ahistorical and static" (Merchant 1992, 31). The research programmes of which we are a part can flow from these associations, as reflected in the two recent special issues of *The Journal of Environmental Education* on gender and environmental education (Volumes 48(1) and 49(4)). The articles demonstrated commitments to different gendered methodologies – ranging from the personal to the political and drawing on a range of theoretical resources including ecofeminism, poststructuralist, materialist and other feminisms, intersectionality, Foucauldian analysis, assemblages, masculinity and environmental justice – and how the different perspectives of such approaches invigorate environmental education scholarship and show what is possible when gendered methodologies are adopted. However, gendered methodologies still remain on the margins of most environmental education research programmes, as the editors of these special issues note (Gough et al. 2017).

What ecofeminism told us in the past was that systematic oppression of a multiply categorised, naturalised, otherness was indelibly linked with systems of patriarchal, hegemonic and colonising thinking that serves to oppress the agency of bodies actively constituted into a category 'nature' and a category of 'women' to their mutual disadvantage. This remains a profound insight. A genealogical examination of the ecofeminist writings of the 1980s and 1990s reveals how such hierarchy of favouritism – dominion over nature, domination over women – has underpinned the accelerating disaster that befalls us all in the present. We must recollect that older ecofeminism contains many insights relevant to new materialist feminism. Yes, the critiques of ecofeminist essentialism are valid (see Gaard 2011). The essentialist dustings were always empirically suspect anyway, as there certainly is no essential, empirically delineated nature to be found, and women were always agentic, and seeking agency, whether this was recognised or not. However, the ecofeminist analysis of the workings of the system against the fabric of life on earth and the struggles of women against this dual domination remains extant and meaningful in material ways. And, on a historical note, it is not wise to abandon powerful insights solely on the basis of critique.

Barad (2012, 49) is quite emphatic on this point, saying:

> Critique is all too often not a deconstructive practice, that is, a practice of reading for the constitutive exclusions of those ideas we cannot do without, but a destructive practice meant to dismiss, to turn aside, to put someone or something down – another scholar, another feminist, a discipline, an approach, et cetera. So, this is a practice of negativity that I think is about subtraction, distancing and othering.

And Barad (2012, 50) herself is always keen for any proposal (theoretical, philosophical, conceptual, practical) to be read "in the sense of it being suggestive, creative and visionary".

It is our view that twentieth-century ecofeminism was certainly visionary. For example, according to Karen Warren (2000, 1), "Nature is a feminist issue, might be the slogan of ecofeminism". The purpose of earlier ecofeminism was to recognise and investigate "the interconnection among all unjustified human domination" (Warren 2000, 2) with a focus on women, because the focus on women (rather than on geography, poverty, race, class, indigeneity etc.) "reveals important interconnected systems of human domination" (2). And, significantly, when human groups are harmed by ecological destruction (inadvertent or deliberate), it is generally women and their children who suffer more material harm than men – a social reality that has been well understood for decades. For example, Warren (2000, 4) quotes from a 1995 United Nations document titled *The World of Women* that highlights how rural women were, at the time, "excluded from traditional rural development programmes that . . . provide training – and from the credit and other institutional support needed for rural development". Overcoming systematised and institutional disadvantage in relation to social development and environmental care experienced by the majority of the world's women remains a work in progress, as evidenced by Goal 5 on Gender Equality of the United Nations (2016) Sustainable Development Goals, and that gender is recognised as "a prerequisite for the achievement of other Goals, including Goal 4", which is concerned with Education (United Nations 2019, 8).

Warren (2000) is but one example of how ecofeminist analysis preconfigured the new materialism. To quote Barad (2012, 54) again:

> Agency is not . . . a property of persons or things; rather, agency is an enactment, a matter of possibilities for reconfiguring entanglements . . . entailed in reconfiguring material-discursive apparatuses of bodily production, including the boundary articulations and *exclusions* that are marked by those practices.

Ecofeminism was always concerned with exclusions, of women's bodies and of "natural", "Othered" bodies from the domains of privilege, and from privilege's rapacious pursuits.

Climate matters [for] ecofeminist new materialism

Barad (2007, 3) has argued that what is 'new' about new materialism is that we come to know that "matter and meaning are not separate entities" (2007, 3). Jamie Mcphie and David Clarke (2018, 5) point out that concepts "matter in very literal ways" and that, "conceiving the world in terms of material – and materiality – allows researchers to think with different conceptions, different stories that matter different consequences. For research is a story and it does matter consequences" (6). As thinking counts, re-collecting the 'eco' into new materialist feminism allows for a larger story to be told, the amnesia forgotten. The term 'eco' is an adjective derived from the Greek *oiko*, meaning house, home, settlement, place of being. In contemporary use, *eco* is combined with nouns and adjectives to indicate an additional consideration of the ecological, the space beyond the wholly human. Eco + feminism produced a sophisticated and complex analysis of human systems of oppression and domination in the later 20th century in order to reconfigure those relations and relationships towards a larger vision of justice and inequality (Warren 2000). Gaard

(2011, 31) sees that what is needed in the 21st century is "an *intersectional*" ecological-feminist approach – one that will frame conditions "in such a way that people can recognize common cause across the boundaries of race, class, gender, sexuality, species, age, ability, nation – and affords a basis for engaged theory, education, and activism" (Gaard 2011, 44). Both ecofeminism and new material feminisms similarly shift the emphasis from humans at the centre and "propose that all manner of bodies, objects and things have agency within a confederation of meaning-making" (Taylor in press).

For Stacy Alaimo (2010), the way forward is through a deeper understanding of "trans-corporeality", which she describes as a way of theorizing the relationship between humanity and the world at large as not being clearly delineated and separate but as fluid. Alaimo (2008, 238) has pointed out that "human corporeality in all its material fleshiness is inseparable from 'nature' or 'environment'", and that body/nature is also intertwined with 'culture'. However, as Greta Gaard (2015, 25) points out, this is another example of amnesia as

> transcorporeality, the physical fact of our co-constituted embodiment with other flows of life, matter, and energy. . . rests on four decades of feminist science studies and ecofeminist perspectives on the human-environment connection, developing knowledge in the study of gender, race, class, age, and public health.

Considering the inherent agency of material bodies is a means for better apprehending the active workings of power (Barad 2003) and for questions regarding causation, resistance and effect (Frost 2011), Blanche Verlie (2017, 2018) has discussed this in terms of climate as entanglement and becoming-climate. Mcphie and Clarke (2018) argue strongly for immanent thinking. All environmental education research is best served when we are able to draw inclusively on all available ideas across ecofeminism and new material feminisms (Gough and Whitehouse 2018). Any form of eco-amnesia in theory and practice is a significant conceptual oversight, particularly when we are facing a climate emergency, which is quite possibly the 'Mother' of all materialities and made all the more frightening because how each and all of us will experience the effects will be a matter of uncertainty. In her complex, historical analysis, Merchant (2016) explores concepts of chance, randomness unruliness and unpredictability and argues that coming to terms with our collective pollution of the atmosphere-ocean system means coming to terms with a world in which "chance and randomness will constitute our new reality" (6).

Viewed temporally, the whole condition of risk 'we' (every material body, large and small) collectively face is the matter of carbon pollution changing the chemistry of our oceans and atmosphere. George Marshall (2014, 2) calls this "the ultimate challenge to our ability to make sense of the world around us". The urgency to make sense is generated through sophisticated scientific modelling, empirical evidence and our lived experiences of what Isabelle Stengers

(2015, 2) says is "the brevity of time to avoid brutal climate change". Driving our dread is the fear that "human action risks being ineffective in a system which is already exhibiting a powerful momentum in an apocalyptic direction" (Knox 2015, 104). As Lawrence Torcello (2017, np) succinctly states, "time is running out and physics is managing the clock". Some recent environmental education research is reconsidering philosophical, theoretical and practical approaches to climate change education and communication (Börebäck and Schweiler 2018; Stapleton 2019; Verlie 2017, 2019; Verlie and CCR 15 2018), exploring the problematics of material entanglements to challenge the hierarchical hegemonies implicit within the climate emergency. For example, Sarah Riggs Stapleton (2019) argues for a climate justice approach to climate change education, through "storying" climate change (Crate 2017) and providing opportunities "to provide needed context, nuance, and personal connection to the problem" (736). Judy Horacek (see Figure 9.1) encapsulates such storying.

Blanche Verlie and CCR 15 (2018, 1) take another approach and "explore cultivating climate action if we decentre the human from our climate

Figure 9.1 Storying climate change.

Source: Cartoon by Judy Horacek ©2019, first published in *The Age*, reprinted with permission.

pedagogies". However, gender is still not generally brought to the forefront, which is particularly dismaying given women and their children are already, *materially*, caught in the grip of rapid environmental change, to wit: "The women of Bangladesh are among the first to face the impacts of climate change, and their suffering is disproportionate. In the cyclone disaster of 1991, for example, 90% of the 140,000 people who died in the country were women" (Environmental Justice Foundation 2017, online).

It is the case that the *material* effects of the climate emergency remain profoundly gendered. Anna Ines Abelenda (2014, 125) remarks that the international order is yet to "seek to bring about just and healthy economies, which form the basis for the realisation of the fulfilment of the full gamut of rights and freedoms, including economic, social and women's rights, rather than merely attempt to make the existing order appear less egregious in its exploitation". And Jeremy Deaton (2018, np) states that, given "the climate crisis is tightly bound with other urgent issues . . . societies cannot effectively deal with the carbon crisis without addressing, racism, xenophobia and economic inequality . . .[and] without tackling pervasive sexism". Such analyses have always formed part of the ecofeminist project. Vandana Shiva (1989, 1991, 1997, 2005) has consistently argued that the nexus between a dominant patriarchy and a dominant hyper-capitalism (and now, a hyper-digital revolution with all its attendant pollution and toxicity problems) is the underlying cause for the rise of global, profit-orientated corporations, the biodiversity crisis, the homogenisation of human cultures and the attendant acceleration of the climate crisis.

For Gaard (2015, 21), the historical focus of the discussion has shifted from being *about* women, particularly as "*victims* of environmental degradation in need of rescue", to more thoughtfully reconsidered "gender as a system structuring power relations" (Gaard 2015, 22). For example, Atieno Mboya Samandari (2016, 25) uses ecofeminist analysis to argue that an inferiorised, ocean/atmosphere "has been used as a seemingly endless waste receptacle for by-products of industrialization which serve the male dominated status quo". Unequal power relations magnify the matter of physical risk. The Global Gender and Climate Alliance (2013) has documented the real and present dangers to women, quoting a World Bank survey finding that 103 out of 140 countries impose legal differences on people solely on the basis of gender. Two-thirds of the world's illiterate adults are women, and normative oppressive relations limit women from acquiring the information and skills necessary to escape or avoid hazards (e.g. swimming or climbing trees to escape rising water levels). Even dress codes imposed on women restricts their mobility in times of immediate disaster, as does their responsibility for the safety of young children (Mann 2015, 1).

Hannah Knox (2015, 92) describes climate as "the new geosocial leviathan", and a major challenge for climate education is "how to encourage people to internalize objective accounts of climate in order that they could see themselves as part of the very dynamic that visualizations of predictive climate models were describing" (95). In a nutshell, Knox's argument is that climate *is* us, which is similar to Verlie's (2017, 569) argument that seeing "[c]limate as entanglement accounts for how climate science works, without conceptualizing humans and the climate as independent entities that preexist their intra-action". If we come to understand the climate as "not primarily a thing but a set of relationships" (Knox, 2015, 103), then it follows that the gendered nature of those relationships also comes to the fore. As Megan Capriccio (2017) writes on the *One Million Women* website: "Are you also passionate about equality across the gender spectrum? Do you generally think that all living things should be treated with a sense of respect and dignity? Then jump on in, you're an ecofeminist!"

Gaard (2015, 23) is very clear when she writes, "make no mistake: women are indeed the ones most severely affected by climate change and natural disasters, but their vulnerability is not innate; rather it is a result of inequities produced through gendered social roles, discrimination, and poverty". The critiqued essentialisms of 20th-century ecofeminism sieved through newer understandings of material relationships and inter-relationality reveal a more nuanced approach as to the value of an eco/logical, new feminist materialism. For whatever its faults, the tenets of ecofeminism were always about care and caring, agency and action. And herein lies the challenge for environmental education scholars and practitioners – we need to take on that leadership and draw attention to the silences around gender and the climate emergency in our research and practice. We have to imagine the possible, predictably unpredictable futures that hinge on our current actions in a real and material world; interpreting the prevalent graphs, models, visualisations and media reports into tangible realms of experience and agency, while knowing matter and meaning are inextricable in our educational practices.

The climate justice approach to climate change education discussed by Stapleton (2019) is a good beginning – but it needs a more robust gender dimension. As the recent report to the United Nations General Assembly on the *Implementation of education for sustainable development in the framework of the 2030 Agenda for Sustainable Development* (2019, 9) noted.

> At present, Sustainable Development Goal 5, "Achieve gender equality and empower all women and girls" is far from being a reality. The key message of UNESCO taken forward through its actions, programmes and strategy is consequently clear: "women and men must enjoy equal opportunities, choices, capabilities, power and knowledge as equal citizens. Equipping girls and boys, women and men with the knowledge, values, attitudes and skills to tackle gender disparities is a precondition to building a sustainable future for all".

For Claire Brault (2017, 27) "imagining climate futures needs feminist insight" and she cautions that we all have to be aware and critical of certain "hegemonic assumptions and epistemological procedures that animate many of today's debates" (5).

For the more optimistic practitioners and researchers, Naomi Klein (2016, online) argues that, while "climate change acts as an accelerant to many of our social ills – inequality, wars, racism . . . it can also be an accelerant for the opposite".

Moving forward

We have previously written about an apparent amnesia around women and bringing feminist perspectives into environmental education research (see, for example, Gough (1994, 1999, 2013), Gough and Whitehouse (2003, 2018)), about posthumanism, more-than-human, biopolitics and environmental education (Gough 2015, 2017, 2019; Gough and Gough 2016, 2017) and about climate change education (Stevenson et al. 2017, Whitehouse 2017, Whitehouse et al. 2019). In this chapter, we bring together an argument that amnesia can be overcome through recollection and reinvigoration – that the past and the present can be drawn together to provide meaningful impetus for research. And, we have argued elsewhere, "environmental education researchers should draw more inclusively on international sociological and cultural thinking across ecofeminism and material discursive analyses" (Gough and Whitehouse 2018, 344). Unfortunately, ecofeminist and related feminisms have not had much influence on standardised school curriculum, even as our socio-environmental crises would logically make the need for such teaching more urgent.

The current guiding document for schooling in Australia is the *Melbourne Declaration on Educational Goals for Young Australians* (MCEETYA 2008). In the Preamble to this Declaration, it states (MCEETYA 2008, 5),

> Complex environmental, social and economic pressures such as climate change that extend beyond national borders pose unprecedented challenges, requiring countries to work together in new ways. To meet these challenges, Australians must be able to engage with scientific concepts and principles and approach problem solving in new and creative ways.

These challenges are translated in Goal 2 as "All young Australians become successful learners, confident and creative individuals, and active and informed citizens . . . work for the common good, in particular sustaining and improving natural and social environments" (MCEETYA 2008, 8–9). However, the only action related to the environment included in the Declaration document is "a focus on environmental sustainability will be integrated across the curriculum" (MCEETYA 2008, 14), which was part of the recommendations or framing what has become the Australian Curriculum, with a cross-curriculum priority of Sustainability (ACARA 2019). In addition, while the Preamble intimates that science understanding is important in meeting complex challenges such as climate change the Science curriculum does little to educate young Australians to meet these challenges. Climate change is only referred to three times in the Science curriculum, all at Year 10 level (ACARA 2019) where the content involves:

- considering the role of science in identifying and explaining the causes of climate change,
- considering how computer modelling has improved knowledge and predictability of phenomena such as climate change and atmospheric pollution, and
- considering the scientific knowledge used in discussions relating to climate change.

None of these content statements provide opportunities to engage with "ecological, economic, and political sustainability by developing theory that responds to the racialized . . . and gendered climate crises of the Anthropocene" (Gaard 2017, 126). (The term 'Anthropocene' does not yet gain a mention in the national Australian Curriculum.) Given these deliberate silences (Whitehouse et al. 2019), the research challenge for, and within, all formal educational institutions is to transform the normative definitions and practices of environmental or sustainability education to address the material crises of a destabilised climate. And similarly, community education in the form of lifelong learning has a strong place in "the promotion of sustainable ways of living; challenging traditional views on development and the exclusively technical approach to reversing climate change; awareness raising on the need to have a gender perspective; and disseminating local, indigenous and traditional knowledge" (Agostino 2011, online).

It is our conclusion that more deeply conceptualised forms of feminism are needed now more than ever. *The Global Gender Gap Report 2017* (World Economic Forum 2017) benchmarked gender parity in 144 countries and came to the conclusion that, at present rates of social change, global gender equality was still over 200 years away. This is not a good sign for a humankind facing a planetary accounting in the next decade when the absence of gender parity is analysed as being largely responsible for the climate crisis. Sandra Harding (2015, 79) admonishes:

> The failure to address women's issues directly in development contexts not only damages women's chances for flourishing and for equality; it also renders it impossible to achieve the eradication of poverty and advance of other, noneconomic kinds of flourishing that supposedly have been the goals of development projects for more than six decades.

The concept of climate justice is gaining international traction through NGOs such as *The Climate Reality Project* (see www.climaterealityproject.org/) and through the United Nations Framework Convention on Climate Change (UNFCCC). At COP 23 in Bonn, Germany, November 2017, a Gender and Climate Change Gender Action Plan (GAP) was proposed noting that "a gender-responsive climate policy continues to require further strengthening in all activities concerning adaptation, mitigation and related means of implementation . . . as well as decision making on the implementation of climate policies" (UNFCCC 2017, 1). The GAP aims to increase the number of women who are active in climate decision making, and train up policy makers on bringing gender equality into climate funding programmes, and engage grassroots women's organisations for local and global climate action. How this plays out in practice is another question, but the aims at least are clear. Hilda Heine (2017, online), the president of the Marshall Islands, pointed out that women who face "a multitude of structural barriers . . . aren't making enough of the decisions, and the decisions aren't yet doing enough for women". A climate justice approach to climate change education and research has the potential to bring together the personal and the political and to foreground "how climate, climate knowledge and climate knowers comerge through intra-action" (entanglement) (Verlie 2017, 569). Recent United Nations (2019) and UNESCO (2019) documents related to education for sustainable development do give much greater prominence to gender. For example, UNESCO (2019, Annex page 3) is currently exploring education for sustainable development beyond 2019 and makes particular mention of gender together with climate change:

> ESD is an instrument to achieve all the SDGs, and each of the SDGs comes with specific gendered challenges. ESD takes on a cross-disciplinary and systemic approach that enables the question of gender equality to be linked to the various issues of sustainable development. There is, for example, a gendered facet of vulnerability to hazards induced by climate change. . . . It should therefore become a priority to provide women with access to ESD. In this regard, ESD actively promotes gender equality, and creates conditions and strategies that empower women.

Gender justice can be further promulgated in environmental education research and practice drawing on new materialist, new feminist materialism, ecofeminist and more-than-human concepts and approaches, as all have their conceptual strengths. We need to make haste towards full realisation of the conditions and extent of the calamity of the climate emergency. This is not a time in which to condone amnesia. This is a time for recollection and reconnection – a time to remember our materialities, our corporealities, our natures, and our intersecting and desiring feminisms. Diffusing boundaries in order to achieve a great whole, we have to get our heads around a present future, and clearly there is much work to be done, and many spaces for new thinking, new research and resistance actions are opening up for environmental education research and praxis in a rather frightening hurry. Cecilia Åsberg (2017, 198) argues we must truly consider "what kinds of ethics and critiques, arts and sciences, politics and methods, can account for the changes in spatial and temporal scales introduced by climate change, species extinction, the life sciences, or the emergence of the politics of Life itself".

It is reassuring to see examples of environmental education research emerging that are "making connections" (Gaard (2017, 126) found in the two recent special issues of *The Journal of Environmental Education* (volumes 48(1) and 49(4)), and in the writings of environmental education researchers such as Karen Malone (2016a, 2016b) and Teresa Lloro-Bidart (2017), apart from our own work. What is important in this recent writing, and which needs to be continued in developing a more robust feminism, is the dismantling of the nature-culture binary, the disruption of anthropocentric views, the new ways of encountering more-than-human worlds and the assertion of more political standpoints. As part of this we need to pursue more-than-human perspectives in environmental education research, including "learning how to open the presences of otherness and how to form relationship of mutuality with others we can never fully know" (Gruenewald 2003, 40) because "[t]he more-than-human, if it is to be meaningful as a perspective, makes us confront again and again the relatedness of all entities" (Probyn 2016, 163). This is not what is found in the Foundation to Year 10 Australian Curriculum (ACARA 2019) but it should be there (and in every other curriculum) if we are to achieve gender and climate justice. Given the existential risks we face, can we afford to wait for curriculum to catch up? Such is our continuing challenge as environmental educators and researchers.

Acknowledgement

We would like to sincerely thank the anonymous reviewers together with David Clarke, Jamie Mcphie and Alan Reid for their constructive suggestions for strengthening our arguments. We are very grateful for this feedback.

Notes on contributors

Annette Gough is Professor of Science and Environmental Education in the School of Global, Urban and Social Studies and the School of Education at RMIT University in Melbourne, Australia. Her teaching, research and publications focus on curriculum policy and development in science and environmental education, feminist, more-thanhuman, critical and post-structuralist research, research methodologies, and, most recently, child centered disaster risk reduction and resilience education. She is a past president and life fellow of the Australian Association for Environmental Education and co-editor of the Springer book series International Explorations in Outdoor and Environmental Education.

Hilary Whitehouse is Associate Professor, Deputy Dean of the Graduate Research School, Fellow of the Cairns Institute, and a member of the University's Sustainable Development Working Group at James Cook University in Cairns, Australia. Her teaching, research and publications focus on science education, environmental education, education for sustainability, and research education - particularly change education, new ecofeminism and feminist new materialism. She is an executive editor of the Journal of

Environmental Education, a member of the international advisory board for the Australian Journal of Environmental Education, and a life member of the Australian Association for Environmental Education.

References

Abelenda, Anna Ines. 2014. "A Feminist Perspective on the Post-2015 Development Agenda Background: The Notion of Development at the Crossroads." *Equal Rights Review 13*: 117–128.

Agostino, Ana. 2011. "Gender Equality, Climate Change and Education for Sustainability." *Adult Education and Development 76*. www.dvv-international.de/en/adult-education-and-development/editions/aed-762011/

Alaimo, Stacy. 2008. "Trans-Corporeal Feminisms and the Ethical Space of Nature." In *Material Feminisms*, edited by Stacy Alaimo and Susan J. Hekman, 237–264. Bloomington, IN: Indiana University Press.

Alaimo, Stacy. 2010. *Bodily Natures: Science, Environment, and the Material Self.* Bloomington, IN: Indiana University Press.

Alaimo, Stacy and Susan J. Hekman. 2008. "Introduction: Emerging Models of Materiality in Feminist Theory." In *Material Feminisms*, edited by Stacy Alaimo and Susan J. Hekman, 1–19. Bloomington, IN: Indiana University Press.

Åsberg, Cecilia. 2017. "Feminist Posthumanities in the Anthropocene: Forays into the Postnatural." *Journal of Posthuman Studies 1*(2): 185–204.

Australian Curriculum and Assessment and Reporting Authority (ACARA). 2019. *The Australian Curriculum.* www.australiancurriculum.edu.au/f-10-curriculum/cross-curriculum-priorities/sustainability/

Barad, Karen. 2003. "Posthumanist Performativity: Toward an Understanding of How Matter Comes to Matter." *SIGNS: Journal of Women in Culture and Society 28*(3): 801–831.

Barad, Karen. 2007. *Meeting the Universe Halfway: Quantum Physics and the Entanglement of Matter and Meaning.* London, UK: Duke University Press. doi:10.1215/9780822388128

Barad, Karen. 2012. "Matter Feels, Converses, Suffers, Desires, Yearns and Remembers Interview with Karen Barad." In *New Materialism: Interviews and Cartographies*, edited by Rick Dolphijn and Iris van der Tuin, 48–70. Ann Arbor, MI: Open Humanities Press. doi: 10.3998/ohp.11515701.0001.001

Börebäck. Maria K. and Elias Schweiler. 2018. "Elaborating Environmental Communication within 'Posthuman' Theory." *Journal for the Philosophical Study of Education 3*: 1–30.

Brault, Claire. 2017. "Feminist Imaginations in a Heated Climate: Parody, Idiocy and Climatological Possibilities." *Catalyst: Feminism, Theory, Technoscience 3*(2): 1–33. doi: 10.28968/cftt.v3i2.123.g276

Capriccio, Megan. 2017, February 13. "What is Ecofeminism?" *One Million Women.* www.1millionwomen.com.au/blog/what-ecofeminism/

Crate, Susie. 2017. "Storying Climate Change." *Anthropology News 58*(2): e64–e69. doi:10.1111/AN.410

Davies, Bronwyn. 2018. "Ethics and the New Materialism: A Brief Genealogy of the 'Post Philosophies in the Social Sciences." *Discourse: Studies in the Cultural Politics of Education 39*(1): 113–127.

Deaton, Jeremy. 2018, February 2. "Why Climate Deniers Target Women. Women Who Work on Climate Change Regularly Endure Sexist Attacks." *Nexus Media.* https://nexusmedianews.com/why-climate-deniers-target-women-f64776ee7c5b

d'Eaubonne, Francoise. 1974. *Le Feminisme ou La Mort.* Paris, France: Pierre Horay.

Environmental Justice Foundation. 2017. *Climate Displacement in Bangladesh.* https://ejfoundation.org/reports/climate-displacement-in-bangladesh

Frost, Samantha. 2011. "The Implications of New Materialism for Feminist Epistemology." In *Feminist Epistemology and the Philosophy of Science: Power in Knowledge,* edited by Heidi E. Grasswick, 69–83. Dordrecht: Springer.

Gaard, Greta. 2011. "Ecofeminism Revisited: Rejecting Essentialism and Re-Placing Species in a Material Feminist Environmentalism." *Feminist Formations 23*(2): 26–53. doi:10.1353/ff.2011.0017

Gaard, Greta. 2015. "Ecofeminism and Climate Change." *Women's Studies International Forum 49*: 20–33. doi:10.1016/j.wsif.2015.02.004

Gaard, Greta. 2017. "Posthumanism, Ecofeminism, and Inter-Species Relations." In *Routledge Handbook of Gender and Environment,* edited by Sherilyn MacGregor, 115–129. Abingdon, UK: Routledge.

Global Gender and Climate Alliance (GGCA). 2013. *Overview of Linkages between Gender and Climate Change.* https://issuu.com/undp/docs/pb1_asiapacific_capacity_final

Gough, Annette. 1994. "Fathoming the Fathers in Environmental Education: A Feminist Poststructuralist Analysis." Unpublished doctoral dissertation. Geelong, Victoria: Deakin University.

Gough, Annette. 1999. "Recognising Women in Environmental Education Pedagogy and Research: Toward an Ecofeminist Poststructuralist Perspective." *Environmental Education Research 5*(2): 143–161. doi:10.1080/1350462990050202.

Gough, Annette. 2013. "Researching Differently: Generating a Gender Agenda for Research in Environmental Education." In *International Handbook of Research on Environmental Education,* edited by Robert B. Stevenson, Michael Brody, Justin Dillon and Arjen Wals, 375–383. New York, NY: Routledge. doi:10.4324/9780203813331.ch35

Gough, Annette. 2015. "Resisting Becoming a Glomus Body within Posthuman Theorizing: Mondialisation and Embodied Agency in Educational Research." In *Posthumanism and Educational Research,* edited by Nathan Snaza and John Weaver, 254–275. New York: Routledge.

Gough, Annette. 2017. "Searching for a Crack to Let Environment Light in: Ecological Biopolitics and Education for Sustainable Development Discourses." *Cultural Studies of Science Education 12*(4): 889–905. doi:10.1007/s11422-017-9839-8

Gough, Annette. 2019. Symbiopolitics, Sustainability, and Science Studies: How to Engage with Alien Oceans. *Cultural Studies <=> Critical Methodologies.* doi:10.1177/1532708619883314

Gough, Annette and Noel Gough. 2016. "The Denaturation of Environmental Education: Exploring the Role of Ecotechnologies." *Australian Journal of Environmental Education 32*(1): 30–41. doi:10.1017/aee.2015.34

Gough, Annette and Noel Gough. 2017. "Beyond Cyborg Subjectivities: Becoming-Posthuman Educational Researchers." *Educational Philosophy and Theory 49*(11): 1112–1124. doi:10.1080/00131857.2016.1174099

Gough, Annette, Constance Russell and Hilary Whitehouse. 2017. "Moving Gender from Margin to Center in Environmental Education." *The Journal of Environmental Education 48*(1): 1–4. doi:10.1080/00958964.2016.1252306

Gough, Annette and Hilary Whitehouse. 2003. "The 'Nature' of Environmental Education from a Feminist Poststructuralist Viewpoint." *Canadian Journal of Environmental Education 8*: 31–43.

Gough, Annette and Hilary Whitehouse. 2018. "New Vintages and New Bottles: The 'Nature' of Environmental Education from New Material Feminist and Ecofeminist Viewpoints." *The Journal of Environmental Education 49*(4): 336–349. doi:10.1080/00958964.2017.1409186

Gruenewald, David. 2003. At Home with the Other: Reclaiming the Ecological Roots of Development and Literacy. *The Journal of Environmental Education* 35(1): 33–43. doi: 10.1080/00958960309600593

Harding, Sandra. 2015. *Objectivity & Diversity: Another Logic of Scientific Research.* Chicago: The University of Chicago Press.

Heine, Hilda. 2017, November 15. "Global Climate Action Must be Gender Equal." *The Guardian.* www.theguardian.com/environment/2017/nov/15/global- climate-action-must-be-gender-equal

hooks, bell. 2000. *Feminism is for Everybody: Passionate Politics.* Cambridge, MA: South End Press.

Klein, Naomi. 2016. "Let Them Drown: The Violence of Othering in a Warming World." *London Review of Books 38*(11): 11–14.

Knox, Hannah. 2015. "Thinking Like a Climate." *Distinktion: Journal of Social Theory* 16(1): 91–109. doi:10.1080/1600910X.2015.1022565

Lenz Taguchi, Hillevi. 2013. "Images of Thinking in Feminist Materialisms: Onto-logical Divergences and the Production of Researcher Subjectivities." *International Journal of Qualitative Studies in Education* 26(6): 706–716. doi:10.1080/095183 98.2013.788759

Lloro-Bidart, Teresa. 2017. "A Feminist Posthumanist Political Ecology of Education for Theorizing Human-Animal Relations/Relationships." *Environmental Education Research* 23(1): 111–130. doi:10.1080/13504622.2015.1135419

MacGregor, Sherilyn. (Ed.). 2017. *Routledge Handbook of Gender and Environment.* Abingdon, UK: Routledge.

MacIntyre, Alasdair. 1984. *After Virtue.* Notre Dame, IN: Notre Dame University Press.

Malone, Karen. 2016a. "Reconsidering Children's Encounters with Nature and Place Using Posthumanism." *Australian Journal of Environmental Education* 32(1): 42–56. doi:10.1017/aee.2015.48

Malone, Karen. 2016b. "Posthumanist Approaches to Theorizing Children's Human-Nature Relations." In *Space, Place, and Environment,* edited by Tracey Skelton, Karen Nairn and Peter Kraftl, 185–206. Singapore: Springer.

Mann, Tracy. 2015, December 31. *COP21 Insights: Climate Wise Women.* http://gender-climate.org/resource/cop21-insights-climate-wise-women/

Marshall, George. 2014. *Don't Even Think About It: Why Our Brains are Wired to Ignore Climate Change.* New York, London: Bloomsbury.

Mcphie, Jamie and David A. G. Clarke. 2018. "Nature Matters: Diffracting a Keystone Concept of Environmental Education Research – Just for Kicks." *Environmental Education Research.* doi: 10.1080/13504622.2018.1531387

Merchant, Carolyn. 1990. "Ecofeminism and Feminist Theory." In *Reweaving the World: The Emergence of Ecofeminism,* edited by Irene Diamond and Gloria Feman Orenstein, 100–105. San Francisco, CA: Sierra Club Books.

Merchant, Carolyn. 1992. *Radical Ecology: The Search for a Livable World.* New York, NY: Routledge.

Merchant, Carolyn. 2003. *Reinventing Eden: The Fate of Nature in Western Culture.* New York, NY: Routledge.

Merchant, Carolyn. 2016. *Autonomous Nature: Problems of Prediction and Control from Ancient Times to the Scientific Revolution.* New York, NY: Routledge.

Merrick, Helen. 2017. "Naturecultures and Feminist Materialism." In *Routledge Handbook of Gender and Environment,* edited by Sherilyn MacGregor, 101–114. Abingdon, UK: Routledge.

Ministerial Council on Education Employment Training and Youth Affairs (MCEETYA). 2008. *Melbourne Declaration on Educational Goals for Young Australians*. www. curriculum.edu.au/verve/_resources/National_Declaration_on_the_Educational_ Goals_for_Young_Australians.pdf

Phillips, Mary and Nick Rumens. (Eds.). 2016a. *Contemporary Perspectives on Ecofeminism*. Abingdon, UK: Routledge.

Phillips, Mary and Nick Rumens. 2016b. "Introducing Contemporary Ecofeminism." In *Contemporary Perspectives on Ecofeminism*, edited by Mary Phillips and Nick Rumens, 1–16. Abingdon, UK: Routledge.

Probyn, Elspeth. 2016. *Eating the Ocean*. Durham, NC: Duke University Press.

Ruether, Rosemary. 1975. *New Woman/New Earth: Sexist Ideologies and Human Liberation*. New York, NY: The Seabury Press.

Samandari, Atieno Mboya. 2016, March. "An Ecofeminist Perspective on the Climate Change Regime." Paper presented to the 9th Annual Feminist Legal Theory Conference, University of Baltimore School of Law. Baltimore, MD. https://law.ubalt. edu/centers/caf/2016_conference/Samandari_Atieno.pdf

Sandilands, Catriona. 2017. "Feminism and Biopolitics: A cyborg Account." In *Routledge Handbook of Gender and Environment*, edited by Sherilyn MacGregor, 229–238. Abingdon, UK: Routledge.

Seager, Joni. 1993. *Earth Follies: Coming to Feminist Terms with the Global Environmental Crisis*. New York, NY: Routledge.

Shiva, Vandana. 1989. *Staying Alive: Women, Ecology and Development*. London, UK: Zed Books.

Shiva, Vandana. 1991. "Biodiversity, Biotechnology and Profits." In *Biodiversity: Social & Ecological Perspectives*, edited by Vandana Shiva, 43–58. London, UK: Zed Books/World Rainforest Movement.

Shiva, Vandana. 1997. *Biopiracy: The Plunder of Nature and Knowledge*. Boston, MA: South End Press.

Shiva, Vandana. 2005. *Earth Democracy: Justice, Sustainability, and Peace*. Boston, MA: South End Press.

Stapleton, Sarah R. 2019. "A Case for Climate Justice Education: American Youth Connecting to Intragenerational Climate Injustice in Bangladesh." *Environmental Education Research* 25(5): 732–750. doi:10.1080/13504622.2018.1472220

Stengers, Isabelle. 2015. *In Catastrophic Times. Resisting the Coming Barbarism* (trans. Andrew Goffey). London, UK: Open Humanities Press. http://openhumanities press.org/books/download/Stengers_2015_In- Catastrophic-Times.pdf

Stevens, Lara, Peta Tait and Denise Varney. (Eds.). 2017a. *Feminist Ecologies: Changing Environments in the Anthropocene*. Cham: Palgrave Macmillan.

Stevens, Lara, Peta Tait and Denise Varney. 2017b. "Introduction: 'Street-Fighters and Philosophers': Traversing Ecofeminisms." In *Feminist Ecologies: Changing Environments in the Anthropocene*, edited by Lara Stevens, Peta Tait and Denise Varney, 1–22. Cham: Palgrave Macmillan.

Stevenson, Robert B., Jennifer Nicholls and Hilary Whitehouse. 2017. "What is Climate Change Education?" *Curriculum Perspectives* 37(1): 67–71.

Sturgeon, Noël. 2009. Environmentalism in Popular Culture: Gender, Race, Sexuality, and the Politics of the Natural. Tucson, AZ: University of Arizona Press.

Taylor, Carol A. in press. "Diffracting the Curriculum: Putting 'New' Material Feminist Theory to Work to Reconfigure Knowledge-Making Practices in Undergraduate Higher Education." In *Turning Feminist Theory into Practice: Enacting Material*

Change in Education, edited by Kate Scantlebury, Carol A. Taylor and Anna Lund. Dordrecht: Sense Publishers.

Taylor, Carol A. and Gabrielle Ivinson. 2013. "Material Feminisms: New Directions for Education." *Gender and Education* 25(6): 665–670. doi:10.1080/09540253.2013.834617

Thompson, Charis. 2006. "Back to Nature? Resurrecting Ecofeminism after Poststructuralist and Third-Wave Feminisms." *Isis* 97(3): 505–512.

Torcello, Lawrence. 2017, April 30. "Yes, I am a Climate Alarmist. Global Warming is a Crime Against Humanity." *The Guardian*. www.theguardian.com/commentisfree/2017/apr/29/climate-alarmist-global-warming-crime-humanity

Twine, Richard. 2010. "Intersectional Disgust? Animals and (Eco)Feminism." *Feminism & Psychology* 20(3): 397–406.

UNESCO. 2019. SDG 4 – Education 2030. Part II. Education for Sustainable Development Beyond 2019. *Executive Board 206EX/6.11.* https://unesdoc.unesco.org/ark:/48223/pf0000366797

United Nations. 2016. *Sustainable Development Goals: 17 Goals to Transform Our World.* www.un.org/sustainabledevelopment/sustainable-development-goals/

United Nations. 2019. *Implementation of Education for Sustainable Development in the Framework of the 2030 Agenda for Sustainable Development.* https://undocs.org/A/74/258

United Nations Framework Convention on Climate Change (UNFCCC). 2017. *Gender and Climate Change – Establishment of a Gender Action Plan.* https://unfccc.int/resource/docs/2017/sbi/eng/l29.pdf

Verlie, Blanche. 2017. "Rethinking Climate Education: Climate as Entanglement." *Educational Studies* 53(6): 560–572.

Verlie, Blanche. 2019. "Bearing Worlds: Learning to Live-with Climate Change." *Environmental Education Research* 25(5): 751–766.

Verlie, Blanche and CCR 15. 2018. "From Action to Intra-Action? Agency, Identity and 'Goals' in a Relational Approach to Climate Change Education." *Environmental Education Research*. doi: 10.1080/13504622.2018.1497147

Warren, Karen J. 1993. "Introduction to Ecofeminism." In *Environmental Philosophy: From Animal Rights to Radical Ecology*, edited by Michael E. Zimmerman, J. Baird Callicott, George Sessions, Karen J. Warren and John Clark, 253–267. New York, NY: Prentice-Hall.

Warren, Karen J. (Ed.). 1994. *Ecological Feminism.* New York, NY: Routledge.

Warren, Karen J. 2000. *Ecofeminist Philosophy: A Western Perspective on What It is and Why It Matters.* Boulder, CO: Rowman and Littlefield Publishers.

Warren, Karen J. and Jim Cheney. 1991. "Ecological Feminism and Ecosystem Ecology." *Hypatia* 6(1): 179–197.

Whitehouse, Hilary. 2017. "Point and Counterpoint: Climate Change Education." *Curriculum Perspectives* 37(1): 63–65.

Whitehouse, Hilary, Larraine Joy Larri and Angela Colliver. 2019, September 19. "Ever Wondered What Our Curriculum Teaches Kids About Climate Change? The Answer is "Not Much"." *The Conversation*. https://theconversation.com/ever-wondered-what-our-curriculum-teaches-kids-about-climate-change-the-answer-is-not-much-123272

World Economic Forum. 2017. *The Global Gender Gap Report 2017.* www.weforum.org/reports/the-global-gender-gap-report-2017

10 Reconceiving nature, gender and sustainability

Annette Gough

Gough, A. (2020). Symbiopolitics, sustainability and science studies: How to engage with alien oceans. *Cultural Studies* <=> *Critical Methodologies*, 20(3), 272–282. doi:10.1177/1532708619883314

Abstract

This chapter explores the implications for inquiries in sustainability education of Helmreich discussions of how human biocultural practices scramble nature and culture, life forms and forms of life, and his ethos of acceptance of ambiguous boundaries and transformative linkages with others. The silences in Helmreich's arguments around gender and sustainability through looking at Probyn and *Merchant* the possibilities for a more-than-human scientific inquiry curriculum, are discussed, as is how science studies offer an image of a more-than-human anthropology that leads us to reconceive evolution, nature, gender and sustainability, to which educational researchers need to pay attention.

This chapter takes as its starting point Stefan Helmreich's (2009) book *Alien Ocean* but puts it into conversation with Elspeth Probyn's (2016) *Eating the Ocean* and other works, such as Carolyn Merchant's (2016) *Autonomous Nature,* and Steve Mentz's (2012) article "After sustainability", as background to rethinking possibilities for inquiries towards a more-than-human scientific inquiry curriculum. Helmreich is an anthropologist and his book documents an anthropological journey with American marine scientists and how revolutions in genomics, bioinformatics and remote sensing have changed the orientation of marine scientists' interests towards microbes as well as informing new thinking about the origins of life, climate change, biotechnology and extraterrestrial life. Probyn has written an anthropology of food: firstly, as a feminist and queer perspective on fisheries, and secondly as an ethnographic exploration into the food politics of fisheries in multiple locations: "There is no innocent place in which to escape the food politics of human-fish entanglement" (p. 5). Merchant's book is not about oceans, but it considers changing conceptions of nature and the implications of these for sustainability. Mentz

DOI: 10.4324/9781003390930-12

troubles what he sees as the dominant pastoral nostalgia within the sustainability discourse and argues for "blue cultural studies", a challenge that I consider later in this chapter. But first I discuss the implications for how we understand evolution and the origins of life emerging from Helmreich's work.

Trees and rhizomes

In 1735, the Swedish botanist Carl Linnaeus proposed a system for putting species into groups based on physical similarities "believed to be markers of immutable essences"; however "it was not obvious which similarities – of, say, reproductive organs, locomotion, or habits – should claim priority in classification" (Helmreich, 2009, p. 76). Such taxonomy "is a rigid human construct that is forced on top of the cacophonous uncertainty of the real wild world" (Ashby, 2016, n.p.). However, according to Ashby, "taxonomic thinking is really important for how we understand the world and our place in it, as each of these terms come with implicit information about how they relate to other groups. It neatly puts the world into boxes". Several recent books (particularly Helmreich (2009), Probyn (2016) and Merchant (2016)) have drawn attention to how the sharp distinctions of Linnaean classification are no longer defensible. Even Neil Shubin (2008), the paleontologist who, with colleagues, in April 2006, discovered Tiktaalik, the evolutionary missing link between fish and land-living animals, helps to destabilise Linnaeus' boxes by providing evidence that "the world was an evolving, changing, developing place and ongoing change came from within matter, as opposed to a world determined by a transcendent deity or a pure Form" (Merchant, 2016, p. 10). However, I am not so sure that Shubin would subscribe to the notions of autonomous nature discussed by Merchant (2016).

In the mid-1800s, Richard Owen (1849) argued in his classic monograph, *On the Nature of Limbs,* that there were "exceptional similarities among creatures as different as frogs and people" (Shubin, 2008, p. 30, emphasis in original) and that "there is a fundamental design in the skeleton of all animals" (Shubin, 2008, p. 32). Owen attributed this plan to the Creator, just as Linnaeus believed that his Systema Naturae "would reveal the divine order of God's creation" (Ashby, 2016, n.p.). Nevertheless, "hierarchical taxonomies tell us a lot about an animal's evolutionary history as by their nature they show what came from what" (Ashby, 2016, n.p.), which Charles Darwin (1859), in his landmark work *On the Origin of Species,* explained in his theory of evolution by natural selection that enabled us to make precise determinations of the ancestry of humans. Shubin argues, "deep similarities exist between creatures living today and those long deceased – ancient worms, living sponges, and various kinds of fish" (2008, p. 181), thereby bringing together Owen's observations and Darwin's theory. However, unlike Owen and Linnaeus, Darwin rejected the divine order of God's creation and instead argued for the lineal descent of beings in the final section of his conclusion (Darwin, 1859, Chapter 14):

> Authors of the highest eminence seem to be fully satisfied with the view that each species has been independently created. To my mind it

accords better with what we know of the laws impressed on matter by the Creator, that the production and extinction of the past and present inhabitants of the world should have been due to secondary causes, like those determining the birth and death of the individual. When I view all beings not as special creations, but as the lineal descendants of some few beings which lived long before the first bed of the Silurian system was deposited, they seem to me to become ennobled.

Darwin drew on genealogy as practiced in the Victorian era and the notion of a branching family tree as a "natural system", "genealogical in its arrange-ment", to make it clear to track the divergence of species "which he felt were distinct and coherent, if mutable, biological forms" (Helmreich, 2009, p. 76). For Darwin, trees traced heredity, a concept figured at the time through the metaphor of blood and bloodlines. This metaphor was replaced by genes after the rediscovery of Mendel's work, and genes provided the substance linking together all living things on earth traced by the tree of life and the currency of genealogy.

As a nonreligious person, I was surprised that Helmreich raised Christian-ity as an elephant in the room in the context of marine microbiology and questioned whether Christianity was a force against environmental conscious-ness in America. He found that intelligent design, creationism and Christianity were not far from the surface for some of the marine microbiologists who see making discoveries as "like seeing the handiwork of the Creator" (p. 237). For example, one scientist, who also does not accept Stephen Jay Gould's (1997) nonoverlapping magisteria, said, "Investigations into the evolution of hyper-thermophiles are perfectly consistent with views of the world as a Creation in which God is continually present, coeval with the evolutionary becoming of reality itself" (pp. 237–238). However, for others,

[t]he implications of this new microbiology for evolution and the ori-gin of life provoke devout Christians and "intelligent design" advocates quite as much as – perhaps more than – the ape ancestry of humans: it would seem to imply "the uselessness of a Creator God," as Pasteur once put it.

(Strick, 2009, p. 950)

This position seems to go with one scientist's belief that humans have "some-thing that makes us special", whereas the microbes are "little living machines" (Helmreich, 2009, p. 238). However, while he does recognize that both humans and microbes have metabolism, he objectifies only the microbes, a position far from recognizing the entanglement of human and more-than-human life.

In Chapter 2 of *Alien Ocean*, drawing on marine microbiology and the work of scientists in marine spaces, Helmreich (2009) discusses dissolving the tree of life because what counts as biological life is changing, as is the relation between "life forms" and "forms of life", and what counts as native or alien, familial or other. For Helmreich, "forms of life" are "those cultural, social, symbolic, and pragmatic ways of thinking and action that organize human communities . . . my usage emphasizes a plurality of forms of life – scientific, religious, economic, ethical – as well as their uneven, contested, and overlapping character" (p. 6) and "life forms" are

> those embodied bits of vitality called *organisms*, variously apprehended as ranged into species (durable but changeable genealogical kinds) or as sorted into types occupying spaces of physical, metabolic, or ecological possibility (photosynthesizers, deep-sea dwellers) . . . I also mean the relations of creatures with one another, following here those biologists who regard organisms as inextricably situated in ecologies.
>
> (p. 6)

These different relationships between organisms lead Helmreich to assert that there cannot be a tree relating to all organisms as the sequencing of complete genomes of bacteria, eukarya and archaea have shown that microbes are mosaic of acquired genes and that a "rhizomatic, reticulated representation as an alternative to the linearity of the tree diagram" (p. 83), which has biological resonances with the work of Deleuze and Guattari (1987).

In the next section, I focus on the implications of the changes in what counts as biological life for our understanding of biopolitics, and what Helmreich calls *symbiopolitics*, and for how nature is conceived. I then discuss the silences in Helmreich's arguments related to sustainability, gender and new materialisms before considering how we can conceptualize a more-than-human scientific inquiry curriculum, which is very different from the current image of scientific inquiry common in Western curricula. I begin by situating the significance of oceans in these discussions.

Oceans

Oceans cover 71 percent of the Earth's surface and contain over 90 percent of the biosphere (DeLoughrey, 2007, p. 20), yet we know very little about life in these marine environments, even though, as Darwin (1859) argued and Shubin (2008) provided evidence for, land-based animals (including humans) evolved from fish. For Helmreich (2009), the alien ocean is a strange place on the border of human consciousness, "a zone suspended between places . . . a medium . . . to which humans connect only through chains of negotiation, translation, and transduction" (p. 249). In particular, Helmreich examines how the microbes and extremophiles in this alien ocean are changing our conceptions about the origins of life, kinship and our social, political and cultural relationships with the ocean and its resources.

Humans have a range of responses to the marine environment, including seeing an "alien ocean" (Helmreich, 2009), "earth's last frontier" (Helmreich, 2009, p. 102), a food source or, as Probyn (2016) phrases it, "eating the ocean". Alternatively, the marine environment can be seen as a place of wonder, "a feared vestige of primordial chaos, host to demonic monsters" (Helmreich, 2009, p. 15), a place of human-wildlife conflict (Crossley et al., 2014), a place for the origin of human life (Shubin, 2008), a place to be feared (thalassophobia) or where one feels vulnerable (Mentz, 2012), a place that is "awesomely beautiful and terrifying" (Helmreich, 2009, p. 15), a source of "materials for pharmaceutical and industrial products" (Helmreich, 2009, p. 12) or a place for fun in the sun.

Interdisciplinary writers have alerted us to their concerns about the ocean over the years, including Rachel Carson (1951, 1955), and John Steinbeck (1951/2011) with Ed Ricketts. Helmreich (2009, p. 46) notes that Carson's work articulated an ecological linkage between life forms and forms of life. More recently David Attenborough (in Carrington, 2017) noted in discussing the series *Blue Planet*:

> For years we thought the oceans were so vast and the inhabitants so infinitely numerous that nothing we could do could have an effect upon them. But now we know that was wrong. . . It is now clear our actions are having a significant impact on the world's oceans. [They] are under threat now as never before in human history. Many people believe the oceans have reached a crisis point.

The oceans might be at a crisis point, but for Steve Mentz (2012), blue oceans are "an already present, partly explored environment for postsustainability thinking" (p. 568) and "a model for how to live in our world today, when landed life increasingly resembles conditions at sea" (p. 590). Mentz's argument is consistent with both Helmreich (2009) and Probyn (2016), who, while agreeing that the oceans are in crisis, see the ocean as no longer a simple system but one that is complex and alien: "when the relation of life forms to forms of life, of *bio-* to whatever word it is prefixing, is unsteady, when flux crosses form" (Helmreich, 2009, p. 282, emphasis in original), or, providing "another take on what the oceanic more-than-human might afford" in considering human relations with nonhumans (Probyn, 2016, p. 163).

What is clear from these writings is that the ocean is a site for re-thinking the more-than-human, and our changing concepts of nature. As Helmreich (2009, p. 282) clearly asserts,

> In a seascape swimming with ambiguous ancestors, organisms out of place, and submarine cyborgs, it is worth asking whether the microbial ocean, as it becomes a site in which links between life forms and forms of life are redrawn, might also key us in to transformations around the boundaries of the *human* and the biotic substances and connections through which this is being imagined.

Mentz (2012, p. 587), drawing on Earle (1996), also argues that "[a]s the global climate becomes increasingly unstable, we have begun to recognize that planet-sized ecological questions are really about the ocean". This perspective is reflected in Sustainable Development Goal 14 – Conserve and sustainably use the oceans, seas and marine resources (United Nations, 2016) – which states,

> The world's oceans – their temperature, chemistry, currents and life – drive global systems that make the Earth habitable for humankind.
>
> Our rainwater, drinking water, weather, climate, coastlines, much of our food, and even the oxygen in the air we breathe, are all ultimately provided and regulated by the sea. Throughout history, oceans and seas have been vital conduits for trade and transportation.
>
> Careful management of this essential global resource is a key feature of a sustainable future.

Although some aspects of this goal can be found in school science and geography curricula – such as the water cycle and trade and transport – other aspects (as discussed later) are ignored, particularly those related to forms of life, climate, food, gender and biopolitics, which are discussed by Helmreich (2009) and Probyn (2016), albeit in very different ways.

Biopolitics and symbiopolitics

Practices which organize human life forms according to the social imperatives of forms of life are what Foucault (1978) named *biopolitics*, but Helmreich remixes this articulation:

> The biopolitics of our contemporary world comes in more varied forms; genomics and transgenics are not eugenics, and they enable different biopolitical constellations, ones not so neatly organized around genealogy and birth, or, for that matter, through human bodies.
>
> (p. 101)

Helmreich extends biopolitics in *symbiopolitics*: "the governance of relations among entangled living things" (p. 15). His notion of symbiopolitics involves the organisms that live in symbiosis with bacteria, as what Haraway (2008) has called companion species, as well as *stranger species*, in an association that "recognizes novel kinds of networked agents, human and nonhuman in the drama of the sciences" (p. 24). These stranger species include extremophiles, such as deep sea vent microbes that thrive at extremely high temperatures which are now being brought to the market as enzymes to make biochemical reactions run hotter and faster: "These microbes are hyperlinked not just to other organisms through gene transfer but also to new kinds of biotechnological science, capital, politics" (Helmreich, 2009, p. 100). Thus, "human biocultural practices flow into the putatively natural zone of the ocean, scrambling nature and culture, life forms and forms of life" (Helmreich, 2009, p. 13). In Chapter 4 Helmreich discusses a particular example of blue-green capitalism as it is manifested in the biopolitics surrounding indigenous biodiversity, native species, marine biotechnology and bioprospecting in Hawai'i. Such an example relates to sustainability and it is discussed later in the chapter.

In Hawai'i and elsewhere, marine biologists are struggling to classify many of the microbes and extremophiles found in the alien ocean because they exhibit evidence of lateral gene transfer, shuffling "genes back and forth with their contemporaries, an activity mixing up their own and others' genealogies", making it "extremely difficult to arrive at a root for the tree of life" (p. 82). Indeed, Helmreich asserts that "we are seeing the rise of an informatically inflected bare life that is increasingly agenealogical, molecular, and modular" (p. 101) because, as Haraway (in Jones and Haraway, 2006, p. 243) suggests, bare life is "never 'bare', never just itself". Helmreich thus suggests:

> If *sex* was the pivotal point of classical biopolitics, tying together individuals and populations as well as citizens and states, *transfer* may well be the practice through which new biopolitical links – between persons and patents, polymorphisms and politics – will be forged.
>
> (p. 101, emphasis in original)

Indeed, that lateral gene transfer is occurring adds to the notions of human-ocean-fish entanglements, discussed by Probyn (2016). Probyn also draws on Helmreich's athwart theory but expands it. For Helmreich, athwart theory asks for

> an empirical itinerary of associations and relations, a travelogue which, to draw on the nautical meaning of athwart, moves sideways, tracing the contingent, drifting and bobbing, real-time, and often unexpected connections of which social action is constituted, which mixes up things and their descriptions.
>
> (p. 23)

To Helmreich's use of this term, Probyn adds Eve Sedgwick's understanding that to work athwart is "to be within the spheres of 'continuing moment, movement, motive – recurrent, eddying, *troublant*'" (p. 19, emphasis in original). Probyn argues that "athwart" closely describes her book: travelling "sideways across different bodies of disciplinary knowledge, across the scales of the intimate and the public, of poems, literature, government reports" as well as to various geographic locations and different communities, addressing questions such as "What, in practice, does it mean to be athwart the ocean and her more-than-human dependent inhabitants?" (p. 19).

This chapter in many ways works athwart the writings by Helmreich (2009), Probyn (2016), and other related authors as I explore their potential relevance to how we conceive of 'inquiry' in the image of science. This requires changing our understandings about concepts of nature, sustainability and the more-than-human. However, before engaging these implications, I discuss the changing understandings about concepts of nature and sustainability, the relevance of new materialisms and the importance of gender in such discussions.

Changing concepts of nature

Helmreich (2009) is not alone in recognizing that what counts as biological life is changing and that human and more-than-human life is very much entangled. For example, although she does not mention oceans, Carolyn Merchant (2016) discusses new concepts of nature based on the idea of *autonomous nature*: "the nature at the root of the new chaos and complexity paradigm in which humans and nonhuman nature must exist together and thrive" (p. 161), and she concludes that

> [n]ature becomes postnature in ways that so thoroughly blur any human/ nature differences as to make a single interactive, mutually influential, and mutually interdependent post-human-nature . . . a new relationship between humanity and nature based on the idea of autonomous nature.
>
> (p. 161)

Merchant uses the terms 'nonhuman' and 'posthuman'; however, Probyn (2016) argues that, while these terms and 'more-than-human' "are generative in that they seek to shake up any assumptions that we might have had about what conjoins and what separates us, not to mention what that profoundly confusing 'us' might be", she prefers 'more-than-human' because it is "ontologically and materially relational, and opens up new epistemologies as it narrows the diverse and shifting relations between and among humans, and the many different aspects of that are so much more-than-human" (p. 110).

Other scholars have attempted to redefine nature in relationship to human culture and human society. These include Bruno Latour's (1993) concept of natures-cultures as an interactive human/nature system: "The very notion of culture is an artifact created by bracketing Nature off. Cultures – different or universal – do not exist any more than Nature does. There are only natures-cultures, and these offer the only possible basis for comparison" (p. 104). Kate Soper (1995), from the perspective of discourses about nature and concerns about the meaning of 'nature' and 'non-human', questions whether there is such a thing as non-human nature and argues that nature as 'other' encompasses everything that is not human while recognizing that we also see ourselves being within a wider understanding of nature. "'Nature' in this is both that which we are not *and* that which we are within" (p. 21, emphasis in original).

Katherine Hayles (1999) discusses posthuman nature from the perspective of cybernetics and chaos and complexity theories and argues for a new relationship between intelligent humans and intelligent machines:

> Although some current versions of the posthuman point toward the antihuman and the apocalyptic, we can craft others that will be conducive to the long-range survival of humans and of the other life-forms, biological and artificial, with whom we share the planet and ourselves.
>
> (p. 291)

Hayles' work is relevant to Helmreich's (2009) discussion of submarine cyborgs. In Chapter 6 he alludes to Donna Haraway's (1991) cyborg manifesto argument "that cybernetics, far from re-imagining organisms as rigid machines, opened up possibilities for recoding out bodies and selves, for short-circuiting the idea that a fixed 'nature' dictated our personal and political destinies", and notes that his submarine cyborg "also calls attention to the difference between media (e.g. seawater and air) that must first be bridged in order for boundaries to be blurred" (p. 214). For Helmreich, the research submersible Alvin, as an "assemblage of the sub and its encapsulated scientists[,] is clearly a cyborg, a combination of the organic and machine kept in tune and on track through visual, audio, and (human) metabolic feedback" (p. 213). Helmreich discusses cyborgs and cybernetics; however, he does not make the shift in terminology to 'nonhuman', 'posthuman' or 'more-than-human', although, as noted previously, he asserts that "human biocultural practices flow into the putatively natural zone of the ocean, scrambling nature and culture, life forms and forms of life" (p. 13).

These changing conceptions of nature all draw attention to the entanglement of human and more-than-human life, which has implications for how we teach about forms of life and life forms in science education, which I discuss in later sections. They are also relevant to discussions about new materialism. For example, Karen Barad (2007) argues:

> Bodies do not simply take their places in the world. They are not simply situated in, or located in, particular environments. Rather "environments" and "bodies" are intra-actively co-constituted. Bodies ("human", "environmental," or otherwise) are integral "parts" of, or dynamic reconfigurings of, what is.
>
> (p. 170)

This entanglement of the world and everything in it leads Barad to argue that there are no separate "objects" with boundaries in nature, but there are identifiable "phenomena", which are the "*ontological* inseparability of agentially intra-acting components" and "basic units of reality", where "'intra-action' *signifies the mutual constitution of entangled agencies*" (p. 33, emphasis in original). Similarly, for Sonu and Snaza (2015, p. 259) new materialism is "a subset of the posthumanist drift in the fields of philosophy, biology, and the human sciences – attempts to rethink human subjectivity so that it accounts for its relationship with non-human affect and force". These issues are discussed further later in the chapter.

Sustainability

The entanglement of human and more-than-human life also has implications for how we think about sustainability, while recognizing that the term is problematic. I have discussed the concepts, debates and challenges associated with

Table 10.1 The 17 Sustainable Development Goals and the Three Pillars of Sustainable Development

Economic Development	Environmental Protection	Social Development
1. No Poverty	2. End Hunger	1. No Poverty
2. End Hunger	6. Clean Water and	2. End Hunger
4. Quality Education	Sanitation	3. Good Health and
7. Affordable and	7. Affordable and Clean	Well-being
Clean Energy	Energy	4. Quality Education
8. Decent Work and	12. Responsible Con-	5. Gender Equality
Economic Growth	sumption and	6. Clean Water and
9. Industry, Inno-	Production	Sanitation
vation and	13. Climate Action	7. Affordable and Clean
Infrastructure	14. Life below Water	Energy
11. Sustainable Cities	15. Life on Land	8. Decent Work and Eco-
and Communities		nomic Growth
12. Responsible		9. Industry, Innovation
Consumption and		and Infrastructure
Production		10. Reduce Inequalities
13. Climate Action		11. Sustainable Cities and
17. Partnerships for the		Communities
Goals		12. Responsible Consump-
		tion and Production
		13. Climate Action
		16. Peace, Justice and
		Strong Institutions
		17. Partnerships for the
		Goals

sustainable development elsewhere (Gough, 2018) as well as how sustainability and sustainable development tend to be conflated in the literature. I will not repeat this discussion here but will continue to conflate the terms because the authors I am working with also do so.

A recommendation from the 2012 United Nations Conference on Sustainable Development was for a set of sustainable development goals (SDGs) to be developed, to supersede the Millennium Development Goals (MDGs) from 2016. Rather than providing general statements about sustainable development, the 17 Sustainable Development Goals (United Nations, 2016) provide targets for transforming the world by 2030, and they can be grouped under the three pillars of environmental, social and economic (see Table 10.1). These seventeen goals and their associated targets are not new. Their origins can be traced back to *Agenda 21* (United Nations, 1993), the outcomes document from the 1992 United Nations Conference on Environment and Development, which integrated environmental, social and economic concerns. However, while it is the first time that a single assessment matrix for sustainable development had been devised and accepted, the set of goals, as Probyn (2016, p. 7) points out, "covers so much that it becomes nigh on meaningless. It may conjure feel-good effects, but materially it continues to be intractable".

It is therefore not surprising that Mentz (2012, p. 586) has declared that "the era of sustainability is over" as a result of our shared cultural narratives of the pastoral fantasy about stasis, "an imaginary world in which we can trust whatever happened yesterday will keep happening tomorrow". He continues, "The ecological crisis we live in challenges our appetites for change. We must learn to love disruption, including the disruption of human lives by non-human forces. . . . After sustainability, we need dynamic narratives about our relation to the biosphere" (p. 587). This call for new relationships is echoed by Merchant's (2016) discussion of partnership ethics:

By merging anthropocentric with ecocentric ethics – the ethic that includes all of nature within it – we can develop an integrated, interactive ethic based on partnership between the human and the nonhuman worlds. . . . *A partnership ethic holds that the greatest good for the human and nonhuman communities is in their mutual living interdependence*

(p. 162, emphasis in original)

Merchant's partnership ethic is further discussed below in the context of the possibilities for a new image of scientific inquiry in sustainability education.

It is apparent from Table 14.1, and recent discussions around the Anthropocene, that there has been a shift to a human/posthuman focus and a reduced concern about environment protection and conservation. For example, the concept of the Anthropocene has been critiqued as universalist in its intention to install a "white liberal human subject of the Enlightenment" (Caluya, 2014, p. 32) at its center, which alludes to a larger problem of a gendered and cultural amnesia. In a similar vein, Stacy Alaimo (2012, p. 559) asks a key question: "how is it that environmentalism as a social movement became so smoothly co-opted and institutionalized as sustainability?" Probyn (2016) explains how corporate sustainability schemes have gone from being a marginal concern to being big business which has eliminated multiple actors from key discussions in an effort to develop a simplistic cultural and social ecosystem. However, she argues that there are multiple actors in any sustainability discussions: food politics "compels us to understand how entangled we are as consumers in the geopolitical, economic, cultural, and structural intricacies of the fishing industries" (Probyn, 2016, p. 4) in a complex entanglement.

Complex entanglements are also evident in Helmreich's discussions of the biopolitics of 'blue-green capitalism' in Hawai'i, where marine scientists are trying to use local marine microorganisms, such as cyanobacteria (blue-green algae), as resources for industrial and pharmaceutical biotechnology. The term 'blue-green capitalism' incorporates blue as "the freedom of the open ocean and sky-high speculation, and green for biological productivity and ecological sustainability" (p. 26). The story involves tracing the transfer of biological and intellectual property between academia and industry, and the resistance from Native Hawaiians "concerned that 'bioprospecting' the Islands' waters reinvigorates colonial practices through which they were alienated from their territory" (p. 26). Adding to the entanglements are the introduced species in Hawaiian waters: visible algae that densely entangle scientific and everyday interactions with sea creatures and "show that the taxonomy and politics of alien and native species are slippery" (p. 26).

As evidenced by both Helmreich (2009) and Probyn (2016), biopolitics are very much caught up in achieving sustainability, particularly in marine contexts, as is gender: "women are differentially affected by the man-made disaster of the Anthropocene, just as they are excluded in the righteous framing of the men who will save the environment" (Probyn, 2016, p. 104). However, gender and environment are mutually constituted – in forms of entanglement that "produce worlds that we don't yet, will never, fully know" (Probyn, 2016, p. 111).

Gender

Although Helmreich (2009) notes that "[v]olumes could be written about how *Alvin*'s cyborg manifest has been marked by gender and sexuality" (p. 215), he actually says very little on either topic beyond noting that males worried about how they could urinate discreetly into their "pee bottles" on the sub if women were present, that women felt that they had to become "sexless" so no one will notice that they are different: "the gender-neutral body was understood in the culture of mid-century American oceanography, as in early space travel, to be male" (p. 216), and that "American oceanographic infrastructure blocked women from working on navy ships" during the cold war (p. 208). He does not expand on sexuality, but he is strong on racial discrimination: "Marine biology in the United States, while fairly international, is pretty white" (p. 202). He draws attention to "the various ethnic and racial histories entangled in American ocean science . . . [and] . . . ideologies of nature through which 'lakes, creeks, and oceans' call forth curiosity in the first place" (p. 203) to account for the absence of underrepresented minorities, but he does not mention women much, except to note that half the science crew on his trip were female, "though there are fewer in senior ranks" (p. 240). This oversight is notable given the decades of writings about the absence of women in science – a mystery that deepens when he has three references to Evelyn Fox Keller's work in his bibliography (including her 1992 book on language, gender and science), but none in his index, although she was one of the early writers in this space.

In contrast with Helmreich, Probyn's focus is very much on gender. According to Probyn (2016, p. 126), gender is "athwart history and experience", and it requires that we hone "the arts of noticing the entwined relations of humans and other species across multiple non-nesting scales" (Tsing, 2014, p. 233). However, "gender and sexuality, as well as ethnicity and class, have been squeezed out of current debates on the Anthropocene, climate change, and the more-than-human" (Probyn, 2016, p. 12). Two recent special issues of *The Journal of Environmental Education* (volumes 48(1) and 49(4)) have highlighted the relationship between gender and environmental education and the need to move gender from margin to center in environmental education. As the editors of these issues wrote:

> Greater attention to the dynamics of gender in the field – including intersections with race, class, sexuality, ability, body size, and animality in the contexts of colonization, neoliberalism, globalization, and anthropocentrism – is needed if we are to understand and respond effectively to the complex environmental and social issues we face.
>
> (Gough et al., 2017, p. 1)

Although gender is a silence in Helmreich's book, achieving gender equality is a Sustainable Development Goal (United Nations, 2016), and it should not be a silence in sustainability education or a more-than-human scientific inquiry in sustainability education because gender and more-than-human can enhance each other, with gender more particularly helping "to disrupt the somewhat flat equation of the more-than-human" (Probyn, 2016, p. 113).

Sustainability education

I have been concerned about the challenges facing sustainability (or environmental) education being incorporated into the curriculum for some time (see, for example, Gough, 1997, 2002, 2016, 2017a, 2017b, 2018), despite recognition of the need for it dating back to the 1970 United States Environmental Education Act and the Stockholm Declaration from the 1972 United Nations Conference on the Human Environment (United Nations, 1972).

The marine environment is internationally recognized through the annual World Oceans Day on 8 June, which started in 2008. In 2018 the theme was "Clean Our Ocean", which highlighted the complex entanglement of humans, oceans, ecosystems, technology and tourism, to name a few components (Guterres, 2018). UNESCO (2017) has also published learning objectives for Education for Sustainable Development as part of the Global Education 2030 Agenda, which is associated with the UN Sustainable Development Goals. The learning objectives specifically associated with Goal 14 (Life below water) are:

1. The learner understands basic marine ecology, ecosystems, predator-prey relationships, etc.
2. The learner understands the connection of many people to the sea and the life it holds, including the sea's role as a provider of food, jobs and exciting opportunities.
3. The learner knows the basic premise of climate change and the role of the oceans in moderating our climate.
4. The learner understands threats to ocean systems such as pollution and overfishing and recognizes and can explain the relative fragility of many ocean ecosystems including coral reefs and hypoxic dead zones.

5. The learner knows about opportunities for the sustainable use of living marine resources. (p. 37)

Despite the attention being given to marine and oceanic issues internationally, such topics are difficult to find in the curriculum of most countries. For example, Castle et al. (2010, p. 428) commented, "Although there is no statutory requirement to cover marine and coastal topics in education in England, the perceived lack of coastal and marine topics in the formal education system has been the subject of criticism". Indeed, the lack of marine topics in the current national curriculum in England is more than perceived – 'marine' does not appear in the Science curriculum (Department of Education, 2015), and it appears only once in the Geography curriculum (Department of Education, 2013). 'Marine' also only appears once in the British Columbia, Canada K-7 Science curriculum (British Columbia Ministry of Education, 2005, p. 301). Ocean does get 30 mentions in their Science curriculum (British Columbia Ministry of Education, 2005) and 4 in the across curriculum document (British Columbia Ministry of Education, 2008/2009); however, these tend to be related to oceans as a geological feature and mapping them rather than a broader notion of ocean literacy (Ocean Literacy Network, 2005/2013). In New Zealand, although using the nautilus, a marine animal, as the symbol for the curriculum, the current version of the curriculum does not include the words marine or ocean (New Zealand Ministry of Education, 2007). More recently, Berchez et al. (2016, p. 151) noted that in the Brazilian context, "Marine and coastal environment-education activities in Brazil are still scarce and need to receive effective support from such integrated networks as ReBentos". The provisional secondary 1–3 Science curriculum for Hong Kong (Education Bureau, 2017) contains two mentions of marine: visit some marine parks and conduct a project on the conservation of marine animals in Hong Kong (e.g. corals, sharks, Green Turtle and Chinese White Dolphin). There is also little encouragement to study the marine environment in the Australian Curriculum for Years Foundation to 10 (ACARA, 2018). 'Marine' only appears once in the Science curriculum, at Year 10 level as an Earth and Space Sciences understanding: "examining the factors that drive the deep ocean currents, their role in regulating global climate, and their effects on marine life". This is also the only reference to 'ocean' in the whole curriculum. 'Marine' appears in the Year 2 Humanities and Social Sciences curriculum content elaboration:

> making generalisations from data showing patterns and relationships (for example, the relationship between the distance of places and the frequency of visits to them; between rubbish in the school and eating areas; **between marine animals and where human rubbish may go**; between climate zones and clothing or housing).
>
> (my emphasis)

There are no other references to 'marine' or 'ocean' in the curriculum content. Such sparse opportunities (studying deep ocean currents and the effect of human rubbish on marine animals) provide little opportunity or encouragement for discussing the bigger issues raised in SDG 14 and its associated learning objectives (UNESCO, 2017), nor any of the entanglements that are the focus of Helmreich and Probyn's research. "Careful management of this essential global resource is a key feature of a sustainable future" (United Nations, 2016) may be the goal, but students are not being educated to be able to participate in its achievement. We need to learn from Helmreich and Probyn's problematizing of our presumptions about how to study oceans, the great 'other'. Social sciences can show science educators how to think differently about studying the more-than-human relations in our environment and develop new ways of thinking about inquiry and our other educational practices.

Engaging with alien oceans: Toward more-than-human scientific inquiry in education

More-than-human inquiries include "learning how to open the presences of otherness and how to form relationship of mutuality with others we can never fully know" (Gruenewald, 2003, p. 40). This is not what is found in the Foundation to Year 10 Australian Curriculum (ACARA, 2018). The only mentions of evolution and natural selection are in two of the Year 10 Science content descriptions and their elaborations – one in Science Understanding and the other in Science as Human Endeavour – both of which are grounded in a very traditional conceptualization of evolution. The first is: "The theory of evolution by natural selection explains the diversity of living things and is supported by a range of scientific evidence" in the Year 10 Science Understanding Biological Sciences content (ACARA, 2018). The second is: "Scientific understanding, including models and theories, is contestable and is refined over time through a process of review by the scientific community" in the Year 10 Science as Human Endeavour content (ACARA, 2018), with relevant elaboration: "considering the role of different sources of evidence including biochemical, anatomical and fossil evidence for evolution by natural selection". While there is the possibility for approaching scientific inquiry differently through this second content description, it is still grounded in a traditional conception of natural selection, whereas this view has been dissolved by Helmreich.

A more-than-human curriculum would need to take into account the dissolving of the evolutionary tree that relates to all organisms because, as previously discussed, the sequencing of complete genomes of bacteria, eukarya and archaea has shown that microbes are the mosaic of acquired genes, and that lateral gene transfer is occurring blurs boundaries. This goes far beyond what is in the current curriculum.

More-than-human scientific inquiries would need to stop simplifying the sea and accept that oceans are complex entanglements. They would also need to include "pedagogies inspired by posthumanist and new materialist ontologies [that] are situational encounters made up of entanglements and interweavings, conjoint actions and political ecologies, entanglements that are alive, vibrant, and powerful" (Sonu and Snaza, 2015, p. 274). Probyn (2016, p. 16) provides a clear example of entanglements and interweavings, which relates to the Year 2 Humanities and Social Sciences curriculum content elaboration noted previously (studying patterns and relationships between marine animals and where human rubbish may go) (ACARA, 2018), when she writes: "Fish eat the microplastics used in daily skin care; humans eat the fish and the microplastics; and fish and human bodies intermingle". Studying the ocean and our entanglements with it means examining the more-than-human assemblages of fish, institutional power, gender and class relations, and technology.

One existing framework that warrants consideration in this context is Merchant's (2016) partnership ethic. This contains five precepts for a human community in a sustainable partnership with what she calls a nonhuman community – which is in a particular place, a place in which connections to the larger world are recognized through economic and ecological exchanges:

- Equity between the human and nonhuman communities.
- Moral consideration for both humans and other species.
- Respect for both cultural diversity and biodiversity.
- Inclusion of women, minorities, and nonhuman nature in the code of ethical accountability.
- An ecologically sound management that is consistent with the continued health of both the human and the nonhuman communities.

To these we would probably need to add overt recognition of the more-than-human assemblages discussed by Probyn (2016) and Helmreich (2009). But they are a start for a new approach to scientific inquiry that, instead of seeking simplicity and certainty, recognizes that we live in a complex world of assemblages that we can never fully know.

Conclusion

The three books that have provided the underpinnings for this athwart-essaying journey have much to contribute to our understanding of the changing nature of life forms and forms of life, and how we should think about them. Much more could be said about all of these troublings, and many others, and their relevance for scientific inquiry in sustainability education while we keep in mind that "[t]he more-than-human, if it is to be meaningful as a perspective, makes us confront again and again the relatedness of all entities" (Probyn, 2016, p. 163). This perspective brings us to the development of very different inquiry approaches.

Biography

Annette Gough is Professor of Science and Environmental Education in the School of Education and the School of Global, Urban and Social Studies at RMIT University, Melbourne, Australia, and a Life Fellow of the Australian Association for Environmental Education (since 1992). She has over 130 publications (books, chapters and articles and curriculum materials) including Green Schools Globally: Stories of Impact on Education for Sustainable Development (forthcoming from Springer). Her research interests span environmental, sustainability and science education, research methodologies, posthuman and gender studies.

References

Alaimo, S. (2012). Sustainability this, sustainability that: New materialisms, post-humanism and unknown futures. *PMLA*, *127*(3), 558–564.

Ashby, J. (2016, 30 November). The absurdity of natural history – or, why humans are 'fish'. *The Conversation*. Retrieved from https://theconversation.com/the-absurdity-of-natural-history-or-why-humans-are-fish-69384

Australian Curriculum and Assessment and Reporting Authority (ACARA). (2018). *The Australian Curriculum. 8.1.* Retrieved from www.australiancurriculum.edu au/

Barad, K. (2007). *Meeting the universe halfway: Quantum physics and the entanglement of matter and meaning.* London: Duke University Press. https://doi.org/10.1215/9780822388128

Berchez, F.A.S., Ghilardi-Lopes, N.P., Correia, M.D., Sovierzoski, H.H., de Pedrini, A.G., Ursi, S., Kremer, L.P., de Almeida, R., Schaeffer-Novelli, Y. Marques, V., & Brotto, D.S. (2016). Marine and coastal environmental education in the context of global climate changes – synthesis and subsidies for RioBentos (Coastal benthic habitats monitoring network). *Brazilian Journal of Oceanography*, *64*(sp2), 137–156. doi.org/10.1590/S1679–875920160932064sp2

British Columbia Ministry of Education. (2005). *Science K to 7: Integrated resource package 2005.* Revised edition. Retrieved from www2.gov.bc.ca/gov/content/education-training/k-12/teach/curriculum/english/sciences/science-k-to-7

British Columbia Ministry of Education. (2008/2009). *The environmental learning & experience curriculum maps: Environment & sustainability across BC's K-12 curricula.* Retrieved from www2.gov.bc.ca/assets/gov/education/kindergarten-to-grade-12/teach/teaching-tools/environmental-learning/ele_maps.pdf

Caluya, G. (2014). Fragments for a postcolonial critique of the Anthropocene: Invasion biology and environmental security. In J. Frawley & I. McCalman (Eds.), *Rethinking invasion ecologies from the environmental humanities* (pp. 31–44). Abingdon, UK: Routledge.

Carrington, D. (2017, 5 December). Oceans under greatest threat in history, warns Sir David Attenborough. *The Guardian*. Retrieved from www.theguardian.com/environment/2017/dec/05/oceans-under-greatest-threat-in-history-warns-sir-david-attenborough

Carson, R. (1951). *The sea around us.* New York: Oxford University Press.

Carson, R. (1955). *The edge of the sea.* Boston: Houghton Mifflin.

Castle, Z., Fletcher, S., & McKinley, E. (2010). Coastal and marine education in schools: Constraints and opportunities created by the curriculum, schools and teachers in England. *Ocean Yearbook, 24*, 425–444. doi.org/10.1163/22116001-90000066

Crossley, R., Collins, C.M., Sutton, S.G., & Huveneers, C. (2014). Public perception and understanding of shark attack mitigation measures in Australia. *Human Dimensions of Wildlife, 19*(2), 154–165, DOI: 10.1080/10871209.2014.844289

Darwin, C. (1859). *On the origin of species.* London: John Murray. Retrieved from www.gutenberg.org/files/1228/1228-h/1228-h.htm#link2H_4_0020

Deleuze, G., & Guattari, F. (1987). *A thousand plateaus: Capitalism and schizophrenia.* Brian Massumi, Trans. Minneapolis, MN: University of Minnesota Press.

DeLoughrey, E. (2007). *Routes and roots: Navigating Caribbean and Pacific island literatures.* Honolulu: University of Hawai'i Press.

Department of Education. (2013). *National curriculum in England: geography programmes of study.* Retrieved from www.gov.uk/government/publications/national-curriculum-in-england-geography-programmes-of-study/national-curriculum-in-england-geography-programmes-of-study

Department of Education. (2015). *National curriculum in England: Science programmes of study.* Retrieved from www.gov.uk/government/publications/national-curriculum-in-england-science-programmes-of-study/national-curriculum-in-england-science-programmes-of-study

Earle, S. (1996). *Sea change: A message of the oceans.* New York, NY: Ballantine.

Education Bureau, Hong Kong Special Administrative Region Government (HKSARG). (2017). Supplement to the science education key learning area curriculum guide, science (secondary 1–3). *Provisional Final Draft.* Retrieved 1 June 2017 from www.edb.gov.hk/attachment/en/curriculum-development/kla/science-edu/Science(S1-3)_supp_20170302_prov_final_draft.pdf

Foucault, M. (1978). *The history of sexuality, volume one, an introduction.* Robert Hurley, Trans. New York: Pantheon.

Gough, A. (1997). *Education and the environment: Policy, trends and the problems of marginalisation.* Australian Education Review Series No.39. Melbourne, Victoria: Australian Council for Educational Research.

Gough, A. (2002). Mutualism: A different agenda for science and environmental education. *International Journal of Science Education, 24*(11), 1201–1215. doi: 10.1080/09500690210136611

Gough, A. (2016). Tensions around the teaching of environmental sustainability in schools. In T. Barkatsas & A. Bertram (Eds.), *Global learning in the 21st century* (pp. 83–102). Rotterdam: Sense Publishers.

Gough, A. (2017a). Searching for a crack to let environmental light in: Ecological biopolitics and international education for sustainable development discourses. *Cultural Studies of Science Education, 12*(4), 889–905. doi.org/10.1007/s11422-017-9839-8.

Gough, A. (2017b). Educating for the marine environment: Challenges for schools and scientists. *Marine Pollution Bulletin 124*(2): 633–638. doi.org/10.1016/j.marpolbul.2017.06.069

Gough, A. (2018). Sustainable development and global citizenship education: Challenging imperatives. In I. Davies, L-C. Ho, D. Kiwan, C. Peck, A. Peterson, E. Sant, & Y. Waghid (Eds.), *The Palgrave handbook of global citizenship and education* (pp. 295–312). London: Palgrave Macmillan. doi.org/10.1057/978-1-137-59733-5_19

Gough, A., Russell, C., & Whitehouse, H. (2017). Moving gender from margin to center in environmental education. *The Journal of Environmental Education, 48*(1), 1–4. doi.org/10.1080/00958964.2016.1252306

Gould, S.J. (1997, March). Nonoverlapping magisterial. *Natural History, 106,* 16–22.

Gruenewald, D. (2003). At home with the other: Reclaiming the ecological roots of development and literacy. *The Journal of Environmental Education, 35*(1), 33–43.

Guterres, A. (2018, 8 June). Secretary-general's message. *World Oceans Day.* Retrieved from www.un.org/en/events/oceansday/message.shtml

Haraway, D.J. (1991). *Simians, cyborgs, and women: The reinvention of nature.* London, England: Free Association Books.

Haraway, D.J. (2008). *When species meet.* Minneapolis, MN: University of Minnesota Press.

Hayles, K. (1999). *How we became posthuman: Virtual bodies in cybernetics, literature, and informatics.* Chicago: University of Chicago Press. https://doi.org/10.7208/chicago/9780226321394.001.0001

Helmreich, S. (2009). *Alien ocean: Anthropological voyages in microbial seas.* Berkeley: University of California Press.

Jones, C., & Haraway, D. (2006). Zoon. In C.A. Jones (Ed.), *Sensorium: Embodied experience, technology, and contemporary art* (pp. 241–243). Cambridge, MA: MIT Press.

Keller, E.F. (1992). Secrets of life, secrets of death: Essays on language, gender and science. New York, London: Routledge.

Latour, B. (1993). *We have never been modern.* Cambridge, MA: Harvard University Press.

Mentz, S. (2012). After sustainability. *PMLA, 127*(3), 586–592.

Merchant, C. (2016). *Autonomous nature: Problems of prediction and control from ancient times to the scientific revolution.* New York, NY: Routledge.

New Zealand Ministry of Education. (2007). *The New Zealand curriculum.* Wellington: Ministry of Education.

Ocean Literacy Network. (2005/2013). *Ocean literacy: The essential principles and fundamental concepts of ocean sciences for learners of all ages version 2.* First published June 2005, revised March 2013. Washington, DC: National Oceanic and Atmospheric Administration. Retrieved from www.oceanliteracy.net

Owen, R. (1849). *On the nature of limbs.* London: John Van Voorst.

Probyn, E. (2016). *Eating the ocean.* Durham: Duke University Press.

Shubin, N. (2008). *Your inner fish: A journey into the 3.5-billion-year history of the human body.* New York, NY: Pantheon.

Sonu, D., & Snaza, N. (2015). The fragility of ecological pedagogy: Elementary social studies standards and possibilities of new materialism. *Journal of Curriculum and Pedagogy, 12*(3), 258–277. doi:10.1080/15505170.2015.1103671

Soper, K. (1995). *What is nature? Culture, politics, and the non-human.* Oxford: Blackwell.

Steinbeck, J. (1951/2011). *The log from the sea of Cortez.* London: Penguin Classics.

Strick, J. (2009). Review of *alien ocean: Anthropological voyages in microbial* seas by Stefan Helmreich. *Isis, 100*(4), 949–950.

Tsing, A.L. (2014). Strathern beyond the human: Testimony of a spore. *Theory, Culture and Society, 31*(2–3), 221–241.

UNESCO. (2017). *Education for sustainable development goals: Learning objectives.* Paris: UNESCO.

United Nations. (1972). *Declaration of the United Nations conference on the human environment.* Retrieved 24 October 2014 from www.unep.org/Documents. Multilingual/Default.asp?documentid=97&articleid=1503

United Nations. (1993). *Agenda 21: Earth summit: The United Nations programme of action from Rio.* Retrieved from https://sustainabledevelopment.un.org/content/documents/Agenda21.pdf

United Nations. (2016). *Sustainable development goals: 17 goals to transform our world.* Retrieved from www.un.org/sustainabledevelopment/sustainable-development-goals/

United Nations Educational, Scientific, and Cultural Organization. (2017). *Education for sustainable development goals.* Learning objectives. Paris, France: Author.

Section III

Moving beyond feminisms and gender

The chapters in this section represent 20 years of my writings about and beyond feminisms and gender. It includes discussions of queer(y)ing gender and other marginalised voices in environmental education research, embodying cyborg subjectivities in writing about the body of an environmental education researcher's cancer experiences, and being playful through introducing humour into environmental education by constructing Camp Wilde and writing with a feminist and environmentalist cartoonist. The section concludes with my article on education in the Anthropocene, which takes us back into discussions of the more-than-human started in Chapter 10.

Chapter 11 provides an umbrella argument for several of the subsequent chapters. When I was working on my doctoral thesis, I became increasingly aware that women were one of many marginalised groups in environmental and environmental education discourses and broadened my focus to a more social justice agenda which was incorporated into the set of five principles for research and practice in environmental education that emerged from my research (see Chapter 1). Following my gender agenda chapter in the section Connie Russell co-edited for the *International Handbook of Research on Environmental Education* (Gough 2013, see Chapter 5), Connie invited me to write a chapter on voices from the margins for an environmental education reader she was co-editing; however, this book was never published. When Joshua Russell called for chapters for his book on *Queer Ecopedagogies* (2021a) I proposed a revision of this earlier chapter for inclusion. He agreed, and I made some major revisions to the earlier draft to take account of notions of assemblages and intersectionality that had emerged in more recent times.

Chapter 12 dates from 2003 when Noel Gough and I collaborated with a wonderful cast of friends (Warren Sellers, Mary Aswell Doll, Peter Appelbaum and Sophia Appelbaum) in Camp Wilde to "probe the ways in which heteronormativity configures ignorance in environmental education" (Gough et al. 2003, 48). This article also appeared in the book *Queer Ecopedagogies*, because Joshua saw Camp Wilde as providing

an opportunity to truly engage with queer thought, being, and engagement with others. It is a space to explore the possibilities of a critical

DOI: 10.4324/9781003390930-13

pedagogy rooted in love and desire (Santavicca et al. 2019), a crucially important element of reconnecting learning with emotion (Russell and Oakley 2016).

(Russell 2021b, 7)

Although heterosexual scholars might not be best suited to adopt queer perspectives, we endeavoured to demonstrate that, as scholarly allies, we took seriously the concern that we not speak for queer people in EE but rather that we could interrupt heteronormativities in the field, as Russell (2021b, 12) acknowledges.

Inventing Camp Wilde was also prompted by our chance encounter with Elizabeth Vallance's (1980) observations concerning the dearth of humour in educational discourses. We intended Camp Wilde to be a play space: to play, with ourselves and others, in exploring our understandings of people and environments. Chapter 13 fits into a similar category. When *Environmental Education Research* journal called for papers on humour and environmental education I immediately thought of Judy Horacek, a local feminist environmentalist cartoonist, and approached her to be involved in a writing project that became Chapter 13. Judy was keen for this to be a collaboration rather than an interview, so we constructed a duoethnography drawing on critical and new materialist feminist theory where we each responded to some of her cartoons related to four themes (cartoons, feminism, the environment and formal environmental education) to discuss the generativity of cartoons in environmental education teaching and research.

Chapter 14 has its origins in my experiences of breast cancer in 2001. My way of coping with my cancer journey was to write, and so I did (Gough 2002, 2003, 2004), culminating in this final piece. This chapter began as a presentation at the 2003 "Body Modification: Changing Bodies, Changing Selves" conference held at Macquarie University and was published in a special issue of *Women's Studies* that brought together papers from that conference. I saw my experiences of breast cancer surgery and reconstruction as an embodied cyborg subjectivity and drew on both Michel Foucault's (1981, 1988, 1991) notions of surveillance, power and discipline as well as Rosi Braidotti's (2002) Deleuzean notion of becoming-woman/animal/machine and Donna Haraway's (1985) cyborg manifesto to write about how "the binaries of human/animal and human/machine were dissolving in me".

My writings about embodying cyborg subjectivities and dissolving binaries were part of my concerns with needing to see humans as part of nature not separate from it, as discussed in earlier chapters. If the language had existed when I was doing this writing, I would have drawn on notions of more-than-human scientific inquiries, entanglements and posthumanist and new materialist ontologies that were discussed in Chapter 10. These notions become increasingly important as we look forward in Section IV.

Before moving on I do need to clarify and correct something I wrote in the article that is Chapter 14. In re-seeking permission from Deena Metzger to reproduce her photograph (Figure 14.2) I shared the article with her, and in response she drew attention to the contradiction between what I quote from

Elizabeth Van Schaick (1998, 3) about Metzger's breast scar tattoo and her own reasoning. I quoted from Van Schaik:

> The emphasis which Western culture has put on women's appearance leads Metzger to view her illness as ugliness, to feel shame, to apologize for the offensiveness of her appearance, to attempt to hide it. . . . The photograph represents a respite from previous periods of revulsion toward the breast cancer body, and Metzger chooses to emphasise her accomplishment.

Metzger (personal communication) drew attention to the reality obvious in the photo and its accompanying poem:

> I am no longer afraid of mirrors where I see the sign of the amazon, the one who shoots arrows.
> There was a fine red line across my chest where a knife entered, but now a branch winds about the scar and travels from arm to heart.
> Green leaves cover the branch, grapes hang there and a bird appears.
> What grows in me now is vital and does not cause me harm. I think the bird is singing.
> I have relinquished some of the scars.
> I have designed my chest with the care given to an illuminated manuscript.
> I am no longer ashamed to make love. Love is a battle I can win.
> I have the body of a warrior who does not kill or wound.
> On the book of my body, I have permanently inscribed a tree.

She did not feel shame or revulsion. Instead, as she wrote in *Tree* (Metzger 1997, p. 89), "I claim my body, such as it has become, as my own".

In her Preface to the book, Metzger (1997, xii) brings together her concerns with personal healing from illness and healing of the violated planet, and thus our interests coincided much more than I thought.

The final chapter in this section focuses on understandings of and contestations around the concept of the Anthropocene, the changing role of education in society and what education in the Anthropocene could include, particularly from multiple feminist approaches and perspectives because women are so affected by climate change and other environmental crises. It provides an appropriate end to this section and a segue to the final chapter on looking forward in gender and environmental education research.

References

Braidotti, Rosi. 2002. *Metamorphoses: Towards a Materialist Theory of Becoming*. Cambridge: Polity Press.

Foucault, Michel. 1981. *The History of Sexuality, Vol. One, An Introduction*. Trans. Robert Hurley. Harmondsworth: Pelican.

Foucault, Michel. 1988. *The Care of the Self. The History of Sexuality, Vol. Three*. Trans. Robert Hurley. New York: Vintage Books.

Foucault, Michel. 1991. *Discipline and Punish: The Birth of the Prison*. Trans. Alan Sheridan. Harmondsworth: Pelican.

Gough, Annette. 2002. "I looked at the Place Where My Breast Used to be. . ." *WISENET Journal* 61: 22–24.

Gough, Annette. 2003. "Embodying a Mine Site: Enacting Cyborg Curriculum." *Journal of Curriculum Theorizing* 19: 33–47.

Gough, Annette. 2004. "Blurring Boundaries: Embodying Cyborg Subjectivity and Methodology." In *Educational Research: Difference and Diversity*, edited by Heather Piper and Ian Stronach, 113–127. Burlington, VT: Ashgate.

Gough, Annette. 2013. "Researching Differently: Generating a Gender Agenda for Research in Environmental Education." In *International Handbook of Research on Environmental Education*, edited by Robert B. Stevenson, Michael Brody, Justin Dillon and Arjen Wals, 375–383. New York: Routledge for the American Educational Research Association.

Gough, Noel, Annette Gough, Peter Appelbaum, Sophia Appelbaum, Mary Aswell Doll, and Warren Sellers. 2003. "Tales from Camp Wilde: Queer(y)ing Environmental Education Research." *Canadian Journal of Environmental Education* 8: 44–66.

Haraway, Donna. 1985. "A Manifesto for Cyborgs." *Socialist Review* 80: 65–107.

Metzger, Deena. 1997. *Tree*. Berkeley: North Atlantic Books.

Russell, Joshua, ed. 2021a. *Queer Ecopedagogies: Explorations in Nature, Sexuality, and Education*. Cham: Springer.

Russell, Joshua. 2021b. "Introduction: Why Queer Ecopedagogies?" In *Ecopedagogies: Explorations in Nature, Sexuality, and Education*, edited by Joshua Russell, 1–16. Cham: Springer.

Vallance, Elizabeth. 1980. "A Deadpan Look at Humor in Curriculum Discourse (or, the Serious Versus the Solemn in Education)." *Curriculum Inquiry* 10(2): 179–189.

Van Schaick, Elizabeth. 1998, Fall. "Palimpset of Breast: Representation of Breast Cancer in the Work of Deena Metzger and Jo Spence." *Schuylkill: A Creative and Critical Review from Temple University* 2(1). www.temple.edu/gradmag/fall98/schaick.htm

11 Listening to voices from the margins

Transforming environmental education

Annette Gough

Gough, A. (2021). Listening to voices from the margins: Transforming environmental education. In J. Russell (Ed.), *Queer ecopedagogies: Explorations in nature, sexuality, and education* (pp. 161–181). Springer.

Abstract

In the past, women, queer, Indigenous and disabled people, as well as race, class and body size issues, have been overlooked in most environmental education programs through being subsumed into the notion of "universalised people", the "norm". However, people in each of these marginalised groups has a distinctive contribution to make to environmental education, as a form of anti-oppressive resistance, which needs to be foregrounded. In this chapter I problematize the relative silencing of queer theory and theorizing in environmental education and discuss the contributions that queer(y)ing perspectives bring to environmental education. In particular, I discuss some possibilities for new directions for environmental education where queer ecopedagogies and research, intersectionality and assemblages are used to listen to, and work with, "queer" voices.

Introduction

"The personal is political" was a definitive slogan of second-wave feminism in the 1970s (Nicholson 1981; Rogan and Budgeon 2018). It highlighted how "interpersonal interactions could be assessed according to the ideals of the public world" (Nicholson 1981, p. 91), and seen as oppressing women. The slogan encapsulated the belief that "If social norms and practices were what lay behind problems in the personal sphere, then removing such problems might require a social or political movement" (Nicholson 1981, p. 92). Frances Rogan and Shelley Budgeon (2018) trace the slogan to its origins in a 1970 paper of the same title by Carol Hanisch (in Firestone and Koedt 1970), and argue that Hanisch's paper still resonates with contemporary critiques of gender relations through its threads of "four interrelated categories: power, the private/public dichotomy, political action, and subjectivity" (Rogan and Budgeon 2018, p. 2). Rogan and Budgeon (2018) argue that these conceptual

DOI: 10.4324/9781003390930-14

threads are still relevant because of the top-down oppressive power of neoliberal discourses on women, and an ongoing concern with the production of gendered subjects, and that "the personal is still political" (p. 14).

While the focus of this book is queer (not feminist) ecopedagogies, I believe that it is important to engage with feminism, related marginalised theories, and posthumanism for several reasons. Firstly, as Alison Jaggar and Paula Rothenberg Struhl argue, women's oppression "provides a conceptual model for understanding all other forms of oppression" (1978, p. 84). So, if "the personal is still political" then understanding from feminist theories and practices are still relevant for understanding other forms of oppression. Secondly, as Jesse Bazzul and Nicholas Santavicca (2017) argue, drawing on Judith Butler (2011), we should try to "keep sex/gender and sexuality together as much as possible simply to signify that both sex and gender are always socially situated, discursively constituted terms that emerge in context with sexualities" (p. 65) and as they "can be central to how students come to creative modes of being in dynamic environments" (p. 56). They go on to argue that "a commitment to talking of sex/gender and sexuality is what gives queering its subversive power" (p. 57), which is particularly relevant for transgender studies.

The bringing together of sex and gender is apparent in the United Nations' (2016) Sustainable Development Goals, particularly Goal 5: Achieve gender equality and empower all women and girls, and campaigns such as the UNESCO Bangkok (2019) #ColourMeIn campaign to raise awareness for International Day Against Homophobia, Transphobia and Biphobia (17 May) as part of its commitment "to preventing and addressing school-related gender-based violence as part of its priority to make schools safe, inclusive and health promoting for all learners." However, while gender equality is mentioned in the recent UNESCO (2019) *Framework for the implementation of Education for Sustainable Development (ESD) beyond 2019* because it is one of the United Nations' Sustainable Development Goals (United Nations 2016), sex and sexuality are not (and neither are other forms of oppression discussed in this chapter).[1]

A third reason to engage with feminism here is the need to disrupt oppressive dualisms such as the segregating of queer from nature and the natural. As Stacy Alaimo (2010, pp. 51–52) notes,

> Much queer theory has bracketed, expelled, or distanced the volatile categories of nature and the natural, situating queer desire into an entirely social, and very human, habitat. This now compulsory sort of segregation of queer from nature is hardly appealing to those who seek queer green places, or, in other words, an environmentalism allied with gay affirmation, and a gay politics that is also environmentalist.

However, this needs to be tempered with an understanding that there is no essential nature. For example, Jamie Mcphie and David Clarke (2018) "think with the material turn about the concept nature due to its significant performativity in its role within environmental education and research" and "initiate a pluralistic thought experiment that purposefully diffracts nature into eight performances, to see what it does" (p. 1). Their performances of nature incorporate racism, classism, and androcentrism as well as colonialism, homogenisation, and mass extinction, which are other forms of oppression that relate back to "the personal is political." They reject combining nature and culture (such as in natures-cultures (Latour 1993) or natureculture (Haraway 2003, 2016)), or overcoming "Cartesian dualism with a piece of sticky tape holding the subject and object together in 'interaction' or 'connection'" (Mcphie and Clarke 2018, p. 14) and instead argue, drawing on Karen Barad (2003), that their performativity in creating new concepts of nature, "is in effect a mode of becoming of the world." For them "the matter of concern is the political mattering of the ontologically suffused concept nature" (p. 4). They explore eight conceptions of nature at play in environmental education and encourage readers to consider "their use of 'nature' in the practice of both research and pedagogy, so as to ponder the effects of their use of it" (p. 11).

Interestingly, while Mcphie and Clarke (2018) argue for an "environmental literacy that attends to what a concept is capable of, what a concept can do, and perhaps even what a concept can prevent, *post-nature*" (p. 1),[2] others are seeing value in playing with posthuman assemblages (Bazzul and Santavicca 2017) and with engaging intersectionality and entanglements (C. Russell 2018, 2019). As Connie Russell explains, "Intersectional analyses help us better understand power and oppression, and the complex ways that race, class, gender, sexuality, ability, size, and other identities interconnect and are experienced" (2018, online). Similarly, Sumi Cho, Kimberlé Crenshaw, and Leslie McCall (2013) describe intersectionality studies as "a gathering place for open-ended investigations" of "overlapping and conflicting dynamics" of oppression (p. 788). Connie Russell (2018, 2019), Teresa Lloro-Bidart and Michael Finewood (2018), and Corey Lee Wrenn (2019) argue for intersectionality to encompass species other than humans. Bazzul and Santavicca (2017) use Gilles Deleuze and Félix Guattari's (1988) notion of assemblage, "a heterogeneous grouping of material, discursive, and affective entities and forces" (p. 57), because "working with assemblages can help educational communities realize that all identities, plants, animals, and discourse, etc., are entangled together – so much so that faith in the fixity of singularities becomes impossible" (p. 59).

In this way, intersectionality and assemblages, together with queer ecopedagogy, can be seen as a successor to Kevin Kumashiro's (2000) anti-oppressive education. Anti-oppressive education calls attention to the ways various forms of oppression marginalize and harm individuals and groups. Anti-oppressive education together with an expanded notion of Others, intersectionality and assemblages are discussed below as a strategy for queer(y)ing the taken-for-granted in environmental education.

Background

Environmental education is very much concerned with environmental problems as perceived from so-called mainstream perspectives – particularly white, English speaking, middle class, able, heterosexual, and male. Yet it continues to sit on the margins of most education policy, and it frequently ignores sexuality, gender, and other oppressions, even when these groups are central to the formation and resolution of environmental problems. There is not only the issue of whose voices are heard in environmental education research and practice, but also how we can enable other voices to be heard, without being patronising or tokenistic.

In the past, women, Indigenous, queer and disabled people, and race, class and body size issues have been overlooked in most environmental education programs through being subsumed into the notion of 'universalised people', or the 'norm'. However, each of these groups (together with people who experience other sources of oppressions, such as race, indigeneity, culture, language, age and religion) has a distinctive contribution to make to environmental education, as a form of anti-oppressive resistance, which needs to be foregrounded. In this chapter, I problematise the relative silencing of queer theory and theorising in environmental education and discuss the contributions marginalised voices bring to environmental education. As part of this, I discuss some possibilities for new directions when queer ecopedagogies and research methodologies are used in environmental education to listen to, and work with, the voices from the margins to address social justice issues. As Connie Russell et al. (2013, p. 35) argue, "if we are committed to both environmental education and social justice, surely reproducing oppression of any sort. . . is problematic."

Social justice is now on the international agenda. *The Future We Want*, the outcomes document adopted at the Rio + 20 United Nations Conference on Sustainable Development (2012), reaffirmed the need for "promotion of social equity, and protection of the environment, while enhancing gender equality and women's empowerment, and equal opportunities for all, and the protection, survival and development of children to their full potential, including through education" (Paragraph 11). *The Future We Want* also went beyond gender equality, emphasising that it is "the responsibilities of all States, in conformity with the Charter of the United Nations, to respect, protect and promote human rights and fundamental freedoms for all, without distinction of any kind to race, colour, sex, language or religion, political or other opinion, national or social origin, property, birth, disability or other status" (Paragraph 9). This is an important development for queer(y)ing environmental education.

More recently, the Sustainable Development Goals (United Nations 2016) reflected a similar position in the Goal 5 on Gender Equality and Women's Empowerment target, which focuses on ending all forms of discrimination against all women and girls, and in the Goal 16 target on Peace, Justice, and Social Institutions through its focus on ensuring responsive, inclusive, participatory and representative decision-making at all levels. However, neither goal mentions other forms of gender discrimination, nor is there acknowledgement of the other marginalised groups in 'normalised' environmental education discourses.[3]

If the common vision of the Sustainable Development Goals is to be achieved, then everyone needs to be able to participate in environmental education. However, as I discuss in this chapter, there are many marginalised groups in society who are not being given the opportunity to participate in environmental education and have their voices heard. As the Sierra Youth Coalition (n.d.) argues,

> Historically, the leadership of the mainstream environmental movement – including SYC – has tended to be mostly white, and mostly people of affluence. This leaves out some of the groups of people most affected by environmental degradation. . . Furthermore, racism, sexism, classism, transphobia, ableism and heterosexism (among other things) are just as harmful to our human environment as is its physical degradation.

Similarly, Liz Newbury (2003, p. 205) notes, environmental education continues to "divide groups of people into strong/able/male and Other." The need to change (decenter) the perspectives from which we think and perceive (and act) is central to anti-oppressive environmental education and queer ecopedagogy. Michel Foucault's work is particularly relevant when considering voices from the margins given his concern with the relationship between power and knowledge and the marginalisation and silencing of knowledge that results from the exercise of power. As he argued in *Discipline and Punish* (Foucault 1975/1991, p. 27), "there is no power relation without the correlative constitution of a field of knowledge, nor any knowledge that does not presuppose and constitute at the same time power relations". Voices from the margins, as discussed by Naomi Maina-Okori et al. (2018), are voices that have no power and are not seen as possessing legitimate knowledge. Sandra Harding (1991, pp. 269–270) further elaborates this point when she argues,

> it *does* make a difference who says what and when. When people speak from the opposite sides of power relations, the perspective from the lives of the less powerful can provide a more objective view than the perspective from the lives of the more powerful.

In an education context, "Teachers perpetuate ideology through curriculum (content), pedagogy, and everyday actions" (SUNY-Oswego 2010, online) and the challenge is to ensure that they perpetuate anti-oppressive ideology, not oppression. As I have argued previously, "what is needed is a transformative process in which new empirical and theoretical resources are opened up to reveal new purposes and subjects for inquiry" (Gough 2013, p. 375). Thus, it is important to queer environmental education "to denaturalize coherent selves, to resist the narrow logic of binaries, and to dislodge the sense of safety that comes with 'really knowing'" (Fifield and Letts 2014, p. 405).

Towards queer ecopedagogy

Joshua Russell (2013, p. 13) describes queer ecopedagogy in this way:

> Queer ecopedagogy promotes embodied attentiveness and reflection on being or feeling queer in the world, as well as various personal and political commitments for engaging with dominant "political, bureaucratic, or ideological structures" (van Manen 1997, p. 154) that oppress and silence a wide range of beings, not just LGBTQ-identifying humans.

Queer ecopedagogy calls attention to the ways various forms of oppression marginalise and harm individuals and groups. This is using "queer" to mean more than just "making strange', and more as "an evolving term of disruption" (Bazzul and Santavicca 2017, p. 57), in ways such that "Educators create a queer(ing) response by exposing the rhetorical strategies and irrational assumptions inherent in curriculum and pedagogy" (Santavicca et al. 2019, p. 290).

As noted previously, a precursor to queer ecopedagogy was what Kumashiro (2000) called anti-oppressive education, which "must involve learning to be unsatisfied with what is being learned, said, and known" and "to participate in the ongoing, never-completed construction of knowledge. . . [asking] 'what is not said?' and then [going] to places that have, until now, been foreclosed". He also argued that it is learning about Others[4] (those groups who are traditionally marginalised in society) and "rethinking who one is by seeing the Other as an 'equal' but on different terms" (Kumashiro 2000, p. 45). He did not extend this to beings other than humans. In the two decades since Kumashiro developed his concept there has been growing understanding of the need to look beyond the human in environmental education (see, for example, Russell (2013) and Flessas and Zimmerman (2019)), and this is where a queer ecopedagogy can make a contribution. However, before moving to beyond the human, it is still important to focus on the Othering in human society.

These Other groups include the following dominant/marginalised positions on each dimension of oppression (also see Figure 13.1):

- Gender (male/female)
- Sexuality (heterosexual/LBGTQIA+)
- Class (rich, educated/poor, uneducated)
- Ability (able/disabled)
- Religion (Christian/non-Christian)
- Body discrimination (thin/fat)
- Race (Anglo-Amero-Europeans/people of color)
- Age (youthful/very young, old)
- Language (English/other languages)

Such a listing exemplifies the ambiguous and ambivalent tensions that are inherent in deconstructing the binaries in master narratives that position the first named in the binary as normal and natural (male, heterosexual, rich, educated, able-bodied, Christian, thin, Anglo-Amero-European, youthful and English speaking) as dismantling the binaries can undermine identities that rest upon binary oppositions, and some groups, such as LGBTIA+, inherently dismantle the male/female gender binary. While this is not the place to pursue identity politics, it is important to recognise that "[i]dentity politics pursues recognition, re-naming, and re-representation for people who claim shared histories, cultures, and concerns that have been denied voice and self-representation in dominant cultures" (Fifield and Letts 2014, p. 395), and that, when working toward queer ecopedagogy, to remember that it is easy to get caught up in binary thinking about binaries and their simultaneous affordances and limitations.

In this vein, several recent articles are relevant (see, for example, Bazzul and Santavicca (2017), Knaier (2019), Santavicca et al. (2019)). Bazzul and Santavicca (2017) powerfully argue that "sex/gender and sexuality are phenomena that emerge from complex material and discursive entities – which are simultaneously biological/cultural, individual/collective, non-human/

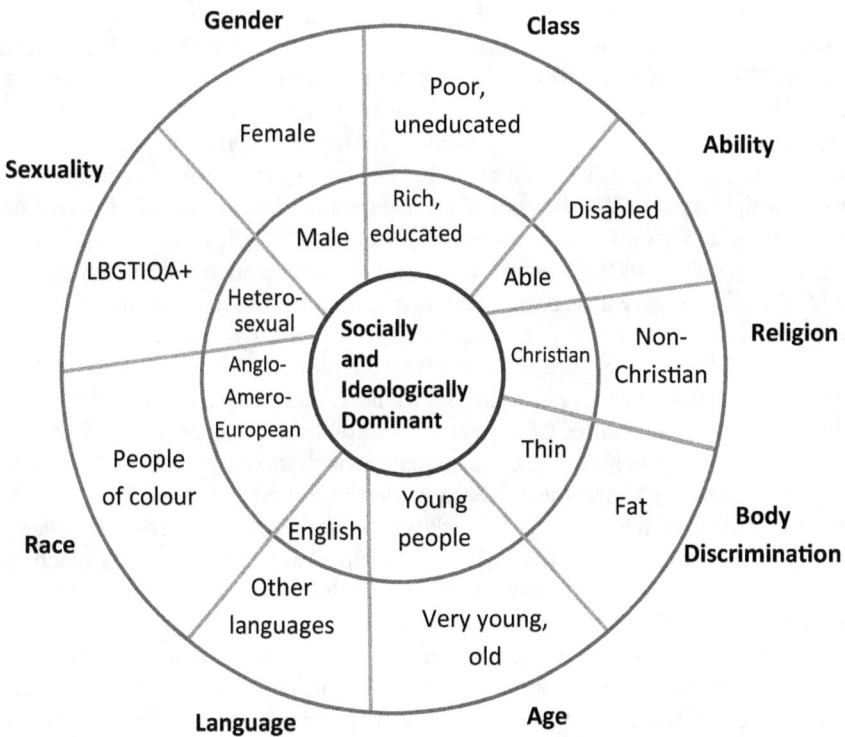

Figure 11.1 Oppression Circle (based on diagram from State University of New York – Oswego, 2010).

human" (p. 56). Similarly, reaching across the species divide in intersectionality analysis (Deckha 2008; Ellsworth 1989) helps to disrupt binaries.

Elizabeth Ellsworth (1989) raises questions, grounded in her classroom experience of critical pedagogy, which are relevant when considering queer ecopedagogy. Of particular relevance are the questions:

- what is critical pedagogy's relationship as a rationalist discourse to the ways that rational competence has been and continues to be constructed as a series of exclusions – of white women, of women and men of color, of nature, of aesthetics?
- what happens to a pedagogy dependent upon the practice of "dialogue" in classrooms when all voices present do not speak, are not heard as they are intended, do not carry equal legitimacy, are not and can never be fully rational? (p. 399)

She also queries the contradictory, partial, and irreducible knowledges that are present in the classroom and the partial knowledges that are created through "terror and loathing of any difference". While recognising that these questions are reflective of the anti-oppressive education foci (gender, race and class) of their time, they are still relevant questions for marginalised voices of all kinds and complement more recent work by, for example, Kumashiro (2000, 2001), Humes (2008), Sykes (2011), Gough (2013), Gray and Mitten (2018), Haluza-Delay (2013), Russell et al. (2013) and Shava (2013), the authors in a special issue of the *Canadian Journal of Environmental Education* on marginalised voices (Russell et al. 2013b) and the authors of articles in the two special issues of *The Journal of Environmental Education* that focused on gender and environmental education (Gough et al. 2017; Russell et al. 2018).

In responding to Ellsworth's questions in the context of environmental education, I am very aware that, just as it is undesirable to be romantic about the environment, so it is meaningless to be romantic about breaking down the barriers between groups from different gender and sexuality backgrounds, or assuming that such a task is simple. Melissa Lucashenko (1994), for example, sees the complexities of interrelationships between marginalised groups and argues that it is meaningless to equate black women's powerlessness with white women's powerlessness, because white women are often complicit in the situation that has resulted in black women becoming powerless. Similarly, Kumashiro (2001) notes that while anti-oppressive approaches to teaching and research can operate in ways that challenge some forms of oppression, they can also – albeit unintentionally or invisibly – comply with others.

In the context of environmental education, I interpret the task of queer ecopedagogy as being to interrupt oppression in its broadest anti-oppressive sense, such as the sexism of silencing and marginalising non-male voices in discussions of environmental education issues – or the silencing and marginalising of queer voices. Just as Valerie Brown and Margaret Switzer (1991) draw attention to the gender bias in the discourses of sustainable development, there is also a need to draw attention to the queer and other discrimination and bias in the discourses of both sustainable development and environmental education. Like Brown and Switzer argued for developing women's as well as men's awareness of the extent of women's exclusion from environmental decision making, this will involve developing everyone's awareness of the extent of "Others'" exclusion from environmental decision making. Such a strategy is anti-oppressive because it subverts hegemonic structures.

In the following sections of this chapter, I discuss six of the above nine marginalised groups in society. While recognising that all nine dimensions interact, for the purposes of this chapter I have focused on these six as body discrimination (beyond consideration of fat and thin), race, class and ability are often closely associated with gender and sexuality discrimination than age and language. I have previously critiqued the dominance of the English language in environmental education discourses (Greenall Gough 1993) and issues around age (Gough 1999) so I will not repeat those arguments here. These sections take up the challenge from Connie Russell and Leesa Fawcett (2013), who note the small amount of attention that has been given to gender, class, disability, sexuality, race and body size in environmental education research – and address some of the gaps and silences in environmental education pedagogy and research around these three groups, placing them with a context of queer ecopedagogy.

Intersectionality between the marginalised groups means that it is difficult to consider a group in isolation – for example, as noted previously, Bazzul and Santavicca (2017) believe that sex/gender and sexuality should be kept together as much as possible to signify that both are always socially situated. Others have also argued that it is often difficult to look at individual oppressed groups in isolation; social and environmental justice matters are often intrinsically linked and benefit from intersectional analysis, which "involves the concurrent analyses of multiple, intersecting (and interacting) sources of subordination/oppression" (Denis 2008, p. 677). Russell et al. (2013, p. 36), for example, note, "obesity discourse has been shown to be sexist, heterosexist, classist, racist and ableist" and Sykes (2011, p. 1) discusses "how students form embodied subjectivities within particular constellations of ableism, heterosexism, racism and body discrimination." Still, each of the following sections foregrounds one group while recognising that there are significant intersections.

Gender

The need to achieve active involvement of women in environmental decision making has been prominent in United Nations documents since the 1992 United Nations Conference on Environment and Development whose outcomes document, *Agenda 21* (United Nations 1993), had a chapter devoted to global action for women towards sustainable and equitable development. Recommended actions included "Programmes to eliminate persistent negative images, stereotypes, attitudes and prejudices against women through changes in socialization patterns, the media, advertising, and formal and non-formal education" (United Nations 1993, Paragraph 24.3(i)). This is continued in

The Future We Want (United Nations 2012), where there is a section on gender equality and women's empowerment in the framework for action and follow up. Specifically, here it is recognised that,

> although progress in gender equality has been made in some areas, the potential of women to engage in, contribute to and benefit from sustainable development as leaders, participants and agents of change has not been fully realised due, inter alia, to persistent social, economic, and political inequalities
>
> (Paragraph 237).

The fifth of the seventeen Sustainable Development Goals (United Nations 2016) also focuses on promoting gender equality and empowering women.

Despite this strong focus on the role of women there is a paucity of research in environmental education in this area. As I have argued elsewhere, "environmental education research has rarely addressed areas of different women's experiences and knowledges, which means that many useful insights have not been pursued" (Gough 2013, p. 376), and that, throughout its history, environmental education has been dominated by universalised (masculine and other dominant) perspectives. That chapter also reviewed feminist and gender research in environmental education that had been undertaken up until 2010, but the list is not long and a recent update (Gough and Whitehouse 2019) did not add much, except for the two previously mentioned recent special issues of *The Journal of Environmental Education* on gender and environmental education, and the recent extensive edited collection, *The Palgrave International Handbook of Women and Outdoor Learning* (Gray and Mitten 2018).

A gender dimension is needed in environmental education so that we have less partial and less distorted stories about environmental experiences; ones that include women's voices, remembering that, "[t]o make sense of any actual woman's life or the gender relations in any culture, analyses must begin in real, historic women's lives, and these will be women of particular races, classes, cultures, and sexualities" (Harding 1991, p. 151). This gender agenda needs to also strengthen its subversive power by voicing "a commitment to talking of sex/gender and sexuality [because this] is what gives queering its subversive power" (Bazzul and Santavicca 2017, p. 57), as Knaier (2019) clearly articulates. Such action will avoid the essentialising of women as a category and recognise the gender identities of queer, trans* and XY women. These aspects are further discussed in other chapters in this book.

Sexuality

There have been silences around sexuality and queer theory and theorising in environmental education research, which is surprising given the volume of writing around queering education since the mid-1990s. The first two articles in this space were Russell et al. (2002) followed by Gough et al. (2003) – but their calls to queer environmental education fell on deaf ears until Joshua Russell's (2013) article on queer ecopedagogy. Russell et al discussed the possibilities for queering environmental education through problematising heteronormativity and essentialised identities, and Gough et al fabricated a fictional location, Camp Wilde, and stories that enabled them to expose the facticity of the field's heternormative constructedness. More recently, Bazzul and Santavicca (2017) employed the concept of assemblages to discuss how our world is always in a state of queer becoming, and Santavicca et al. (2019) revisited Gough et al.'s (2003) Camp Wilde.

Promoting a queer perspective that troubles heteronormativity within environmental education is an important anti-oppressive agenda. As Joshua Russell (2013, p. 24) writes,

> A queer ecopedagogy seeks out the margins in our educational endeavours, exploring uniqueness and diversity among ourselves, each other and the more-than-human world. . . much of the work we engage in as environmental educators needs queering – to explore the (dis)orientations of our technologically-mediated bodies, to seek out and include 'othered' voices and to prop up those who fall 'out of line' with exclusionary and oppressive models of education and politics.

Heterosexuality as an oppression has been discussed more within outdoor education contexts, particularly the macho ethos of the practices (see for example, Gray and Mitten 2018; Kidd and Mason 2019; Mitten 1997; Musinsky 2019; Newbury 2003, Russell et al. 2002), which is relevant for environmental educators because of their shared experiential component. Noel Gough et al. (2003) and Blake Flessas and Timothy Zimmerman (2019) are rare voices in environmental education to draw attention to its heteronormativity. In a recent edited collection of 62 chapters, *The Palgrave International Handbook of Women and Outdoor Learning* (Gray and Mitten 2018), three chapters (Argus 2018; Hauk 2018; Mitten 2018) discussed lesbian and queer experiences. In one of these, Denise Mitten (2018, p. 22) argues the importance of problematising heteronormativity and confronting homophobia:

> Homophobia works through silence and invisibility. By talking with other women and men about how lesbian baiting and other forms of homophobia, transphobia, and the like work to restrict the lives of all women, men, and agender people, doors open for people to be themselves and contribute in many ways to OLEs [outdoor learning experiences] and beyond.

The notion of opening doors is important as part of listening to voices from the margins in a queer ecopedagogy, and expanding queering's subversive power by opening up new empirical and theoretical resources.

Body Discrimination

As Michel Foucault made so clear, body discrimination is another form of oppression or making people feel queer in the world. Firstly, "Discourse transmits and produces power; it reinforces it but it also undermines and exposes it, renders it fragile and makes it possible to thwart it" (Foucault 1976/1981, p. 101), and power is shown on a subject's body because events are 'written' on the body: they shape the way we perform, or act out, our bodily selves (Foucault 1975/1991). This is made clear by several of the authors in Gray and Mitten (2018) who discuss female body image and gender identity in outdoor education contexts.

Another area where there is body discrimination is the area of fat studies. Marilyn Wann (2009, pp. xxi–xxii), for example, argues,

> Like feminist studies, queer studies, and disability studies, which consider gender, sexuality, or functional difference, fat studies can show us who we are via the lens of weight. Fat studies can offer an analysis that is in solidarity with resistance to other forms of oppression by offering a new and unique view of alienation.

While fat studies may seem to some to be irrelevant to environmental education, it is important to remember that size is no different from race, class, gender or sexuality in terms of discourses being marginalised by the dominant discourses in society. As Esther Rothblum (2011, p. 173) argues, "Fat studies scholars ask why we oppress people who are fat and who benefits from that oppression. In that regard, fat studies is similar to academic disciplines that focus on race, ethnicity, gender or age". Each of us brings our own perspective to environmental issues and each needs to be valued and not allowed to be dismissed by the dominant discourses, as Elena Levy-Navarro (2009, p. 16) argues,

> Our histories, however, must never allow this dominant culture to define us. . . As one Latina writer observes, she lives in two worlds with two very different understandings of her "fat" body. Where the one would insist that she is "obese," the other understands her to be "bien cuidada" or well-cared for.

In their article on fat pedagogy and environmental education, Russell et al. (2013) assert, "environmental education is not immune to obesity discourse, although it is mostly used in a throwaway line about childhood obesity being one of the dire consequences of lessening contact with nature" (p. 34), and argue that "[w]e need to consider how bodies are turned into political sites of privilege and oppression through (self) regulation, disciplining, and degradation in mainstream Western society in general and in environmental education in particular" (p. 28). An example of the latter suggests tackling childhood obesity through stealth interventions that "heighten the social desirability to be environmentally friendly" (Skouteris et al. 2013, p. 2), consistent with actions to prevent climate change: "[s]etting goals to achieve a more sustainable society offers a unique opportunity to reduce levels of obesity" (p. 2), which is an argument very different from that being advanced by Russell et al. (2013), and is a good example of a researcher not recognising their own positionality and privilege.

Body discrimination often intersects with gender/sexuality and class and affects access to environmental educational opportunities. It is very much part of being made to feel queer in the world so a queer ecopedagogy needs to promote embodied attentiveness and reflection on the silencing of people who are seen as, and made to feel, different, asking who benefits from such oppression and working to confront it.

Class

Eradicating poverty is the first of the Sustainable Development Goals (United Nations 2016) and it continues as "the greatest global challenge facing the world today" (United Nations 2012, Paragraph 2). Poverty is among the principal sources of environmental damage in developing countries, while the countries themselves are often facing huge environmental challenges from environmental and health hazards, climate change and threats to their resources and biodiversity (Masron and Subramaniam 2019). As poor people are often also uneducated, they have fewer opportunities to voice their concerns and to participate in some environmental activities, neither are their human rights respected.

Poverty impacts on children's play too in terms of the quality of their outdoor spaces and their types of play (Thomson and Philo 2004), and Newbury (2003) notes how outdoor recreation is steeped in cultural capital.

Randolph Haluza-DeLay (2013, p. 394) clearly articulates how environmental educators ignore poverty as an environmental issue, "in so doing environmental educators have reduced the scope of environmental sustainability and missed opportunities to connect with more people and potential allies among a broader reach of civil society organizations and other educators". Poverty is one of many issues that impact on poor people that are being ignored by society at large.

The environmental movement and environmental education have long been recognised as a concern of the middle class rather than lower classes (Bullard 1993; Loomis 2016; Taylor 1996), and the silencing of poor people's voices can be compounded through their intersections with gender, sexuality, race and body discrimination too. But there is a need for queer ecopedagogy where poor people's perspectives are heard and valued, not Othered. Within the Sustainable Development Goals agenda (United Nations 2016) access to education, especially for girls, is seen as a way out of poverty, but other forms of intersecting discrimination are not mentioned, and they need to be addressed. As Harding suggests, we should be moving from:

> . . . *including* others' lives and thoughts in research and scholarly projects to *starting from* their lives to ask research questions, develop theoretical concepts, design research, collect data, and interpret findings . . .

that would provide less partial and distorted accounts of nature and social relations.

(1991, p. 268, emphasis as in original)

Race

A key example of intersectionality and assemblages is the close association of race with class in oppressions in environmental education, because it is often non-Western voices from poor areas that are silenced in the discourses. This was noted since the early days of environmentalism. For example, Peruvian Leopoldo Chiappo (1978, p. 456) questioned whether the inhabitants of the needy South can "accept as valid the way of seeing and interpreting ecological facts adopted by the countries of the super-industrialized, wealthy North", asserting that it is "necessary to reveal the ideology that underlies the attitude of dominance." African Godfrey M'Mwereria (1993, p. 2) commented on "a protest against and rejection of co-option, exclusion and marginalisation of Southern countries in international economic, political and cultural spheres, and an affirmation of African alternatives". In a similar vein, Dorceta Taylor (1996, p. 3) saw environmental education as a white middle class package transmitted to other races and social classes:

More often than not, other cultures and perspectives have been excluded, or played marginal or insignificant roles. This being the case, students of color or poor students wanting to learn about the environment have had to divorce themselves from their surroundings and familiar experiences to do so.

Additionally, Erik Loomis (2016, online) commented that "environmentalism's reputation as a rich white social movement will have enormous negative consequences," and Hannah Miller (2018, p. 854) pointed out that white people do not recognise their whiteness "until people of color enter these spaces and expose the oppressive structures through their own personal struggles." Danielle Purifoy (2018) discussed how people of colour generally and women of colour specifically are subject to multiple forms of exclusion, and Aguilar et al. (2017) argued for environmental education to be inclusive of class, race, ethnicity, and all other aspects of difference and identity. Such an intersectional/assemblage approach is very much part of queer ecopedagogy, but, as Miller (2018) indicates, it is not easy. She provides suggestions for developing collective critical consciousness of race in place-based EE programs drawing on critical race theory, but in queer ecopedagogy these need to be extended to also consider the class, gender, sexuality and other oppressions that can associated with race as a discrimination. As discussed below, the task of queer ecopedagogy educators is to draw attention to these discriminations and resist white and other hegemonies.

Ability

Although there is engagement with disability issues in outdoor and experiential education, to date there has not been any writing from an environmental education perspective. Access to the outdoors is an important component of environmental education for learning and experience. Mitch McLarnon (2013, p. 19), for example, argues, "OE needs to be accessible, affordable and distributed with equity". However, Fox and Avramidis (2003) found that outdoor education represents a powerful, albeit underused, tool for reducing disaffection, promoting inclusive practice and decreasing the risk of permanent exclusion for students with emotional and behavioural difficulties, and Brannan et al. (2000) found that disabled youth significantly increased their social interactions as a result of inclusive outdoor programs. Ability can also intersect with race, gender, sexuality, class and body discrimination as barriers to engagement with environmental education.

However, people with disabilities need to be able to access environmental education activities as part of their education, and people without disabilities need to better understand the barriers faced by people with disabilities and other forms of discrimination and work to enrich their experiences.

Decentering the discourses

At the core of a queer ecopedagogy is the need to decenter the dominant discourses in order to achieve genuine change in society. As Audre Lorde (1984, pp. 113–114) argues,

> the master's tools will never dismantle the master's house. They may allow us temporarily to beat him at his own game, but they will never enable us to bring about genuine change. . . . Racism and homophobia are real conditions of all our lives in this place and time. I urge each one of us here to reach down into that deep place of knowledge inside herself and touch that terror and loathing of any difference that lives there. See whose face it wears. Then the personal as the political can begin to illuminate all our choices.

Lorde was writing of the oppressions of that time that were relevant to the forum at which she was speaking – gender, race and homophobia – and there are others that we can now name as being relevant to a queer ecopedagogy that can both dis-orient and re-orient (Russell 2013). Russell et al. (2013) note how fit, able and healthy bodies are privileged, yet these categories need to be deconstructed to disrupt fat oppression, and Harding (1993, p. 18) shares this quest for a more democratic future by arguing, "democratic values, ones that prioritize seeking out criticisms of dominant belief from the perspective of the lives of the least advantaged groups, tend to increase the objectivity of the results of the research".

A first step towards queer ecopedagogy is to be aware of our own identity and to recognise that our knowledge is a partial. From this standpoint we can monitor our use of oppressive discourses and critically examine our pedagogical practices in order to identify opportunities to deconstruct privileged categories (Russell et al. 2013). When recognising our complicity in oppressive practices some people experience feelings of guilt about their privilege, or, as Kelsey and Armstrong (2012, in Russell et al. 2013, p. 38) describe, a group of environmental educators self-reported feeling "angry, discouraged, hopeless, alienated, and ashamed of the contradictions between their knowledge of climate change and their behaviour." Stopping at feeling guilty or ashamed can mean not acting to challenge oppression. Instead we should adopt allied behaviour and work to dismantle any form of oppression from which we receive benefit and take "personal responsibility for the changes we know are need in society" (SYC n.d.). Timothy San Pedro's (2018) use of culturally disruptive pedagogy to create ally behaviours is particularly relevant, as is Katy Swalwell's (2013) use of social justice pedagogy in educating activist allies around race and class issues.

It is also important to be playful – such as Gough et al.'s (2003) fabrication of Camp Wilde – in order to destabilise identities, analyse interconnections of oppressions and issues and work across differences (Russell et al. 2002). Intersectional analysis and assemblages provide opportunities to deconstruct multiple discourses of oppression – such as the able/strong/male discourse of outdoor education or the poor/colored/fat categorisations in obesity discourses.

Russell et al. (2013) also suggest that expansion of discussions of embodiment are generative in their environmental education practices, particularly starting with the food that goes into their bodies and its interconnectedness with issues such as ethics, social justice, globalisation, climate change and fat studies.

In queer ecopedagogy it is also important to remember the distinction between speaking *for* someone and speaking *with* someone – and to provide opportunities for oppressed voices to be heard. Queer ecopedagogy is an ongoing challenge to undo the stereotypes and prejudices we have learned over time. As Martin et al. (2018, p. 305) recommend,

> People of colour, those with a lesbian/gay/bisexual/queer/questioning orientation, transgender people, and people with a low socioeconomic status need to be acknowledged, understood, and represented in AE [adventure education] literature and programming to cultivate inclusive and authentic social environments.

Queer ecopedagogy can also move beyond the human as part of a posthuman pedagogy where a pedagogic event is an entanglement that provides for modes of becoming that can foster greater entanglement that is "attuned to basic principles of social justice" (Braidotti 2013, p. 11). Such entanglements encompass intersectionality and assemblages that encourage questioning of assumptions and critical engagement with transformative environmental education. For example, Gough (2020) discusses a developing a more-than-human curriculum for environmental education which includes "learning how to open the presences of otherness and how to form relationship of mutuality with others we can never fully know" (Gruenewald 2003, p. 40), recognises equity between human and non-human communities, and argues that, instead of seeking simplicity and certainty, we need to recognise that "we live in a complex world of assemblages that we can never fully know" (p. 9).

Flessas and Zimmerman (2019, p. 96) assert that "queering pedagogy must engage with intersectionality and share aspects of critical pedagogy, queer theory, cultural studies, and ecofeminism" and acknowledge Russell's (2013) recognition of the need to engage with otherness and reflect on *all* oppressed beings. For them, a queer, critical, environmental education practice requires the naming of oppression; designing opportunities for the interrogation of racism, sexism, classism, Western-centric knowledge, and gendered norms, and for the development of big-picture perspectives that include political, economic, social and cultural worldviews; and acknowledging the privileging of certain knowledge and one's own economic situation. In addition, they include addressing complexity, examining assumptions, and questioning "naturalness" as in "human nature", naturalisation of sexed bodies, social gender and sexualities, and views of the perceived naturalness of nonhuman beings. An environmental education practice than follows their suggested techniques is a giant leap forward towards a queer ecopedagogy – it moves beyond a social justice and socially critical pedagogy and embraces the more-than-human, and it is disruptive.

Conclusion

UNESCO (2018, p. 4) sees transformative actions for sustainability by learners as necessitating "a certain level of disruption, with people opting to step outside the safety of the status quo or the 'usual' way of thinking, behaving or living." Unfortunately, this "disruption" is not named as involving any of the dimensions of oppression discussed in this chapter. Nor does it extend beyond an anthropocentric agenda and engage with beings other than human or a posthuman agenda, despite much discussion that "Bodies are both embedded in and embodied as an environment, and are thus forced into a politics and a culture that are governed by an anthropocentric discourse" (Börebäck and Schwieler 2018, p. 9). Queer ecopedagogy, together with the intersectionality and assemblages discussed in this chapter is an important part of the transformation needed from the taken-for-granted in environmental education to a more socially just environmental education that listens to the human and other beings' voices from the margins.

Acknowledgement

I would like to thank Connie Russell for her encouragement to continue writing about marginalised voices in environmental education, Joshua Russell for his thoughtful interrogation of an earlier draft of this chapter, and both for their writings to date in this area that have informed my writing here.

Notes

1 The earlier version of this document, A UNESCO position paper on the future of education for sustainable development (ESD) (UNESCO 2018), did not mention gender, sex or sexuality; a noteworthy oversight.
2 Mcphie and Clarke (2018) do not say what they mean by "post-nature", but it is generally understood to be what will come after what we currently know as life on earth.
3 This is not the place for an extended discussion of the relationship between environmental education and education for sustainability/sustainable development which I have discussed elsewhere (Gough 2016, 2018). For the purposes of this discussion, the two will be taken as synonymous as both will be enhanced by adopting a queer ecopedadagogy.
4 Kumashiro's "Other" are four groups: "students of color, students from under- or unemployed families, students who are female, or male but not stereotypically 'masculine,' and students who are, or are perceived to be, queer", but he believes his proposal could be extended to "other forms of oppression and to other traditionally marginalized groups, such as students with disabilities, students with limited or no English-language proficiency, and students from nonChristian religious backgrounds" (2000, p. 26).

References

Aguilar, O. M., McCann, E. P., & Liddicoat, K. (2017). Inclusive education. In A. Russ, & M. E. Krasny (Eds.), *Urban environmental education review* (pp. 194–201). Ithaca: Comstock Publishing.

Alaimo, S. (2010). Eluding capture: The science, culture, and pleasure of "queer" animals. In C. Mortimer-Sandilands, & B. Erickson (Eds.), *Queer ecologies: Sex, nature, politics, desire* (pp. 51–72). Indianapolis: Indiana University Press.

Argus, S. (2018). LGBTQ girl scouts reflect on their outdoor experiences. In T. Gray, & D. Mitten (Eds.), *The Palgrave international handbook of women and outdoor learning* (pp. 529–543). London: Palgrave Macmillan.

Barad, K. (2003). Posthumanist performativity: Toward an understanding of how matter comes to matter. *Signs: Journal of Women in Culture and Society, 28*(3), 801–831.

Bazzul, J., & Santavicca, N. (2017). Diagramming assemblages of sex/gender and sexuality as environmental education. *The Journal of Environmental Education, 48*(1), 56–66, doi:10.1080/00958964.2016.1249327

Börebäck, M. K., & Schwieler, E. (2018). Elaborating environmental communication within "posthuman" theory. *Journal for the Philosophical Study of Education, III*. www.researchgate.net/publication/325742768_Elaborating_Environmental_Communication_within_Posthuman_Theory

Braidotti, R. (2013). *The posthuman*. Cambridge: Polity Press.

Brannan, S., Arick, J., Fullerton, A., & Harris, J. (2000). Inclusive outdoor programs benefit youth: Recent research on practices and effects. *Camping Magazine, 73*(4), 2–29.

Brown, V. A., & Switzer, M. A. (1991). *Engendering the debate: Women and ecologically sustainable development*. Canberra: Office of the Status of Women, Department of the Prime Minister and Cabinet.

Bullard, R. D. (Ed.). (1993). *Confronting environmental racism: Voices from the grass-roots*. Boston: South End Press.

Butler, J. (2011). *Gender trouble: Feminism and the subversion of identity*. New York: Routledge.

Chiappo, L. (1978). Environmental education and the third world. *Prospects, VIII*(4), 456–465.

Cho, S., Crenshaw, K., & McCall, L. (2013). Toward a field of intersectionality studies: Theory, applications, and praxis. *Signs, 38*(4), 785–810.

Deckha, M. (2008). Intersectionality and posthumanist visions of equality. *Wisconsin Journal of Law, Gender and Society, 23*(2), 249–267.

Deleuze, G., & Guattari, F. (1988). *A thousand plateaus: Capitalism and schizophrenia*. London: Bloomsbury Publishing.

Denis, A. (2008). Review essay: Intersectional analysis: A contribution of feminism to sociology. *International Sociology, 23*(5), 677–694.

Ellsworth, E. (1989). Why doesn't this feel empowering? Working through the repressive myths of critical pedagogy. *Harvard Educational Review, 59*(3), 297–324.

Fifield, S., & Letts, W. (2014). [Re]considering queer theories and science education. *Cultural Studies of Science Education, 9*, 393–407.

Firestone, S., & Koedt, A. (Eds.). (1970). *Notes from the second year: Women's liberation*. New York: Radical Feminism.

Flessas, B. M. R., & Zimmerman, T. D. (2019). Beyond nature talk: Transforming environmental education with critical and queer theories. In W. J. Letts, & S. Fifield (Eds.), *STEM of desire: Queer theories and science education* (pp. 89–107). Leiden: Brill Sense.

Foucault, M. (1975/1991). *Discipline and punish: The birth of the prison*. Trans. by A. Sheridan. Harmondsworth, UK: Pelican.

Foucault, M. (1976/1981). *The history of sexuality, Vol. one, an introduction*. Trans. by R. Hurley. Harmondsworth, UK: Pelican.

Fox, P., & Avramidis, E. (2003). An evaluation of an outdoor education programme for students with emotional and behavioural difficulties. *Emotional and Behavioural Difficulties, 8*(3), 267–283.

Gough, A. (1999). Kids don't like wearing the same jeans as their mums and dads: So whose 'life' should be in significant life experiences research? *Environmental Education Research, 5*(4), 383–394. doi:10.1080/1350462990050404

Gough, A. (2013). Researching differently: Generating a gender agenda for research in environmental education. In J. Dillon, R. Stevenson, M. Brody, & A. Wals (Eds.), *International handbook of research on environmental education* (pp. 375–383). New York, NY: Routledge.

Gough, A. (2016). Tensions around the teaching of environmental sustainability in schools. In T. Barkatsas, & A. Bertram (Eds.), *Global learning in the 21st century* (pp. 83–102). Rotterdam: Sense Publishers.

Gough, A. (2018). Sustainable development and global citizenship education: Challenging imperatives. In I. Davies, L-C. Ho, D. Kiwan, C. Peck, A. Peterson, E. Sant, & Y. Waghid (Eds.), *The Palgrave handbook of global citizenship and education* (pp. 295–312). London: Palgrave.

Gough, A. (2020). Symbiopolitics, sustainability and science studies: How to engage with alien oceans. *Cultural Studies <=> Critical Methodologies, 20*(3), 272–282. doi:10.1177/1532708619883314

Gough, A., Russell, C., & Whitehouse, H. (2017). Moving gender from margin to center in environmental education. *The Journal of Environmental Education, 48*(1), 5–9. doi:10.1080/00958964.2016.1252306

Gough, A., & Whitehouse, H. (2019). Centering gender on the agenda for environmental education research. *The Journal of Environmental Education, 50*(4–6), 332–347. doi:10.1080/00958964.2019.1703622

Gough, N., Gough, A., Appelbaum, P., Appelbaum, S., Doll, M. A., & Sellers, W. (2003). Tales from Camp Wilde: Queer(y)ing environmental education research. *Canadian Journal of Environmental Education, 8*(1), 44–66.

Gray, T., & Mitten, D. (Eds.). (2018). *The Palgrave international handbook of women and outdoor learning.* London: Palgrave Macmillan.

Greenall Gough, A. (1993). Globalizing environmental education: What's language got to do with it? *Journal of Experiential Education, 16*(3), 32–39, 46. doi:10.1177/105382599301600306

Gruenewald, D. (2003). At home with the other: Reclaiming the ecological roots of development and literacy. *The Journal of Environmental Education, 35*, 33–43.

Haluza-DeLay, R. (2013). Educating for environmental justice. In B. Stevenson, M. Brody, J. Dillon, & A. E. J. Wals (Eds.), *International handbook of research on environmental education* (pp. 394–403). New York: Routledge.

Haraway, D. J. (2003). *The companion species manifesto: Dogs, people, and significant otherness.* Chicago: Prickly Paradigm Press.

Haraway, D. J. (2016). *Staying with the trouble: Making kin in the Chtulucene.* Durham: Duke University Press. doi:10.1215/9780822373780

Harding, S. (1991). *Whose science? Whose knowledge? Thinking for women's lives.* Ithaca: Cornell University Press.

Harding, S. (1993). Introduction: Eurocentric scientific illiteracy – a challenge for the world community. In S. Harding (Ed.), *The "racial" economy of science: Toward a democratic future* (pp. 1–21). Bloomington: Indiana University Press.

Hauk, M. (2018). Living lesbian lands and women-led experiential living: Outdoor learning environments for Gaian flourishing. In T. Gray, & D. Mitten (Eds.), *The Palgrave international handbook of women and outdoor learning* (pp. 335–349). London: Palgrave Macmillan.

Humes, B. (2008). Moving toward a liberatory pedagogy for all species: Mapping the need for dialogue between humane and anti-oppressive education. *Green Theory & Praxis: The Journal of Ecopedagogy, 4*(1), 65–85.

Jaggar, A. M., & Struhl, P. R. (1978). *Feminist frameworks: Alternative theoretical accounts of the relations between women and men.* New York: McGraw Hill.

Kelsey, E., & Armstrong, C. (2012). Finding hope in a world of environmental catastrophe. In A. Wals, & P. Corcoran (Eds.), *Learning for sustainability in times of accelerating change* (pp. 187–200). Wageningen, Netherlands: Wageningen Academic Publishing.

Kidd, K. B., & Mason, D. (2019). Camping out: An introduction. In K. B. Kidd, & D. Mason (Eds.), *Queer as camp: Essays on summer, style and sexuality* (pp. 12–29). New York: Fordham University Press.

Knaier, M. L. (2019). What makes girls and boys so desirable? STEM education beyond gender binaries. In W. J. Letts, & S. Fifield (Eds.), *STEM of desire: Queer theories and science education* (pp. 209–221). Leiden: Brill Sense.

Kumashiro, K. (2000). Towards a theory of anti-oppressive education. *Review of Educational Research, 70*(1), 25–53.

Kumashiro, K. (2001). "Posts" perspectives on anti-oppressive education in social studies, English, mathematics, and science classrooms. *Educational Researcher, 30*, 3–12.

Latour, B. (1993). *We have never been modern.* Cambridge: Harvard University Press.

Levy-Navarro, E. (2009). Fattening queer history: Where does fat history go from here? In E. Rothblum, & S. Solovay (Eds.), *The fat studies reader* (pp. 15–22). New York, London: New York University Press.

Lloro-Bidart, T., & Finewood, M. H. (2018). Intersectional feminism for the environmental studies and sciences: Looking inward and outward. *Journal of Environmental Studies and Sciences, 8*, 142–151. doi:10.1007/s13412-018-0468-7

Loomis, E. (2016, 5 December). Towards a working-class environmentalism. *The New Republic.* https://newrepublic.com/article/139132/towards-working-class-environmentalism. Accessed 25 May 2019.

Lorde, A. (1984). *Sister outsider: Essays and speeches.* Berkeley, CA: Crossing Press.

Lucashenko, M. (1994). No other truth? Aboriginal women and Australian feminism. *Social Alternatives, 12*(4), 21–24.

Maina-Okori, N. M., Koushik, J. R., & Wilson, A. (2018). Reimagining intersectionality in environmental and sustainability education: A critical literature review. *The Journal of Environmental Education, 49*(4), 286–296. doi:10.1080/00958964.2017.1364215

Martin, S., Maney, S., & Mitten, D. (2018). Messages about women through representation in adventure education texts and journals. In T. Gray, & D. Mitten (Eds.), *The Palgrave international handbook of women and outdoor learning* (pp. 293–306). London: Palgrave Macmillan.

Masron, T. J., & Subramaniam, S. (2019). Does poverty cause environmental degradation? Evidence from developing countries. *Journal of Poverty, 23*(1), 44–64. doi:10.1080/10875549.2018.1500969

McLarnon, M. (2013). Inclusive outdoor education: Bridging the gap. *Pathways: The Ontario Journal of Outdoor Education.* www.academia.edu/3847210/Inclusive_Outdoor_Education_Bridging_the_Gap

Mcphie, J., & Clarke, D. A. G. (2018). Nature matters: Diffracting a keystone concept of environmental education research – just for kicks. *Environmental Education Research.* doi:10.1080/13504622.2018.1531387

Miller, H. K. (2018). Developing a critical consciousness of race in place-based environmental education: Franco's story. *Environmental Education Research, 24*(6), 845–858. doi:10.1080/13504622.2017.1357802

Mitten, D. (1997). In the light: Sexual diversity on women's outdoor trips. *Journal of Leisurability, 24*(4). http://lin.ca/resource-details/2375. Accessed 26 June 2014.

Mitten, D. (2018). Let's meet at the picnic table at midnight. In T. Gray, & D. Mitten (Eds.), *The Palgrave international handbook of women and outdoor learning* (pp. 19–34). London: Palgrave Macmillan.

M'Mwereria, G. K. (1993). Southern networks for environment and development (SONED) – Africa region. *The Independent Sectors' Network, 26*, 2.

Musinsky, F. (2019). Queer pedagogy at Indian brook Camp. In K. B. Kidd, & D. Mason (Eds.), *Queer as camp: Essays on summer, style and sexuality* (pp. 51–61). New York: Fordham University Press.

Newbury, L. (2003). Will any/body carry that canoe? A geography of the body, ability, and gender. *Canadian Journal of Environmental Education, 8*, 204–216.

Nicholson, L. J. (1981). "The personal is political": An analysis in retrospect. *Social Theory and Practice, 7*(1), 85–98.

Purifoy, D. (2018, 22 June). On the stubborn whiteness of environmentalism. *Inside Higher Ed.* www.insidehighered.com/advice/2018/06/22/how-environmentalism-academe-today-excludes-people-color-opinion?fbclid=IwAR1rPHnQTonouw7 evraWYVVKl-056Wh4vdnJfHH0q3H0XZOuGG-sbAUoTTw. Accessed 29 May 2019.

Rogan, F., & Budgeon, S. (2018). The personal is political: Assessing feminist fundamentals in the digital age. *Social Sciences, 7*, 1–19, 132. doi:10/3390/socsci7080132

Rothblum, E. D. (2011). Fat studies. In J. Cawley (Ed.), *The oxford handbook of the social science of obesity* (pp. 173–183). New York: Oxford University Press.

Russell, C. (2018, 21 November). How might the idea of intersectionality help us learn about our entanglements? *Common World Research Collective Microblog.* http://commonworlds.net/how-might-the-idea-of-intersectionality-help-us-learn-about-our-entanglements/. Accessed 1 December 2018.

Russell, C. (2019). An intersectional approach to teaching and learning about humans and other animals in educational contexts. In T. Lloro-Bidart, & V. S. Banschbach (Eds.), *Animals in environmental education: Interdisciplinary approaches to curriculum and pedagogy* (pp. 35–52). Cham: Palgrave Macmillan.

Russell, C., Cameron, E., Socha, T., & McNinch, H. (2013). "Fatties cause global warming": Fat pedagogy and environmental education. *Canadian Journal of Environmental Education, 18*, 27–45.

Russell, C., & Fawcett, L. (2013). Moving margins in environmental education. In J. Dillon, R. Stevenson, M. Brody, & A. Wals (Eds.), *International handbook of research on environmental education* (pp. 369–374). New York: Routledge.

Russell, C., Fawcett, L., & Oakley, J. (2013). Editorial: Removing margins in environmental education. *Canadian Journal of Environmental Education, 18*, 5–35.

Russell, C., Gough, A., & Whitehouse, H. (2018). Gender and environmental education in the time of #MeToo. *The Journal of Environmental Education, 49*(4), 273–275. doi:10.1080/00958964.2018.1475954

Russell, C. L., Sarick, T., & Kennelly, J. (2002). Queering environmental education. *Canadian Journal of Environmental Education, 7*, 54–66.

Russell, J. (2013). Whose better? [Re]orientating a queer ecopedagogy. *Canadian Journal of Environmental Education, 18*, 11–26.

San Pedro, T. (2018). Abby as Ally: An argument for culturally disruptive pedagogy. *American Educational Research Journal, 55*(6), 1193–1232. doi:10.3102/0002831218773488

Santavicca, N., Bazzul, J., & Witzig, S. (2019). Camping science education: A trip to camp wilde and the queer nature of nature. In W. Letts, & S. Fifield (Eds.), *STEM of desire: Queer theories and science education* (pp. 290–305). Leiden: Brill Sense. doi:10.1163/9789004331068_017

Shava, S. (2013). The representation of indigenous knowledges. In R. B. Stevenson, M. Brody, J. Dillon, & A. E. J. Wals (Eds.), *International handbook of research on environmental education* (pp. 384–393). New York: Routledge.

Sierra Youth Coalition. (n.d.). *Anti-oppression.* www.syc-cjs.org/anti-oppression. Accessed 13 June 2014.

Skouteris, H., Cox, R., Huang, T., Rutherford, L., Edwards, S., & Cutter-Mackenzie, A. (2013). Promoting obesity prevention together with environmental sustainability. *Heath Promotion International.* doi:10.1093/heapro/dat007.

State University of New York (SUNY) – Oswego. (2010). *Circle of oppression.* www.oswego.edu/~prusso1/circle_of_oppression.htm. Accessed 13 June 2014.

Swalwell, K. M. (2013). *Educating activist allies: Social justice pedagogy with the suburban and urban elite.* New York: Taylor and Francis.

Sykes, H. (2011). *Queer bodies: Sexualities, genders & fatness in physical education*. New York: Peter Lang.

Taylor, D. E. (1996). Making multicultural environmental education a REALITY. *Race, Poverty & the Environment, 6*(2–3), 3–6.

Thomson, J. L., & Philo, C. (2004). Playful spaces? A social geography of children's play in Livingston, Scotland. *Children's Geographies, 2*(1), 111–130.

UNESCO. (2018). A UNESCO position paper on the future of education for sustainable development (ESD). *Draft*. https://en.unesco.org/sites/default/files/unesco_position_paper_on_the_future_of_esd_011118.pdf Accessed 19 September 2018.

UNESCO. (2019). *Framework for the implementation of education for sustainable development beyond 2030*. https://unesdoc.unesco.org/ark:/48223/pf0000370215. page=7. Accessed 27 December 2019.

UNESCO Bangkok. (2019, 16 May). *#ColourMeIn art campaign sheds light on inclusion and school safety for LGBTI students*. https://bangkok.unesco.org/content/colourmein-art-campaign-sheds-light-inclusion-and-school-safety-lgbti-students. Accessed 17 May 2018.

United Nations. (1993). *Agenda 21: Earth summit: The United Nations programme of action from Rio*. New York: United Nations. Retrieved on 20 February 2014 from http://sustainabledevelopment.un.org/content/documents/Agenda21.pdf

United Nations. (2012). *The future we want: Outcomes document adopted at Rio + 20*. Rio de Janeiro: United Nations. www.uncsd2012.org/content/documents/727The%20Future%20We%20Want%2019%20June%201230pm.pdf Accessed 27 December 2019.

United Nations. (2016). *Sustainable development goals: 17 goals to transform our world*. www.un.org/sustainabledevelopment/sustainable-development-goals/. Accessed 21 August 2018

Wann, M. (2009). Foreword: Fat studies: An invitation to revolution. In E. Rothblum, & S. Solovay (Eds.), *The fat studies reader* (pp. ix-xxv). New York: New York University Press.

Wrenn, C. L. (2019). *Defining intersectionality*. www.animalsandsociety.org/human-animal-studies/defining-human-animal-studies-an-asi-video-project/defining-intersectionality-with-corey-lee-wrenn/

12 Queer(y)ing environmental education research

*Noel Gough, Annette Gough, Peter Appelbaum,
Sophie Appelbaum, Mary Aswell Doll and
Warren Sellers*

Gough, N., Gough, A., Appelbaum, P., Appelbaum, S., Doll, M.A. & Sellers, W. (2003). Tales from Camp Wilde: Queer(y)ing environmental education research. *Canadian Journal of Environmental Education, 8,* 44–66.

Abstract

This chapter questions the relative silence of queer theory and theorizing in environmental education research. We explore some possibilities for queering environmental education research by fabricating (and inviting colleagues to fabricate) stories of Camp Wilde, a fictional location that helps us to expose the facticity of the field's heteronormative constructedness. These stories suggest alternative ways of (re)presenting and (re)producing both the subjects/objects of our inquiries and our identities as researchers. The contributors draw on a variety of theoretical resources from art history, deconstruction, ecofeminism, literary criticism, popular cultural studies, and feminist poststructuralism to perform an orientation to environmental education research that we hope will never be arrested by its categorization as a "new genre."

Résumé

L'article questionne le silence relatif de la théorie et de la théorisation queer dans le domaine de la recherche en ERE. Nous explorons certaines possibilités d'ouvrir la voie à cette dimension dans la recherche en ERE en inventant (et en proposant à nos collègues d'inventer) les récits du Camp Wilde, lieu fictif qui nous permet d'exposer la facticité propre à la constructivité hétéronormative de ce domaine de recherche. Ces récits suggèrent de nouvelles méthodes pour (re)présenter et (re)produire le sujet et l'objet de notre questionnement ainsi que nos identités en tant que chercheurs. Les collaborateurs font appel à une variété de ressources théoriques, notamment l'histoire de l'art, la déconstruction, l'écoféminisme, la critique littéraire, les études culturelles populaires et le post-structuralisme féministe, afin de créer une orientation nouvelle dans le domaine de la recherche en ERE qui, souhaitons-le, ne sera jamais interrompue parce que reléguée dans la catégorie des "nouveaux genres."

DOI: 10.4324/9781003390930-15

(This paper questions the relative silence of queer theory and theorizing in environmental education research. We explore some possibilities for queering environmental education research by fabricating (and inviting colleagues to fabricate) stories of Camp Wilde, a fictional location that helps us to expose the facticity of the field's heteronormative constructedness. These stories suggest alternative ways of (re)presenting and (re)producing both the subjects/ objects of our inquiries and our identities as researchers. The contributors draw on a variety of theoretical resources from art history, deconstruction, ecofeminism, literary criticism, popular cultural studies, and feminist post-structuralism to perform an orientation to environmental education research that we hope will never be arrested by its categorization as a "new genre.")

The importance of queering earnestness

In recent years, our poststructuralist methodological dispositions (which include attending to whatever is disregarded, muted, repressed, and/or marginalized by dominant cultural discourses and practices) have led us to lament the relative absence of queer theory and theorizing in environmental education research. We agree with Constance Russell et al. (2002) that "queer pedagogy can enrich environmental education theory and practice" (p. 61), and this chapter both affirms their initiative and expands upon it. Our complementary argument is that queer theorizing can enrich environmental education *research*.[1]

We were initially attracted to queer theorizing by its invitation to question the heteronormative desires that animate much educational research, including desires for prediction, control, and "mastery." Like David Jardine (1992), we suspect that technical-scientific discourses limit our capacities to ask questions that do not already presume the possibility of final solutions:

> The language it [technical-scientific discourse] offers is already foreclosed (or, at least, it longs for such foreclosure). It longs for the last word; it longs for . . . a world in which the droning silence of objective presentability finally holds sway over human life. The difficult nature of human life will be solved. We will finally have the curriculum "right" once and for all. . . . Nothing more will need to be said. Obviously, no educational theorist or practitioner would actually claim to want this. But the hesitancy to make such a claim occurs in the same breath that we hear about "having solved just one piece of the puzzle, just one part of the picture. Further research always needs to be done." Such talk, even in its admirable hesitancy . . . does not disrupt the fundamental belief that human life is an objective picture that, however complex, is objectively "there" to be rendered presentable, piece by relentless piece.
>
> (p. 118, emphasis in original)

We recalled Jardine's characterization of educational researchers relentlessly pursuing "objective presentability" when we read Rita Felski's description of nineteenth-century scientists studying human sexual diversity and "deviations" as "earnest Victorian scholars labouring over lists of sexual perversions with the taxonomical zeal of an entomologist examining insects" (quoted in Russell et al., 2002, p. 56). In our experience, many reports of environmental education research similarly conjure images of "droning . . . objective presentability" and "taxonomical zeal."[2]

So we have invented Camp Wilde, an imaginary intellectual space dedicated to alleviating "the irony deficiency that is a hallmark of so many academic texts" (McWilliam, 1999, p. x) by queer(y)ing the earnestness of much environmental education research (and perhaps provoking some subversive laughter). Rather than trying to *represent* queer theory as it might be "applied" to our field, we have tried here to *perform* a queer(y)ing of environmental education research informed by queer theorizing – and both our means of producing this chapter and its final textual form are part of that performance. By "queer(y)ing" – a word formed by embedding a "y" (*why?*) in "queering" – we suggest a mode of questioning inspired by queer theorizing but not necessarily constrained by its extant formulations and contestations.[3] We especially reject any attempt to essentialize "queer," preferring Catherine Mary Dale's (1999) "alternative view of queer as a term productive of positive difference" (p. 3). Positive difference is not structured by negation but "expresses the immanence of the multiple and the one, rather than the eminence of this over that, of one or many, of identity or chaos . . . There is no essential identity nor loss or lack, only affirmation" (p. 3).

In the spirit of Gilles Deleuze and Félix Guattari (1987) we have produced this chapter as a rhizome – a figuration of knowledge as tangled webs of intersections, nodes, and possible pathways, in contradistinction to "arborescent" (treelike) knowledge configured by finite and hierarchically organized roots and branches. To imagine knowledge as a rhizome is to "work against the constraints of authority, regularity, and common sense, and open thought up to creative constructions" (Lather, 1993, p. 680). In a rhizomatic space, there is no one end to inquiry and speculation, no one way of searching – or *re*searching – its limitless possibilities. In Umberto Eco's (1984) words:

> The rhizome is so constructed that every path can be connected with every other one. It has no center, no periphery, no exit, because it is potentially infinite. The space of conjecture is a rhizome space . . . it can be structured but is never structured definitively . . . it is impossible for there to be *a* story.
>
> (pp. 57–58)

We invited several friends to share this "space of conjecture" – to enact textual performances of their own devising that complement our disruptive project.[4] They wrote their texts in response to our 300-word outline, which consisted of little more than the paragraph that begins the next section. We believe that they have helped us to resist foreclosure – to construct Camp Wilde as a conjectural space with "no center, no periphery, no exit." Their contributions can also be read as "data" in a narrative experiment that readers can interpret for themselves.

Welcome to Camp Wilde

Welcome to Camp Wilde. We dedicate this space to the memory of Oscar Wilde because he embodied a mode of subjugated knowledge production that we believe is significant for environmental education research. His works demonstrate that "camp" signifies a more generative mode of being, believing, and behaving than many environmental educators usually associate with the term "camping." In his archaeology of camp posing, Moe Meyer (1994) shows that Wilde undermined the dominant social order of his day not only by being homosexual but also by performing a camp politics and poetics that mocked bourgeois customs, morals, and norms. We suspect that many of his contemporaries were threatened more by his textual inversions and deviations than by his sexual preferences. For example, in "A few maxims for the instruction of the over-educated," Wilde (1989) complains that "the English are always degrading truth into facts. . . . When a truth becomes a fact it loses all its intellectual value" (p. 1203). Against the then-fashionable approaches to literature and art that sought to replicate Nature and Life faithfully, Wilde argued that artifice was more beautiful and more "real." Wilde was dangerous because a deep moral seriousness informed his camp posturing: he was serious about refusing to take himself seriously. His languorous flippancy barely cloaked a scathing irony. When asked to describe the "philosophy" behind *The Importance of Being Earnest* (subtitled *A Trivial Comedy for Serious People*), Wilde replied, "We should treat all trivial things very seriously, and all the serious things of life with sincere and studied triviality" (quoted in Glenn, 2000). At Camp Wilde, we explore how such a paradoxical philosophy might constructively inform environmental education research.[5]

You don't have to be camp (or gay, lesbian, bi-, trans-, or intersexual) to enjoy Camp Wilde, although you might feel more at home here if you didn't think that this was something you needed to question. Of course, queer studies often focus on queer identity, and many queer theorists and researchers explicitly identify themselves as interrogating regimes of normalcy from a "not heterosexual" standpoint. To date, much queer theorizing in education has both interrogated identity and explored relationships between researcher identities and knowledge construction and legitimation (see, e.g., Pinar, 1998). Studies that simultaneously problematize the politics of location and identity, such as Frank Browning's (1996) *A Queer Geography: Journeys Toward a Sexual Self*, and David Bell and Gill Valentine's (1995) *Mapping Desire: Geographies of Sexualities*, have special relevance for environmental education research. But queer theorizing also questions the very idea of normalcy and seeks to dismantle, dislocate, or relocate the boundaries of identity categories (and we identify with that desire). As Patrick Dilley (1999) points out, queered positions are useful but not exclusive starting points for queer theorizing: "anyone can find a queered position (although some might have a better vantage point than others) . . . such a position is not dependent upon one's sexual orientation or predilections, but rather upon one's ability to utilize the (dis)advantages of such a position" (p. 469).

Deborah Britzman (1998) argues that queer theory questions the grounds of identity *and* theory:

> Queer Theory occupies a difficult space between the signifier and the signified, where something queer happens to the signified – to history and bodies – and something queer happens to the signifier – to language and to representation . . . But the "queer," like the "theory," in Queer Theory does not depend on the identity of the theorist or the one who engages with it. Rather, the queer in queer theory anticipates the precariousness of the signified: the limits within its conventions and rules, and the ways these various conventions and rules incite subversive performances, citations, and inconveniences.
>
> (p. 213)

So here at Camp Wilde we want to queer the "normal" signifieds of environmental education research, such as nature-as-an-object-of-knowledge, ecology, body/landscape relations, and the relationships among bodies of knowledges, teachers, and learners. We also want to queer the "normal" signifiers of environmental education research, including the languages and representations with/in which we speak and write environmental education into existence. For example, we suggest that taken-for-granted formulations of purpose such as "the recovery of the ecological imperative" (Bowers, 1993) and formulaic research designs such as those that measure learners' orientations to the Dominant Social Paradigm and the New Ecological Paradigm (Dunlap et al., 2000) are not as straightforward as many environmental educators and researchers assume. To put this another way, we want to probe the ways in which heteronormativity configures ignorance in environmental education research. Jon Wagner (1993) usefully disaggregates *ignorance* into "blank spots" and "blind spots": what we know enough to question but not answer are our blank spots; what we do not know well enough to ask about or care about are our blind spots – areas in which existing theories, methods, and perceptions actually keep us from seeing or imagining objects or phenomena that provoke the curiosity that initiates research (see also Noel Gough, 2002).

Our first guest, Mary Aswell Doll, offers a subversive performance of what she describes elsewhere as "the greening of the imagination" (Doll, 2000), and in so doing strengthens our conviction that Camp Wilde is most aptly named.

Horrible sympathy: nature turned inside out

Mary Aswell Doll
Those who go beneath the surface do so at their own peril.
(Wilde, 1890)

If I were to take the ecological imperative seriously I might do more listening, digging, and sloshing in the mud. Instead of talking of imperatives, with that imperious-sounding intention of classical urgency, I might go in another direction. The alchemists had a saying for how one deepens imagination about lofty, leafy matters. *Opus contra naturam* was the expression they used to mean going in a direction contrary to growth. The gold of material substance is wrought, they wrote, out of their personal dross. Imagine! By concentrating on the nigredo of their own psychic material, these early ecologists saw parallels between the laboratory and the self. They saw that what matters most was not knowledges out there but matter in here, the material of the imagination. It occurred to them that the "gold" of transformation is really found within and that changing inner patterns would have precious outer effects. The growth model with its hefty upward bound to health, happiness, and development needs revisiting, redirecting, bending, turning back, turning around, and queering. Just there, in the dirt, lies another system, hidden perhaps but not Not there.

I speak of a vegetable imagination. Planter societies knew what centering downward entailed. It means going not outward, like the hunter societies with their aim-and-kill approach, but downward, tending soil in such ways as watering, cutting, pruning, pinching, digging, sniffing, and watching. What the earth gives forth is the flesh of the earth blooming in the vine substance of which we all partake. What the older societies taught was watching and learning from the natural cycle of life-death-life. This is conservation of a different order.

I have slept under the stars on an air mattress and a sleeping bag and I have cooked bacon over a campfire, but here I want to suggest what camping in another sense might entail.

It might take itself less seriously, for one. With dreadful seriousness we have literalized our stance on earth matters. And so we talk of "dominant paradigms" and "power/knowledge relationships," as if knowledge is the key issue and dominating is a new ideology.

This dreadful seriousness is deadly. It sees only a human face in the waters of reflection, whereas the cosmos contains so many other life forms in such wide variation. The problem with seriousness is its literalism, unable to think, for instance, as the Buddha thinks when he compares types of people to rocks, sand, or water. Those who are like letters written in running water, he writes, are more evolved not because they are firm in their beliefs or hold solid convictions or believe in pyramid systems but because they listen more and observe what isn't there in the come and go of natural patterns.

Camping in another sense considers the wild more Wildely. The pun, once considered the lowest form of humor, nevertheless can be profound because it sounds two things, two entities, two words, two worlds simultaneously. Wilde's work is punningly serious, as in *The Picture of Dorian Gray*. The novel is about artistry and surfaces. But it also is about the desire for monstrous laws that work as an *opus contra naturam*. Acting as a constant metaphor in the novel is the mythic story of Narcissus, the youth in love with his beauty, which he sees echoed back to him in the surface waters of a pool. Wilde could be talking about his own infatuation with beautiful youths. Or he could be describing the love of images: what one sees beneath surfaces, what lies in the waters of imagination. Instead of an up/down order of things, where fantasies of domination and power swirl, this suggests a different kind of move that privileges smallness and invisibility. Even Darwin is reputed to have added a footnote to a book: Never say higher or lower. He wanted to turn hierarchies around, to study the lowly earthworm – no recourse here to the progress myth. Here is comedy that hears undertones, reverberations, and echoes as a kind of *opus contra naturam*.

Scandalous work, such as Wilde's, disfigures, cherished ideals and so compels revision. When ecologists today talk of conservation and conserving traditions, perhaps that is just another cherished ideal that needs contradicting. Perhaps the revision that is needed is not the powerful ideals of yore but a more humble – and humourous – meditation on earth's humus. As Wilde (1890) puts it, "if the caveman had known how to laugh, history would have been very different" (p. 30).

Camp Wilde's moot court finds Institute for Earth Education chairman guilty of breach of earth charter!

(Authors' note: Camp Wilde's residents and guests frequently use its Moot Court facilities to amuse themselves by simulating "criminal" trials and civil actions. We obtained the following report from Camp Wilde's archives at www.worldwildeweb.net/mootcourt.html.)

Camp Wilde's Moot Court erupted in cheers and laughter today when the jury in the simulated trial of Steve Van Matre handed down its guilty verdict. The founder and self-described "chairman" of the Institute for Earth Education had been charged in absentia with breaching Principle 1.1 of The Earth Charter, which requires humans to "respect Earth and life in all its diversity" (Earth Charter, International Secretariat, 2001, p. 42). Prosecutors argued that Van Matre had failed to comply with this principle by willfully and deliberately limiting the Earth's subject position to that of a heterosexual female, effectively denying Earth's civil rights to freely express its diversity.

The prosecuting team, led by Deakin University law student Kate Allgreen, built its case on Van Matre's own words, citing his editorial contributions to *The Earth Speaks* (Van Matre, 1983a) as evidence that he assumed sexualized identities for both himself and the earth:

Have you listened to the earth?

Yes, the earth speaks, but only to those who can hear with their hearts. It speaks in a thousand, thousand small ways, but like our lovers and families and friends, it often sends its messages without words. For you see, the earth speaks in the language of love. Its voice is in the shape of a new leaf, the feel of a waterworn stone, the color of evening sky, the smell of summer rain, the sound of the night wind. The earth's whispers are everywhere, but only those who have slept with it can respond readily to its call.

. . . falling in love with the earth is one of life's great adventures. It is an affair of the heart like no other; a rapturous experience that remains endlessly repeatable throughout life. This is no fleeting romance, it's an uncommon affair.

(pp. 3–4)

An expert witness for the prosecution, Dr. Sue Curry Jansen, professor of communication studies at Muhlenberg College, testified that on this evidence, Van Matre's standpoint toward the earth was similar to Francis Bacon's, in whose works a nurturing "mother" nature was metaphorically transformed into a more sexualized object – a "bride," "mistress," or "common harlot" (Jansen, 1990, p. 239).

Another witness, semiotician Leon Patrick, testified that elsewhere in *The Earth Speaks* Van Matre (1983b) uses images for the earth that traditionally have passive and/or female connotations, including "vessel" and "ship of life" (p. 61), and that the young people targeted by Earth Education programs would almost certainly interpret terms such as "lovers," "affair," and "romance" to signify conventional (i.e., heterosexual) relationships. Professor Patrick argued that Van Matre's standpoint toward the earth was offensively patronizing and patriarchal, even if his surface rhetoric was that of the new-age "sensitive man." Against Van Matre's romantic claim that "only those who have slept with [the earth] can respond readily to its call," Patrick quoted eminent feminist scholar Donna Haraway's pithy put-down: "I would rather go to bed with a cyborg than a sensitive man. . . . Sensitive men worry me" (quoted in Penley & Ross, 1991, p. 18). Patrick added: "If the earth really could speak, s/he/it might well agree."

Cross-examining this witness, defending counsel Simon Wolfson pointed out that all but four of the authors of the roughly 75 items of prose and poetry in *The Earth Speaks* are male. Since Van Matre (1983c) chose these writings "because each in some way speaks for the earth" (p. vi), does this not imply, asked Wolfson, that he actually positions the earth as male? This suggestion was quickly ridiculed by a number of students from York University who began chanting "stop reading straight!" until brought to order by Judge Russell Hart. Professor Patrick pointed out that Van Matre made matters worse by suggesting that the earth could "speak" only through chiefly male interpreters – or ventriloquists – and was thus positioned not only as passive and female but also as dumb.

Summing up for the defence, Mr. Wolfson argued that Van Matre was guilty only of good intentions, and that positioning the earth as an object of romantic love was no worse than assertions of familial relationship and love, such as Susan Griffin's (1989) declaration that "the earth is my sister, I love her daily grace . . . and how loved I am" (p. 105).

In Ms. Allgreen's final address to the jury, she argued that interpreting the Earth Charter principles at Camp Wilde meant queering the anthropomorphic image of the earth as an object of love and affection – especially if that image is implicitly identified with women, who have historically been oppressed, exploited, and ignored. The feminization of the earth by straight talking men and women, she said, limits the subjective positions available to both individual humans *and* "nature" to those determined by the binary logic of heteronormativity.

The jury took only a few minutes to reach its unanimous guilty verdict. Judge Hart imposed a Community Service Order requiring Van Matre to attend gender equity counselling and to undertake a minimum of 500 hours service as a volunteer guide with Queer(y)ing Nature, an outdoor recreational group in Fredericton, New Brunswick, that is "open to all, yet directed at a queer audience."[6]

Trouble at Camp Wilde

Although we have had a little fun at Steve Van Matre's expense, we hope that readers will appreciate our serious purpose. Once upon a time, we were members of the Institute for Earth Education (IEE) and are on record as seeing merit in its programs (e.g., Noel Gough, 1987; Annette Greenall Gough, 1990). Our disenchantment with IEE began at about the same time that we engaged with Donna Haraway's (1989b, 1991) work on primates and cyborgs, which we read as an invitation to proliferate a shifting multiplicity of standpoints from which to situate our knowledge claims and to question "normal" and "natural" relations of knowledge and power (see, e.g., Annette Gough, 1994; Noel Gough, 1993a, 1993c). Both cyborg and queer subjectivities and corporealities question the normative use of gender-nature affinities (goddess, mother, sister, lover) in producing human relations with nature. Both cultivate suspicion of straight readings of the subjects/objects of environmental education research, because discourses of kinship and community in environmental politics and environmental education often promote principles of care, compassion, and love, which in turn reproduce implicit heteronormative assumptions about identity and relationships. As Catriona Sandilands (1997) writes, "Queers and cyborgs are not easily gendered or natured, and thus represent a new kind of character to inhabit the shifts and fissures of identities in collision or collusion" (p. 19).

For nearly three decades, ecofeminists have been troubling[7] the normative binaries that associate men with culture, reason, and superiority and women with nature, emotion, and subordination. For example, Greta Gaard (1997) argues that "conceptual, symbolic, empirical, and historical linkages between women and nature as they are constructed in Western culture require feminists and environmentalists to address these liberatory efforts together if we are to be successful" (p. 115; see also Plant, 1989; Plumwood, 1993; Warren, 1997a). Haraway's cyborg manifesto has clearly inspired many ecofeminist writers (e.g., Alaimo, 1994; Diamond & Orenstein, 1990; Merchant, 1996; Sandilands, 1997; Warren, 1994), so we initially thought that ecofeminists would feel at home in Camp Wilde, with its focus on queering the normal (read "male") signifieds of environmental education research. But this has not necessarily been the case.

In fact, some of our ecofeminist colleagues are not at all happy with our construction of Camp Wilde, because they see it as a white masculinist project, albeit queer. For example, the provenance of ecofeminism has expanded recently from its earlier concerns with ecological feminism to encompass a recognition "that there are important connections between how one treats women, people of color, and the underclass on one hand and how one treats the nonhuman natural environment on the other" (Warren, 1997a, p. xi).

Ellen O'Loughlin (1993) encapsulates this changed orientation when she writes: "We have to examine how racism, heterosexism, classism, ageism, and sexism are all related to naturism" (p. 148).

Although O'Loughlin mentions heterosexism, sexuality has been a silence in ecofeminism until recently, just as it has been in the environmental and environmental education movements. Thus the "master" of nature in Plumwood's (1993) *Feminism and the Mastery of Nature* is an unmarked category: the heterosexual male. Sandilands' (1997) work is particularly relevant here, especially her arguments for deconstructing assumptions about identity, politics, and their interrelationships if ecofeminism is to continue as a viable and political social movement. In this regard, some feminist scholars, including Sandra Harding (1991), argue that feminist standpoint epistemologies should include a distinctive lesbian feminist position, as well as a heterosexual one, because there is no "essential or typical or preferred 'woman,' from whose typical life feminist standpoint theory require us to start" (p. 250).

This leads us to another group of women who are troubling Camp Wilde. Although some ecofeminists are comfortable with the idea of "queer ecofeminism" (Gaard, 1997; Sandilands, 1997, 1999), an emerging body of lesbian literature troubles "queer politics" by arguing that the strong lesbian feminist political movement which distinguished itself from gay male politics in the 1970s has been submerged by a gay male agenda in the 1990s (Jeffreys, 2003). According to such views, the queer political agenda is damaging to lesbians' interests, to women in general, and to marginalized and vulnerable constituencies of gay men – and, indeed, we should look to lesbians as the vanguard of social change because they are committed to equality and relationships and sex as the basis of social transformation. Thus, Sheila Jeffreys argues that "the word 'queer' is abhorrent to lesbian feminists because it connotes a 'cult of masculinity' especially when linked with the word 'politics'" and that "queer" is "a generic term meaning men and lesbians had to fit into it" (quoted in Myton, 2003, p. 18). The women for whom Jeffreys speaks might be especially troubled by Sandilands' assertion that "queers . . . are not easily gendered." It remains to be seen how such critiques as these might be taken up within queer ecofeminism, but they clearly constitute a further queer(y)ing of environmental education research.

Within ecofeminism we can also discern various shifts of focus. For example, Sandilands (1999) writes of ecofeminism as a quest for democracy rather than for the essentialist woman-based knowledges pursued by some earlier ecofeminist writers. Sandilands' arguments are:

> based on a notion of political subjectivity in which the subject is imperfectly constituted in discourse through the taking-up of multiple subject positions, discursive spaces describing shifting moments of symbolic representation derived from a temporary common understanding. The categories "women" and "nature," in this formulation, appear as common

(and possibly ironic) representations through which democratic politics can progress, rather than as statements about an inherent, oppositional identity.

(p. xx)

Harding (1993) shares this quest for a more democratic future by arguing that "democratic values, ones that prioritize seeking out criticisms of dominant belief from the perspective of the lives of the least advantaged groups, tend to increase the objectivity of the results of the research" (p. 18).

Although much ecofeminist literature asserts the need to consider the empirical connections between women, people of colour, children, the poor and nature (e.g., Warren, 1997b), the spaces created by queer(y)ing environmental education research from (eco)feminist perspectives seem to us to be more generative with respect to "pointing out how better understandings of nature result when scientific projects are linked with and incorporate projects of advancing democracy; [and how] politically regressive societies are likely to produce partial and distorted accounts of the natural and social world" (Harding, 1993, p. ix). Acknowledging the heterosexual basis of Western culture offers us a space for reading nature differently and undertaking more democratic research in environmental education.

Our guest Warren Sellers demonstrates how a queer aesthetic might generate such alternative readings.

Aubrey Beardsley: Camp Wilde's Picturer

Warren Sellers

Aubrey Beardsley is the picturer[8] I associate with Camp Wilde. His images are among the most flagrantly decadent examples of the irony issuing from the *fin de siècle* that melded organic forms into fashioned *objets d'art nouveau*.

According to Charles Bernheimer (2002), *Salome: A Tragedy in One Act* brought Beardsley into Wilde's camp following a pas de deux that saw Beardsley speculating a drawing titled "J'ai baisé ta bouche, Iokanaan" (see Figure 12.1 [Bernheimer, p. 129]) in the inaugural 1893 issue of *The Studio*, which resulted in Wilde arranging a commission for him to illustrate the 1894 Bodley Head edition, which included "The Climax" (Figure 12.2 [Bernheimer, p. 131]).

Writing of Salomania's grip over Europe at the *fin de siècle*, Bernheimer (2002) refers to Bram Dijkstra's thesis that

Salome embodies a male fantasy of woman's inherent perversity. She is a predator whose lust unmans man, a castrating sadist whose victims can best survive her violence by either finding masochistic pleasure in submission or, better, by ridding the world of this purveyor of vice and degeneracy. Misogynist hatred for the Jewish Salome helps prepare the ground, so argues Dijkstra, for the genocidal violence of the twentieth century.

(pp. 104–105)

Figure 12.1 J'ai baisé ta bouche, Iokanaan.[9]

Figure 12.2 The Climax.[10]

Although Bernheimer's project is to unmask Salome's complex roles beyond "male insecurity and anti-feminism" and to show how "she creates overtures to new modes of insight concerning the role of negativity in the psyche and writing" (p. 106), my project is to recognize the symbolic relationship between Salome's climactic gaze and Gaia's climatic concern. I suggest that the imagery in Beardsley's illustrations is a complex graphic representation of both the consequences of collapsing consciousness around modern reductionist science and culture, and potentialities for emergent notions of complexity suggested by James Lovelock's "Gaia" thesis. In his autobiography, Lovelock (2000) writes:

> We now know enough about living organisms and the Earth System to see that we cannot explain them by reductionist science alone. . . . The deepest error of modern biology is the entrenched belief that organisms interact only with other organisms and merely adapt to their material environment.
>
> (p. 390)

My reading of Beardsley's drawings sees a multi-stable figure, a *gestalt* that fluxes through middles of meanings. On the surface there is obsession with desire and dismemberment, analytical separation and examination, and whimsical allusion to sameness and difference. But embodied within, there are also Benoit Mandelbrot's chaotic fractals (Gleick, 1987) (see Figures 12.3 and 12.3a) and Lynn Margulis and Ricardo Guerrero's (1991) complex spirochetes (see Figures 12.4 [Margulis and Guerrero (1991), p. 63] and 12.4a).

Our challenge at Camp Wilde is to unveil some of the decorative irony, which Victoriana exemplifies, to reveal the deft visualization of emerging scientific and social chaos and complexity – to unwrap paradoxical and portentous images of the potential lifelessness ensuing from an obsessive androcentric desire for prizing the world apart.

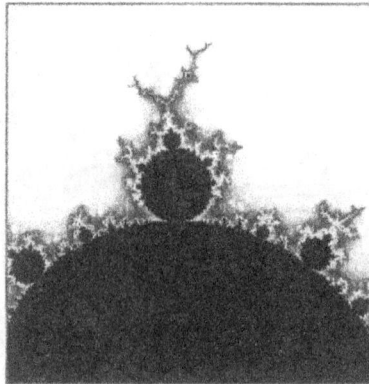

Figure 12.3 Mandelbrot Set.[11]

Figure 12.3a Detail section of Fig 12.2.

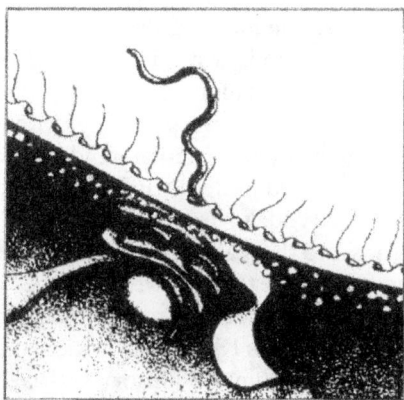

Figure 12.4 Spirochetes.[12]

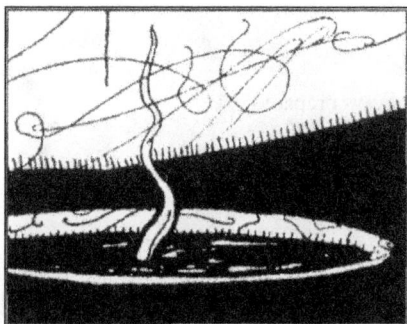

Figure 12.4a Detail section of Fig 12.1.

My reading of *The Climax* reveals the human species' fatal confusion of devolutionary cloning with evolutionary clading. Salome's climax is the nightmare of the disappearing Y chromosome, the spermatozoa of a species oozing back into the eternal primordial potion that is both poison *and* colostrum. Scientism is *analyzing* humanity from existing. As Mary Midgely puts it:

> We have carefully excluded everything non-human from our value system and reduced that system to terms of individual self-interest. We are so mystified – as surely no other set of people would be – about how to recognise the claims of the larger whole that surrounds us – the material world of which we are part. Our moral and physical vocabulary, carefully tailored to the social contract leaves no language in which to recognise the environmental crisis.
>
> (quoted in Lovelock, 2000, p. 390)

This paucity of language is why Camp Wilde needs emergent and generative picturings, which also expose the increasingly corrosive and pervasive illusions of the silvery screen as mainly grand "truth" claims designed to captivate human beings. These claims are the fabrications that Wilde attacked in *The Decay of Lying* by arguing that "art finds her own perfection within, and not outside of herself. She is not to be judged by any external standard of resemblance. She is a veil rather than a mirror" (quoted in Bernheimer, 2002, p. 135).

Bernheimer argues that Wilde's "external standard" was nature, an idea which Beardsley "extends into the realm of the arts. The art of the illustrator . . . need not be subservient to the art of the writer; if the writer veils instead of mirroring nature, so the illustrator veils any resemblance his pictures may have to the external verbal world" (p. 135). This notion of seeing through the veils, looking before-beyond-within the illusory surface, perceiving extensive wholeness, is most revealing – perceiving the need to get over subjectively gazing at the objective, to appreciate becoming whole through complex notions of alluvium, not methods of analysis. The naturally sciential exists *within* picturing being just as well as writing words *about* it.

Different ways with words

The boy scouts are always prepared
To reject him
If they can find him
In their pup-tents
Behind their crackling fires.

(Platizky, 1998)

We share Sellers' distrust of reductionist logocentrism but we also do not want to suggest that verbal modes of representation have any necessary or essential limits. There are many ways of writing other than "straight" prose, and although "queer" inscriptions might sometimes appear to be mere affectations, we should be alert to their interrogative possibilities. For example, the very title of Bronwyn Davies's (2000) *(In)scribing Body/Landscape Relations* demands that readers attempt to decipher not only its words but also its punctuation: the parentheses and backslash invite readers to be suspicious of (and even to disrupt) "normal" relations among and between words, bodies, and landscapes. Davies explores ways in which language – words inscribed in texts and voiced in speech – might trouble (and even collapse) the binaries of landscape and body and their respective "others." For example, she challenges the mind/body binary through collective biography, where participants learn that the mind inhabits not only the brain but the whole body, by writing in a language that recovers the "feeling, poetic body" (p. 168). Her aim is to show bodies in landscape, bodies as landscape (e.g., maternal bodies), and landscapes as extensions of bodies, all being "worked and reworked, scribed and reinscribed" (p. 249). Her writing style seems to be inspired by Hélène Cixous, whose *écriture féminine* inscribes embodied knowledge by using different styles of writing (such as poetry alongside conventional exposition) to fuse experience and subjectivity with analysis.

In a chapter cowritten with Hilary Whitehouse, Davies (2000) re/presents "Australian men talk[ing] about becoming environmentalists" (p. 63) in ways that demonstrate the generativity of poststructuralist approaches to understanding body/landscape relations. Their study explores the take-up of environmental discourses by a small number of men living and working in far north Queensland and analyses the complex relations between the discourse of environmentalism and specific landscapes as they constitute (and are constituted by) these men. The men's talk reveals a boundary between macho/dominant masculinity and more feminine or spiritual or politically correct forms of masculinity. They speak of a stereotype of macho masculinity that they construct as other to themselves, and especially as other to the selves they produce in their talk with Davies and Whitehouse. But some of the men admit to being drawn into this undesirable form of masculinity, and one describes getting caught up in macho talk and patterns of desire through being part of the gay scene:

> When he "came out" as gay, he thought he would find many other men like himself, misfits who had never achieved and did not want to achieve dominant forms of masculinity. To his horror, he discovered that he was as different from other gay men in the rural Queensland city that he moved to, as he had been from heterosexual men.
>
> (p. 72)

Other men in Davies and Whitehouse's study describe adolescent experiences of "mistakenly" expressing masculinity in drunken heroic and dominating ways, and in which they experience themselves as male in relation to female nature. These early experiences were embarrassing to talk about because these men had remade themselves as environmentally caring and profeminist adults. One recalls going out into the bush at the age of 15 and becoming extremely drunk:

> We chose the bush . . . because of the privacy, obviously, because you couldn't be seen. The other thing is, it was like pitting yourself against, you know, you're out there against the environment and you're a man and, I mean, this is an embarrassing confession, that one of the things I did, and I remember doing this, I was really pissed and I dug a hole in the ground and my mates came along and I was rooting the earth . . . and they said, "What are you doing?" and I said, "Oh I'm fucking Mother Earth." I haven't thought of that for twenty years now.
>
> (p. 75)

Davies and Whitehouse find this man's "insight" "interesting": "As a young drunken boy wanting to conquer nature, his act of copulation was one which, he later explained, combined love of nature as well as conquering nature" (p. 75).

Here we must digress. All this talk of rooting and fucking reminds us of Steve Van Matre sleeping with the earth, Mary Doll digging and sloshing in the mud, and especially of Chet Bowers (2002) who returns frequently to "root metaphors" in his arguments for an ecological understanding of curriculum. Each of these authors chooses metaphors that are consistent with our own dispositions to create and conserve organic and evolutionary connections with the earth and one another. But metaphors matter in the literal sense that they have material effects, and even if we cannot *not* think through metaphors, we are responsible for the metaphors we choose to privilege and thus need to be self-critically responsive to the effects of their deployment. So we wonder at what point we need to be suspicious of the materialities we imagine through the metaphors we choose. When and under what circumstances should we remind ourselves that "root metaphor" is a metaphor and does not signify a "real" root? Haraway (1994) asks a difficult but pertinent question for all of us who work with words: "How can metaphor be kept from collapsing into the thing-in-itself?" (p. 60) In other words, how can we resist replicating the worlds we analyze in our own material-semiotic practices? Queer things, metaphors.

Returning to Davies (2000), we note that although only one of the environmentalists in their study identified himself as gay, all of them found a variety of strategies for "troubling the surface of rational dominant masculinity and of coming to (be)long in landscapes in embodied ways" (p. 84). They note that

> "nature" has many meanings, as does "masculinity," and there are many contradictions between them. One way of managing these different meanings is to make discursive and bodily practice specific to particular folds in time and space (such as "the pub" and "Kakadu"). Another way is to merge and meld elements of one discourse and the related set

of practices with other discourses and practices. These men constantly *separate themselves out* from other, lesser men, who are macho exploiters of women and environments. But the individualistic hero image is not easily let go of. Each man escapes from culture and other men in a journey of renewal and return. Each one finds himself vulnerable to the practices and discourses of the culture he finds himself in – vulnerable to becoming "like them."

(p. 85, emphasis in original)

We suggest that the "separating out" to which Davies and Whitehouse refer is continuous with an autonomous queer(y)ing of identity that is "specific to particular folds in time and space," an interpretation that raises generative questions for environmental education research. For example, their analysis suggests that it might be possible to "read" some popular media texts – TV's *The Crocodile Hunter* comes immediately to our minds – not only as banal entertainments but also as complex inscriptions of body/landscape relations. Is Steve Irwin the Liberace of Australian Wilde(r)ness? And in responding to that question what might we learn about our own embodied and locatable knowledges in/of the theatre/landscape we share with him?

Peter Appelbaum and his daughter Sophia demonstrate a similarly deconstructive queer(y)ing of "normal" body/landscape relations in their reading of a popular example of young adult fiction.

The ear, the eye, and the arm: a book review from Camp Wilde

Peter and Sophia Appelbaum

Our family has been reading *The Ear, the Eye and the Arm*, by Nancy Farmer (1995). In this futuristic Zimbabwe, "Dead Man's *Vlei*" is a former toxic waste dump – dense layers of something that used to be called "plastic." People live in Dead Man's *Vlei*. They are almost invisible, blending in with the grayness of the *Vlei* itself. They are part of the *Vlei* as the *Vlei* is part of them. We have been talking about this book as we read it, several chapters a night at bedtime. It strikes me (Peter) that the issue of toxic waste is presented not as a feature of the plot but as a backdrop to important character development.

The main characters have important experiences in the Vlei that help us to understand how they are growing during an experience in which they are outside of their cloistered home for the first time. It is curious that the next sequence of this picaresque novel takes place in a utopian society that has locked modern technology outside of its domain. Here everything initially seems "just right" to the young heroes whose adventure we are lucky to share through reading. In each situation, the environment and technology are not the main features of the story, but the context through which human individuals construct their sense of humanity, ethics, and relationship to the landscape.

It is this perversity of environmental detail, in the denying of centrality, that the centrality of the environmental context is enabled to emerge. Britzman (1996) describes perversity as "pleasure without utility," and it is the specifically non-utilitarian use of the landscape in Farmer's novel that we find to generate, peculiarly, a concern for our own relationship to the environment. Noel Gough (1993b) has written about this phenomenon as well, arguing that in reading and discussing science fiction and cyberpunk literature together, students and teachers can often critically reappraise human relationships with science, technology, and the environment.

In *The Ear, the Eye and the Arm*, mutations caused by environmental devastation also lead to uniquely human manifestations of the changes that might be wrought by such devastation: the terms in the title refer to mutated humans who have perversely heightened abilities of perception. The mutated enhancements are the result of being born in toxic regions of the country. Yet it is through these three people, and by implication, through the changes that humans have introduced into their environment, that the heroes of the novel are able to realize their potential, that the utopian civilization is able to survive, and that the world of magic and science merges into a plot climax that we feel we can't reveal to anyone who has yet to read the book.

A resting (not arrested) place

One way in which we have sought to explore new genres of research in environmental education is to venture beyond our own comfort zones in the production of this chapter. When we invited Mary Doll and Warren Sellers to visit Camp Wilde, we did not know what they would bring with them or how they would perform (in) the "camp" of *their* imaginations. Mary teaches literature and literary criticism, and Warren has worked as a designer, director, producer, consultant, and teacher in the electronic media industries. The modes of inquiry and interpretation through which meanings are produced within their respective traditions of social relationships and organization are different from those with which we are most familiar. We were not particularly surprised that some reviewers of our manuscript had a little difficulty in understanding the implications of their respective contributions, or why they were written in the way they were. One reviewer speculated that s/he was "not well practiced in reading the genre" in which these contributions were written and thus found them "unclear and confusing."

Similarly, when we invited Peter Appelbaum to visit Camp Wilde, we did not expect him to bring his daughter, but we are very pleased that he did. Again, some reviewers did not see any explicit or obvious connection between Peter and Sophia's book review and environmental education research. In our view, this does not mean that such a connection is absent, or that we should try to assimilate the difference between their understandings of Camp Wilde and ours by making the connections that *we* see more explicit. We are suspicious of trying to make the strange familiar and prefer to read each of our guest's contributions as an invitation to work constructively with discourses that appear to be incommensurate without colonizing them.

Although we would prefer not to be defensive, we feel compelled to respond to those reviewers who wanted us to provide "a more clear discussion of queer theory": "As it stands," one wrote, "a less careful reader could come away thinking queer merely referred to the unconventional." We stand by our right to explore how queer theorizing might work, and what it might produce, rather than to explain what it means or what it is. If readers of the *Canadian Journal of Environmental Education* (who we assume to be "careful") want to know what those who claim authoritative status in queer theorizing think it is, we recommend sources such as Suzanne de Castell and Mary Bryson's (1998) "notes toward a queer researcher's manifesto" (p. 249).[10] Because we do not presume to say what queer theory *is*, we also cannot say what it is not, and if our queer(y)ing of heteronormativities in environmental education research therefore looks to some readers like the "merely unconventional" then we accept that risk. To paraphrase Haraway (1989a, p. 307), we are not interested in policing the boundaries between the queer and the unconventional – quite the opposite, we are edified by the traffic.

So, farewell from Camp Wilde. We hope you have enjoyed your visit and that it was not too comfortable. We hope that you return and bring some of your own tales of queer(y)ing environmental education research with you. We have very deliberately eschewed any attempt to provide a straightforward (as it were) account of queer methodology or to present a comprehensive argument for "doing" queer research in environmental education. Rather, we have assembled some of the theoretical resources and cultural materials we had to hand and, with a little help from our friends, performed an orientation to environmental education research that we hope will never be arrested by its categorization as a "new genre."

Acknowledgments

We thank three anonymous reviewers for alerting us to readings of parts of our manuscript that we did not anticipate, which provided us with the opportunity to try to avert such readings. We also thank Connie Russell for her thoughtful editorial mediations and her patience.

Notes on contributors

Noel Gough is Associate Professor in the Faculty of Education, Deakin University, Australia. His research interests include narratives and fictions in educational inquiry, and poststructuralist and postcolonialist analyses of curriculum change, with particular reference to environmental education, science education, internationalization, and inclusivity. He is coeditor (with William Doll) of *Curriculum Visions* (Peter Lang, 2002).

Annette Gough is Associate Professor in the Faculty of Education, Deakin University, Australia. Her research interests are in environmental education, science education, and research methodology, with a particular emphasis on women and poststructuralist perspectives. She is the author of Education and the *Environment: Policy, Trends and the Problems of Marginalisation* (ACER Press, 1997).

Peter Appelbaum is Associate Professor in the Department of Education, Arcadia University, Glenside Pennsylvania, USA.

Sophia Appelbaum is a student at the Project Learn School, Philadelphia, USA.

Mary Aswell Doll teaches in the General Studies program at the Savannah College of Art and Design, Georgia, USA.

Warren Sellers is a doctoral student in curriculum theory at Deakin University, Australia.

Notes

1 We make no categorical distinction between environmental education research and environmental education. We emphasize research because research is what we do. Research is anything that people who call themselves researchers actually do that their peers recognize as research, and thus includes any means by which a discipline or art develops, tests, and renews itself.
2 See Noel Gough (1999) for a critique of examples of environmental education research that can be read as reductio ad absurda of technical-scientific discourse. We first noticed colleagues using the term "queer(y)ing" – and variations such as "que(e)r(y)ing" and "queer-y-ing" – in the mid-1990s (see, e.g., Gibson-Graham, 1997; Nicoll, 1997), but have since found earlier uses (e.g., Sandilands, 1994).
3 All first person plural pronouns ("we," "us," "our") in this chapter refer unequivocally only to the two of us (Noel and Annette Gough). Our guest contributors wrote their own scripts, and we do not presume to speak for them.
4 As Deleuze (1994) notes, paradox is "the passion of philosophy" (p. 227).
5 Queer(y)ing Nature's activities include camping, hiking, cycling, kayaking, skiing, snowshoeing, etc. See www.binetcanada.org/en/mar/play.html, accessed 1 September 2002.
6 We use "troubling" in similar ways to Lather (1996; Lather & Smithies, 1997), to signify that we read terms that are "troubled" as *sous rature* (under erasure), following Derrida's approach to reading deconstructed signifiers as if their meanings were clear and undeconstructable, but with the understanding that this is only a strategy (Derrida, 1985).
7 Beardsley declared his images to "picture" rather than "illustrate": "When he became art editor of *The Yellow Book*, he insisted that the journal's policy allow the artwork to stand on its own rather than illustrate particular contributions" (Bernheimer, 2002, p. 215).
8 Reproduced from http://aleph0.clarku.edu/~djoyce/julia/mandel2.gif (Joyce, 2003).
9 Although we hope our work is consistent with all seven items in de Castell & Bryson's "manifesto," we do not see our own identities as being coterminous with

their characterization of queer researchers. Watch this space for our "notes toward a cyborg researcher's manifesto."
10 Reproduced from Bernheimer (2002, p. 131)
11 Reproduced from http://aleph0.clarku.edu/~djoyce/julia/mandel2.gif
12 Reproduced from the drawing by Christie Lyons in Margulis and Guerrero (1991, p. 63).

References

Alaimo, S. (1994). Cyborg and feminist interventions: Challenges for an environmental feminism. *Feminist Studies, 20*(1), 133–152.
Bell, D. & Valentine, G. (Eds.). (1995). *Mapping desire: Geographies of sexualities.* London: Routledge.
Bernheimer, C. (2002). *Decadent subjects: The idea of decadence in art, literature, philosophy, and culture of the fin de siècle in Europe.* Baltimore, MD: The Johns Hopkins University Press.
Bowers, C.A. (1993). *Critical essays on education, modernity, and the recovery of the ecological imperative.* New York: Teachers College Press.
Bowers, C.A. (2002). Toward a cultural and ecological understanding of curriculum. In W.E. Doll & N. Gough (Eds.), *Curriculum visions* (pp. 75–85). New York: Peter Lang.
Britzman, D. (1996). On becoming a little sex researcher. *Journal of Curriculum Theorizing, 12*(2), 4–11.
Britzman, D. (1998). Is there a queer pedagogy? Or, stop reading straight. In W.F. Pinar (Ed.), *Curriculum: Toward new identities* (pp. 211–231). New York: Garland.
Browning, F. (1996). *A queer geography: Journeys toward a sexual self.* New York: Crown.
Dale, C.M. (1999). A queer supplement: Reading Spinoza after Grosz. *Hypatia, 14*(1), 1–8.
Davies, B. (2000). *(In)scribing body/landscape relations.* Walnut Creek, CA: AltaMira Press.
de Castell, S. & Bryson, M. (1998). From the ridiculous to the sublime: On finding oneself in educational research. In W.F. Pinar (Ed.), *Queer theory in education* (pp. 245–250). Mahwah, NJ: Lawrence Erlbaum Associates.
Deleuze, G. (1994). *Difference and repetition* (Paul Patton, Trans.). New York: Columbia University Press.
Deleuze, G. & Guattari, F. (1987). *A thousand plateaus: Capitalism and schizophrenia* (Brian Massumi, Trans.). Minneapolis: University of Minnesota Press.
Derrida, J. (1985). Letter to a Japanese friend. In D. Wood & R. Bernasconi (Eds.), *Derrida and différance* (pp. 1–5). Warwick: Parousia Press.
Diamond, I. & Orenstein, G.F. (1990). *Reweaving the world: The emergence of ecofeminism.* San Francisco: Sierra Club Books.
Dilley, P. (1999). Queer theory: Under construction. *International Journal of Qualitative Studies in Education, 12*(5), 457–472.
Doll, M.A. (2000). *Like letters in running water: A mythopoetics of curriculum.* Mahwah, NJ: Lawrence Erlbaum Associates.
Dunlap, R.E., Van Liere, K.D., Mertig, A.G. & Jones, R.E. (2000). Measuring endorsement of the new ecological paradigm: A revised NEP scale. *Journal of Social Issues, 56*(3), 425–442.

Earth Charter, International Secretariat. (2001). *The Earth Charter initiative handbook.* San José, Costa Rica: Earth Charter International Secretariat.

Eco, U. (1984). *Postscript to the name of the rose* (William Weaver, Trans.). New York: Harcourt, Brace, and Jovanovich.

Farmer, N. (1995). *The ear, the eye and the arm.* New York: Penguin Putnam.

Gaard, G. (1997). Toward a queer ecofeminism. *Hypatia, 12*(1), 114–137.

Gibson-Graham, J.K. (1997). Queer(y)ing globalization. In H.J. Nast & S. Pile (Eds.), *Places through the body* (pp. 23–41). New York: Routledge.

Gleick, J. (1987). *Chaos: The making of a new science.* New York: Viking.

Glenn, J. (2000). *Hermenaut of the month: Oscar Wilde.* www.hermenaut.com/a163. shtml accessed 5 May 2003.

Gough, A. (1994). *Fathoming the fathers in environmental education: A feminist post-structuralist analysis.* Unpublished PhD dissertation, Faculty of Education, Deakin University, Australia.

Gough, N. (1987). Greening education. In D. Hutton (Ed.), *Green politics in Australia* (pp. 173–202). Sydney: Angus & Robertson.

Gough, N. (1993a). Environmental education, narrative complexity and postmodern science/fiction. *International Journal of Science Education, 15*(5), 607–625.

Gough, N. (1993b). *Laboratories in fiction: Science education and popular media.* Geelong: Deakin University.

Gough, N. (1993c). Neuromancing the stones: Experience, intertextuality, and cyberpunk science fiction. *Journal of Experiential Education, 16*(3), 9–17.

Gough, N. (1999). Rethinking the subject: (De)constructing human agency in environmental education research. *Environmental Education Research, 5*(1), 35–48.

Gough, N. (2002). Ignorance in environmental education research. *Australian Journal of Environmental Education, 18,* 19–26.

Greenall Gough, A. (1990). Environmental education. In K. McRae (Ed.), *Outdoor and environmental education for schools: Diverse purposes and practices* (pp. 41–53). South Melbourne: Macmillan.

Griffin, S. (1989). The earth is my sister. In J. Plaskow & C.P. Christ (Eds.), *Weaving the visions* (pp. 105–110). San Francisco: Harper and Row.

Haraway, D.J. (1989a). Monkeys, aliens, and women: Love, science, and politics at the intersection of feminist theory and colonial discourse. *Women's Studies International Forum, 12*(3), 295–312.

Haraway, D.J. (1989b). *Primate visions: Gender, race, and nature in the world of modern science.* New York: Routledge.

Haraway, D.J. (1991). *Simians, cyborgs, and women: The reinvention of nature.* New York: Routledge.

Haraway, D.J. (1994). A game of cat's cradle: Science studies, feminist theory, cultural studies. *Configurations, 2*(1), 59–71.

Harding, S. (1991). *Whose science? Whose knowledge?* Ithaca: Cornell University Press.

Harding, S. (Ed.). (1993). *The "racial" economy of science: Toward a democratic future.* Bloomington: Indiana University Press.

Jansen, S.C. (1990). Is science a man? New feminist epistemologies and reconstructions of knowledge. *Theory and Society, 19,* 235–246.

Jardine, D.W. (1992). Reflections on education, hermeneutics, and ambiguity: Hermeneutics as a restoring of life to its original difficulty. In W.F. Pinar & W.M. Reynolds (Eds.), *Understanding curriculum as phenomenological and deconstructed text* (pp. 116–127). New York: Teachers College Press.

Jeffreys, S. (2003). *Unpacking queer politics: A lesbian feminist perspective.* Cambridge, MA: Polity Press.

Joyce, D.E. (2003). *Mandelbrot set.* http://aleph0.clarku.edu/~djoyce/julia/man del12.gif accessed 22 May 2003.

Lather, P. (1993). Fertile obsession: Validity after poststructuralism. *The Sociological Quarterly, 34*(4), 673–693.

Lather, P. (1996). Troubling clarity: The politics of accessible language. *Harvard Educational Review, 66*(3), 525–545.

Lather, P. & Smithies, C. (1997). *Troubling the angels: Women living with HIV/AIDS.* Boulder, CO: Westview Press.

Lovelock, J.E. (2000). *Homage to Gaia: The life of an independent scientist.* Oxford: Oxford University Press.

Margulis, L. & Guerrero, R. (1991). Two plus three equal one: Individuals emerge from bacterial communities. In W.I. Thompson (Ed.), *Gaia 2 emergence: The new science of becoming* (pp. 50–67). Hudson, NY: Lindisfarne Press.

McWilliam, E. (1999). *Pedagogical pleasures.* New York: Peter Lang.

Merchant, C. (1996). *Earthcare: Women and the environment.* New York: Routledge.

Meyer, M. (1994). Under the sign of Wilde: An archaeology of posing. In M. Meyer (Ed.), *The politics and poetics of camp* (pp. 75–109). London: Routledge.

Myton, D. (2003). Queering their pitch. *Campus Review Incorporating Education Review, 12*(14), 18.

Nicoll, F. (1997). "Up ya bum"? Queer(y)ing Australian nationalist subjectivity. *Critical InQueeries, 1*(3), 53–57.

O'Loughlin, E. (1993). Questioning sour grapes: Ecofeminism and the United Farm Workers grape boycott. In G. Gaard (Ed.), *Ecofeminism: Women, animals, nature* (pp. 148–160). Philadelphia, PA: Temple University Press.

Penley, C. & Ross, A. (1991). Cyborgs at large: Interview with Donna Haraway. In C. Penley & A. Ross (Eds.), *Technoculture* (pp. 1–20). Minneapolis: University of Minnesota Press.

Pinar, W.F. (1998). *Queer theory in education.* Mahwah, NJ: Lawrence Erlbaum Associates.

Plant, J. (1989). *Healing the wounds: The promise of ecofeminism.* Philadelphia, PA: New Society Publishers.

Platizky, R. (1998). We "were already ticking and didn't even know" [it]: Early AIDS works. In W.F. Pinar (Ed.), *Queer theory in education* (pp. 337–348). Mahwah, NJ: Lawrence Erlbaum Associates.

Plumwood, V. (1993). *Feminism and the mastery of nature.* New York: Routledge.

Russell, C., Sarick, T. & Kennelly, J. (2002). Queering environmental education. *Canadian Journal of Environmental Education, 7*(1), 54–66.

Sandilands, C. (1994). Lavender's green? Some thoughts on queer(y)ing environmental politics. *Undercurrents, 7*(3), 20–24.

Sandilands, C. (1997). Mother earth, the cyborg, and the queer: Ecofeminism and (more) questions of identity. *NWSA Journal, 9*(3), 18–40.

Sandilands, C. (1999). *The good-natured feminist: Ecofeminism and the quest for democracy.* Minneapolis: University of Minnesota Press.

Van Matre, S. (1983a). Earth magic. In S. Van Matre & B. Weiler (Eds.), *The earth speaks: An acclimatization journal* (pp. 3–7). Warrenville, IL: The Institute for Earth Education.

Van Matre, S. (1983b). Earth wisdom. In S. Van Matre & B. Weiler (Eds.), *The earth speaks: An acclimatization journal* (pp. 61–66). Warrenville, IL: The Institute for Earth Education.

Van Matre, S. (1983c). Introduction. In S. Van Matre & B. Weiler (Eds.), *The earth speaks: An acclimatization journal* (pp. v–vi). Warrenville, IL: The Institute for Earth Education.

Wagner, J. (1993). Ignorance in educational research: Or, how can you not know that? *Educational Researcher, 22*(5), 15–23.

Warren, K.J. (1994). *Ecological feminism.* New York: Routledge.

Warren, K.J. (1997a). *Ecofeminism: Women, culture, nature.* Bloomington: Indiana University Press.

Warren, K.J. (1997b). *Taking empirical data seriously, ecofeminism: Women, culture, nature* (pp. 3–20). Bloomington: Indiana University Press.

Wilde, O. (1890). *The picture of Dorian Gray.* New York: Dover.

Wilde, O. (1989). A few maxims for the instruction of the over-educated. In *The complete works of Oscar Wilde* (pp. 1203–1204). New York: Harper and Row.

13 The generativity of feminist and environmental cartoons for environmental education research and teaching

Annette Gough and Judy Horacek

Gough, A., and J. Horacek. 2022. "The generativity of feminist and environmental cartoons for environmental education research and teaching." *Environmental Education Research 29* (4): 500–519. doi:10.1080/13504622.2022.2073332

Abstract

This chapter brings together a feminist environmentalist cartoonist with a feminist environmental educator in an exploration of the generativity of cartoons in environmental education research and teaching. Using duoethnography as a methodology, and drawing on critical and new materialist feminist theory, we explore our personal memories, stories and conversations, as well as discuss the origins and/or significance of particular cultural artefacts (some of Judy's cartoons), to illuminate the reasons for, and influences on, our engagement with cartoons, feminism, the environment and formal environmental education. Drawing on a range of literature around humour and environmental education and feminism, juxtaposed with our conversations and the cartoons, we also seek to identify some possible ways of measuring what impact or influence cartoons about the environment might have once they are in the world, thereby exploring their generativity in environmental education research, theorising and practice.

Introduction

According to the late John Clarke (2017), in his introduction to a collection of Judy Horacek's cartoons,

> Of all the public figures in Australia, the group with whom the people feel a sense of intimacy is the cartoonists. This is interesting because the public doesn't really know the cartoonists; it knows the cartoons. The cartoons resonate against the thoughts and feelings of readers, and the readers thereby contribute to the creative process of understanding the cartoon.
>
> (2–3)

Cartoons are indeed a form of context-dependent humour that is a very familiar part of our lives. We find them in daily newspapers, in magazines, in

DOI: 10.4324/9781003390930-16

collections, and increasingly online and in social media. Readers of newspapers are often familiar with the names of the regular cartoonists and seek them out for their political comments, as do users of social media. Although readership of print newspapers is declining, cartoons continue to be circulated in social media spaces, and most cartoonists have their own webpages and social media profiles. Indeed, social media has probably increased the visibility of some cartoonists given the ease of sharing content online.

In this chapter we explore the generativity of feminist cartoons for environmental education research and teaching through performing duoethnographic methodology and drawing on critical and new materialist feminist theory to draw attention to, and disrupt, some of the conventional thinking around environmental issues. We both position ourselves as feminists, although the feminist theories we draw upon have changed over time, and we share a belief that we need to look at the gender relations that play a significant role in societies' cultural metaphors and the norms of environmental practice (the ideological underpinnings of society), and in the power structures that control and sustain environmental practice (the political-economic underpinnings of society). In particular, our work gives "attention to historical specificity in the production, for women, of subject positions and modes of femininity and their place in the overall network of social power relations" (Weedon 1987, 135) as well as the positioning of the environment. As Giovanna Di Chiro (1987) argues,

> A feminist perspective of environmental education offers a more complete analysis of environmental problems and therefore a better understanding of those problems and their potential solutions. Such an analysis is political, in that it examines how power relations (in, for example, gender, class, race) shape the world in which we live; it asserts that the "polity" (human social world) determines and controls how this social world is and has been historically constructed and organised.
>
> (40)

Environmental cartoons draw attention to power relations and how the 'polity' controls the world, while feminist cartoons draw attention to power relations in the world. By exploring the educational generativity of Judy's cartoons, we hope to "open up new avenues for recognizing the workings of power in the ways we construct our world and its possibilities . . . [and toward] developing more effective social change practices" (Lather 1991, 100). Unsurprisingly then, cartoons are our first theme for discussion.

In her introduction to Judy's first cartoon collection, Dale Spender (1992) wrote of how "the conventional wisdom of a male dominated society has been that women have no sense of humour – therefore they cannot be expected to create comic – or cutting – cartoons" (v). She continued, "so pervasive has been the premise of women's absence of humour, that any suggestion that women can be funny, witty, splendidly subversive or satirical, has itself come to be something of a joke" (v). Thus, she sees Judy's cartoon work as "a fundamental shift in power" (v): "simply by being funny she breaks the established pattern; and by being about authority, by mocking some of the practices of patriarchs, she challenges the very nature of power and the subordination of women" (vi). Feminism, and feminist cartoons and humour form a second theme for this chapter.

A third theme is the relationship between humour and environmental issues. Miriam Kaltenbacher and Stefan Drews (2020), for example, note that "[r]ecently, there has been a growing interest in humorous communication as an unconventional approach to reach people with environmental issues" and that "[s]ome research suggests that the use of humor can be effective when communicating environmental issues, including climate change" (718). An example of such research is the study by Beth Osnes et al. (2019), which found that comedy can help students "to process emotions that allow joy, fun and hope to sustain their commitment to grow as climate communicators" (224). However, we dispute the "recently" because cartoons on environmental issues have been around since at least the 1970s (Starosielski 2011) or even earlier (Murray and Heumann 2007).

The fourth theme is the relationship between humour and formal environmental education. For example, R.L. Garner's (2006) research found that using humour in teaching can have a positive effect on student enjoyment and content retention: "the use of appropriate humor in this study has been shown to enhance the learning environment and has a significantly positive impact on retention of educational materials in a real-world academic setting" (179). More recently, others such as Patrick Chandler et al. (2020) and Kaltenbacher and Drews (2020) also found that humour enhances learning about environmental issues such as climate change.

This chapter explores "the creative process of understanding the [feminist environmentalist] cartoon" (Clarke 2017, 2–3) in environmental education research and teaching across these four themes through a dialogue between an environmentally concerned feminist cartoonist and a feminist environmental educator who sees pedagogical potential in cartoons. It also seeks to identify some possible ways of measuring what impact or influence cartoons about the environment might have once they are in the world and argue for their generativity in environmental education research, theorising and practice.

Methodology

This research is collaborative and dialogic, involving two participants and many cultural artefacts (cartoons). We have adopted duoethnography (Fitzpatrick and Farquhar 2018; Jenssen and Martin 2021; Norris 2008; Norris et al. 2012; Sawyer and Norris 2012; Tlale and Romm 2019) as an appropriate methodology to extend/deepen our interpretations of ourselves engaged in the critical processes of creating and reacting with feminist environmentalist cartoons and associated education practices.[1] As Richard Sawyer and Joe Norris (2012) write:

> A dialogic context in duoethnography is a conversation – not only between people but also between people and their perceptions of cultural artifacts (such as photos) – that generates new meanings.

In duoethnography, two or more researchers work in tandem to dialogically critique and question the meanings they give to social issues and epistemological constructs.

(2)

The narrative that is constructed in a duoethnography is "a scripted conversation with theory blended into the thick description of how two or more individuals experienced the phenomenon they are investigating" (Sawyer and Norris 2012, 77). For us, the phenomenon we are investigating is Judy's cartoons and their generative power for feminism, the environment, and formal environmental education.

The two participants are Judy Horacek and Annette Gough. Judy became a cartoonist in the late 1980s, with cartoons published in a variety of journals, magazines and community publications, and sometimes with regular gigs in major Australian newspapers (see, for example, www.smh.com.au/politics/federal/judy-horacek-20160212-gms7ju.html), as well as on her website (https://horacek.com.au/) and on Facebook (www.facebook.com/judy-horacekcartoons). She published her first book of cartoons in 1992 and nine further volumes of cartoons (1994, 1997, 1998, 1999, 2002, 2007, 2010, 2017, 2019) and created ten children's books, some on her own and some with Mem Fox. For Judy, cartooning fits well with her love of drawing and writing and making people laugh, and it is a wonderful way of having her say about issues that are important to her, such as feminism and the environment. She has recently decided to give up her bi-weekly gig in the major Melbourne newspaper, *The Age*, and the pursuit of work in newspapers, to pursue new collaborations, still with "a whole world to save" (2019, 278). This liminal space, "poised on uncertain ground" (Heilbrun 1999, frontispiece), provides an opportunity to reflect on whether Peter Nicholson (1999) was right when he wrote that Judy's cartoons "are not only our friends today, but they still will be in twenty years' time. They're not ephemeral, they're universal" (x). Annette has been collecting Judy's cartoon books since the first and has used her cartoons many times – to emphasise a point in an argument (Gough 1999b, 2011; Gough and Whitehouse 2020), and as the visual material to support presentations. For Annette, Judy's cartoons capture so much of what needs to be said about both feminist and environmental issues.

As soon as the CFP for this SI came out Annette approached Judy about collaborating on an article. We discussed possibilities by telephone, at which time Judy made it clear that for her, the article should be collaborative, not just an interview. We agreed to proceed with submitting an abstract. Our first task was to exchange responses to "why you do what you do and the way that you do it". Based on these responses, further conversations, and email exchanges, we submitted an abstract for "a dialogue between Judy and Annette about feminist and environmentalist uses of humour, how they are received, and how feminist and environmentalist cartoons may be particularly generative for environmental education research, theorizing and practice, and why". We initially entertained dialogic pedagogy (Archer and Kelen 2015) as a suitable methodology. Annette had previously engaged in collaborative writing (Gough and

Gough 2017) but creating a duoethnographic, scripted conversation fitted our expectations of how we wanted to proceed. In between COVID lockdowns and family crises, over the next few months, we emailed, talked on the telephone, met in person and exchanged writing pieces around our identified themes of cartoons, feminism, environment and education. In this process we revised our thoughts and written pieces many times as we came to know each other better and spurred each other to think other thoughts as we explored personal memories, stories and conversations to illuminate the factors involved in our individual engagement with our themes.

Cartoons

Our first theme was cartoons. We discussed how we came to be interested in cartoons and how we see their potential to change people's minds.

JH: When someone who knew my interests in writing and art suggested cartooning to me, it felt like the perfect thing – a way to have my say about things I felt were important. I was also at that time becoming interested in feminism, suddenly I found I had not only a medium but also content. I felt at that time that cartoons could change peoples' minds – that the medium could offer a different perspective on things that would help them realise what was important, what was wrong – treating half the population as inferior? Destroying forests, the lungs of our planet? Allowing some people to become incredibly rich by keeping others incredibly poor? Cartoons seemed to be a brilliant way of pointing out how these things don't make sense and must be stopped.

AG: I grew up reading comics – the latest Disney comics were a Saturday indulgence as a child, then I moved onto cartoons in newspapers, and gradually came to see their political power. On your latter point, I remember that in your most recent book you reflected, "When I began cartooning, I really did believe that some well-aimed cartoons could change people's minds – that everyone would realise the common sense of things like treating people equally, of not destroying the environment in which we live and on which we depend" (Horacek 2019, 5). Here you remind me of Isabel Galleymore's (2013) argument that "how the comic mode – in its ability to prompt self-identification and self-critique – can offer a corrective function for destructive environmental attitudes" (151).

JH: Obviously, looking at the world as it is now, my cartoons didn't get rid of all the problems. But stepping back from my failure to change the world, have my environmental cartoons had any impact on a smaller scale? They are often reprinted in educational publications and some have become quite well-known. I regularly share cartoons on social media, and it is the ones about the environment that tend to get the most likes and re-shares. But, of course, this latter is a self-selected audience – people follow me because they know I provide a certain sort of content. I have always been aware of the saying "preaching to the converted" and taken comfort in the reply, "The converted need places to pray in too", with the idea that even if the cartoons convince no one new, they help the people already concerned about the environment, bolster their spirits and help them keep fighting. The other applicable phrase is "You have to laugh or you'll cry", which also fits into the bolstering spirits category.

AG: In our introduction we included Spender's (1992) comment about the general belief that women can't be expected to create cutting cartoons because they have no sense of humour. We know women do have a sense of humour, but you mentioned that there is a belief that women cartoonists aren't savage enough in their cartoons. Is this a problem?

JH: I don't think so. While many women do cartoons that aren't vicious, it is abundantly clear that cartoons don't need to be vicious to be effective.

Spender (1992) asserts that women use humour for different ends, "it is one of the strategies for promoting common understandings among women – and for promoting change" by giving "voice to women's views and vision in a succinct and satiric way" (viii). Promoting change in gender relations or relationships with the environment are key messages that are consistent with a (eco) feminist agenda, and environmental education, and feminist cartoons can have a key role in giving voice to these issues.

Feminism

Furthering a feminist agenda is a strong focus in both of our lives and work. We discussed our paths to that place and what we have done in response.

JH: I started drawing cartoons because I wanted to have my say about things that I felt strongly about. The first catalyst was feminism, which I discovered in my mid-twenties. It was only then that I realised clearly that no matter how much women believed they were equal to men and as worthwhile as men, the world wasn't necessarily going to see it that way. So, I started drawing cartoons and calling myself a cartoonist, often specifically "a feminist cartoonist". Along with the territory, for me at least, came concerns with the environment and social justice. It wasn't a change of my central belief system, but a change from thinking "what's the point, we're all going to be blown up anyway" to actually realising that I want to try to change things.

AG: My journey to feminism came through my work in environmental education. Although I had been the sole female (or one of few) in the environmental education field in Australia for much of the 1970s and into the 1980s, for a long time I was unconscious of the absence of female voices in environmental education discourses: an approach that could, perhaps, be caricatured as essentialist and liberal feminist. However, through researching, reading, and writing I increasingly came to recognise that women are one of many marginalised groups absent and/or silenced in the foundational discourses of environmental education, and that multiple subjectivities abound. In the early 1990s I was working on my doctoral thesis, analysing environmental education as a "man-made" subject, and immersing myself in feminist theory. In doing this research and then giving talks about it, I found myself identifying with your feminist cartoons and using them in presentations. My thesis involved feminist poststructuralist analysis (Gough 1994), so Figure 13.1 seemed all about me and my research journey at that time. Another of your cartoons (Figure 13.2) enabled me to talk about social change and equality for women who had been very much in the background in the framing of environmental education (Greenall Gough 1993; Gough 1994, 1997). I used other cartoons from *Unrequited Love N°s 1–100* to illustrate my arguments.

JH: When I started drawing cartoons it was easy to be seen as a feminist cartoonist – all I had to do was put women as main characters in my cartoons. Previously they had only been wives and girlfriends, adjuncts to the male characters. I wasn't the only female cartoonist doing this at the time, there were others as well such as Kaz Cooke (https://kazcooke.com.au/cartoons/). It was a time when women were making huge inroads into all kinds of forms of comedy, and most were using humour for quite political feminist-leaning ends. Women comedians have had more success at making professional inroads into comedy than women

Figure 13.1 Existential dilemma (Horacek 1992, 58), first published in *Australian Society*, Dec-Jan issue, 1989–90.

Figure 13.2 Not a feminist but/Unrequited love No.9 (Horacek 1994, no pagination).

cartoonists; there are very few women who have made a living as a cartoonist. A large part of that is because the profession of cartooning has
actually crumbled around our ears, due to the demise of newspapers
and the ascension of other ways to be funny in visual forms.

AG: But cartoons can still be so powerful to get a message across. The one
between Unrequited Love 70 and 71 is particularly pertinent to discussions of how women are particularly affected by environmental and
social problems – see Figure 13.3. In using the cartoons in this way
I am trying to promote change by drawing attention to the absence of
women in positions of power.

Spender (1990) has argued that the power structure of patriarchy must
be disturbed, and through women undermining language, men are forced to
defend their power in some other way: "If and when sufficient women agree
that they no longer subscribe to the rules and patterns of patriarchy, then
the rules and patterns are likely to be transformed . . . we have the power to
obstruct patriarchy and we can use it" (3–4). Feminist political cartoons are
one way of undermining this language, so they are still needed to draw attention to the inequalities in power in society, because these transformations have
not yet happened. That women have a difficult life in politics, for example, has
been admirably demonstrated recently by events in the Australian parliament
(Palmieri 2021) and by Annabel Crabb (2021) in her Australian Broadcasting

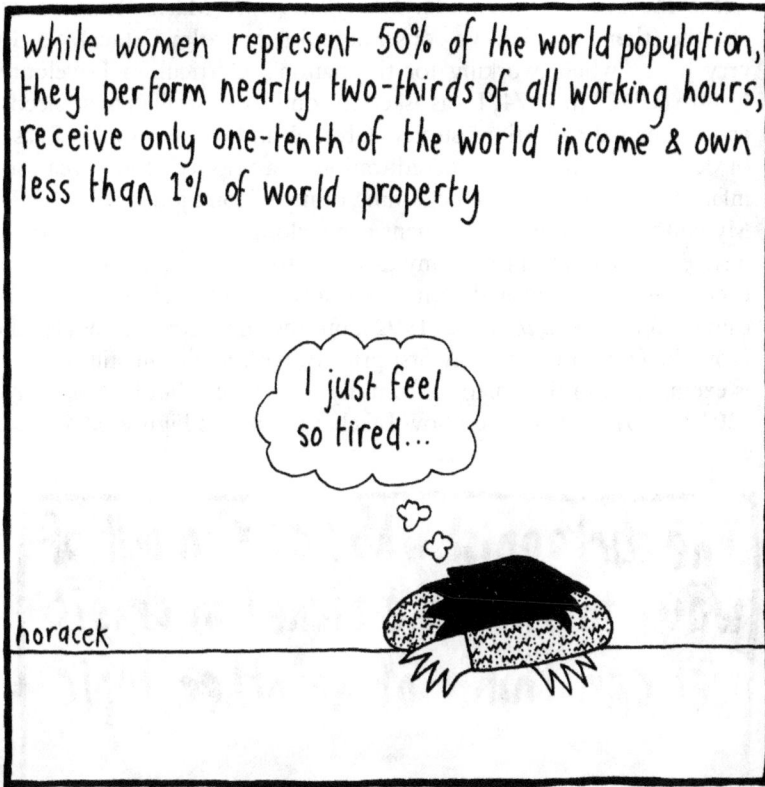

Figure 13.3 Women represent 50% (Horacek 1994, no pagination).

Commission documentary series *Ms Represented*, which, 100 years after Australia elected its very first female parliamentarian, provides a raw and honest account of politics from the female perspective.

The environment

Concern about the state of the environment is another shared theme in our lives and work. We discussed how the environment is such an important focus for us and what we do about it.

AG: I came to a concern for the environment and working in environmental education through a common pathway for early (and probably many current) practitioners and researchers in the field: I completed an undergraduate science degree, majoring in botany and zoology, and a teaching qualification. During the course of my studies, human impact on the environment was often discussed (one of my lecturers wrote *The Last Generation: The End of Survival?* (Martin 1975), another was the (then) well-known conservationist, Malcolm Calder). So, I was very happy when, working for the national Curriculum Development Centre (CDC) in 1974, I was asked to conduct Australia's first "needs for environmental education" survey (Greenall and Womersley 1977) as part of developing environmental education as a priority action area for the infant CDC. This was the beginning of my lifelong journey in the field.

JH: My concern for the environment came along with feminism and social justice, and I wanted to use my cartoons to try to change things. However, there has been a shift in my priorities over the years. Feminism came first in *Life on the Edge* (1992), but the environment was also there. Now the environment is my first priority, with feminism still there. This is exemplified in this cartoon from my most recent book, *Now or Never* (2019, 250), that sums up how I feel at times (see Figure 13.4).

Figure 13.4 No other topic (Horacek 2019, 50), first published in *The Age*, 15 July 2019.

AG: The importance of feminism and social justice in seeking resolution of environmental problems came later for me. Initially I was aware that environmental problems are social problems, as Di Chiro (1987) made clear, but this became more apparent to me in my doctoral research on environmental education as a "man-made subject" (Gough 1994), and both were encapsulated in the set of five principles for research and practice in environmental education that emerged from my research:

- to draw attention to the racism and gender blindness in environmental education and to develop a willingness to listen to silenced voices and to provide opportunities for them to be heard
- to work, individually and collectively, and equally, with other humans and with more-than-human nature rather than separating humans from nature
- to recognise that knowledge is partial, multiple and contradictory
- to develop understandings of the stories of which we are a part and our abilities to deconstruct them
- to recognise resistances to liberatory pedagogy in environmental education and to work with these resistances

JH: Yes, I can recognise many of the things I think are important in these statements. As a very strong recognition of gender blindness, when I first became a cartoonist, there were almost never female main characters in cartoons – women were only adjuncts. Hence my work has always featured women, even if the issue isn't a women's issue. I have also always tried to include people of different races. The subject matter of my cartoons is often the smaller, more silenced voice speaking out against the more powerful force or authority. I also believe that in creating a cartoon, you often make something where people recognise their own experience, and this can be incredibly powerful – 'Oh, it isn't only me who thinks/feels that'. One problem I do have in trying to make cartoons about environmental issues is the temptation to anthropomorphise nature, to make animals and inanimate objects "worthy" of saving because they are just like us. It's a collision of the means that cartooning has at its disposal and a more nuanced political view. Sometimes you just have to have animals talking. I've quite often had the planet itself talking. But I've tried to be strategic about it – using such devices sparingly. Hopefully across the body of my work, it is clear that I don't think human beings are the template for what is most valuable, nor that the planet is only worth saving because we think it is nice.

AG: These principles are consistent with what we would now call feminist new materialism, attempting to 'denature' bodies (human, other, organic and inorganic) as a means to, "establish a radical break with both universalism and dualism as the co-constitutive-ness of cultural discourse and materiality" (Lenz Taguchi 2013, 707). They would also seem to be consistent with your underlying concerns for your cartoons. For example, one of your cartoons that I use frequently in presentations is Figure 13.5, which brings together my two areas of interest of women and environment. Given the huge impact environmental problems, including climate change, are having on women, and that women are taking the lead in environmental protest groups such as Knitting Nannas and School Strike for Climate as well as leading organisations such as the Australian Conservation Foundation, this cartoon communicates the messages beautifully.

AG: Your latest volume of cartoons, *Now or Never* (2019), concludes with a cartoon (see Figure 13.6) that, for me, bookends with Figure 13.5 and continues to visualise your frustration with humans' lack of caring for the earth. It seems to illustrate your change in first priority from feminism and the environment to the environment and humanity.

JH: Global warming and climate change have been a concern for me for some time. For example, back in 2010, the volume, *If You Can't Stand the Heat*, contained many cartoons related to climate change. I wrote, "The issue of climate change is to me, and many people, the most vital and urgent one facing our age" (Horacek 2010, 3) (see Figure 13.7). For my biweekly gig at *The Age*, I did about five cartoons about the environment starting with the words, "The cartoonist who has run out of ways to show the planet in crisis" – Figure 13.6 was the last one. The

Figure 13.5 Babysitter for this (Horacek 1997, 58).

The cartoonist who has run out of ways to show the planet in crisis leaves you with this picture of the earth as the blue-green jewel that is our only home.

And the reminder that it is not actually the planet that is in crisis – it's us.   horacek

Figure 13.6 Blue-green jewel (Horacek 2019, 274), first published in *The Age*, August 2019.

cartoon reflected a very real frustration I had about doing cartoons on the environment, given that I have been doing them for so long, and had tried so many different ways. It is a real representation of how I was feeling, but also was play, as saying I had run out of ways actually led to cartoons.

AG: Around the same time Kate Manzo (2012) was writing of the power of the "geopolitical visions that are communicated in climate change cartoons", while recognising that "the complexities of climate change make its visualisation not only challenging but also necessary" (481). Your various climate change cartoons have become more political and more powerful over time. One in particular that I love is Figure 13.8; it challenges the climate change deniers while showing the facts of rising temperatures. This is one I would use when teaching about climate change.

Figure 13.7 Wrong place wrong time (Horacek 2010, 102), project for Oxfam.

JH: I'm not sure that I would agree that my environmental cartoons have become more political and powerful over time. I'd have to go back and examine them. Often a cartoon is influenced by where it is to be published, what it has been commissioned for. The "wrong place at the wrong time" cartoon (Figure 13.7) was done for an Oxfam report on people in the developing world having to adjust already to the effects of climate change, while richer countries are feeling less effects and also have more of a buffer, given their prosperity. For these cartoons, the subject was a given, I had to find ways to encapsulate the information

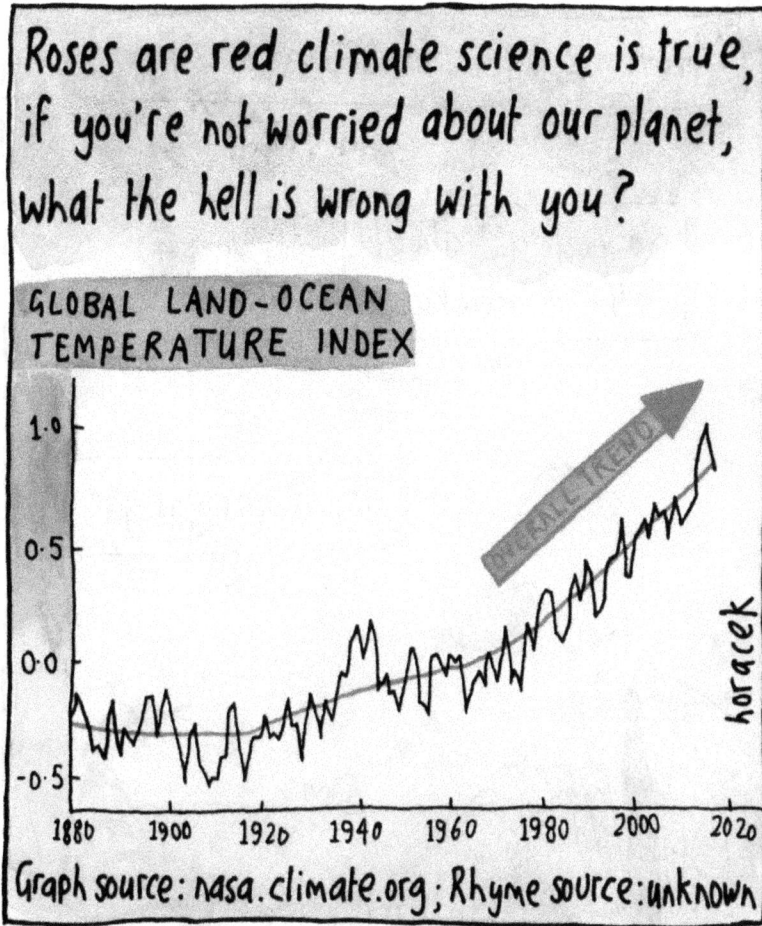

Roses are red, climate science is true,
if you're not worried about our planet,
what the hell is wrong with you?

GLOBAL LAND-OCEAN TEMPERATURE INDEX

OVERALL TREND

horacek

1·0
0·5
0·0
-0·5

1880   1900   1920   1940   1960   1980   2000   2020

Graph source: nasa.climate.org; Rhyme source: unknown

Figure 13.8 Roses are red climate science graph (Horacek 2019, 261), first published in *The Age*, 21 August 2019.

in cartoons. In some ways they are possibly a bit didactic. I think that nowadays the general public's understanding of climate change is much more sophisticated than it was when I first started, when it was often more a case of raising awareness, and so my cartoons have been able to tap into different ways of representation. Using an actual graph in a cartoon? I wouldn't have been able to pull that off thirty years ago.

It was obvious from these discussions that we share a concern to replace the

culture of domination with a world of participatory economics grounded in communalism and social democracy, a world without discrimination

Figure 13.9 Mick Jagger in science class (Horacek 1998, 29), first published in *The Age*, 1990s.

based on race or gender, a world where recognition of mutuality and interdependency would be the dominant ethos, a global ecological vision of how the planet can survive and how everyone on it can have access to peace and well-being.

(hooks 2000, 10)

Judy's concerns are reflected in her cartoons, and Annette's are in her education practices. Both of us operate from a belief that "[c]limate change is rooted in the exploitation and degradation of the planet, peoples, and cultures, which were the foundational principles of colonialism. Rooted in white supremacy, colonialism's impacts on current challenges and solutions to climate change are seldom explored" (Martinez and Irfan 2021, n.p.), and that "ending climate change requires the end of capitalism" (McDuff 2019, n.p.). Cartoons such as Figures 13.3, 13.7 and 13.11 draw attention to the effects of capitalism and colonialism on global warming and women, and provide an opportunity "to engage deeply with and confront historical patterns in concrete pedagogical practices in order to interrupt our own epistemic, political, ethical, and strategic place and categories" (Sund and Pashby 2020, 156) either in classrooms or as part of "the creative process of understanding the [feminist environmentalist] cartoon" (Clarke 2017, 2–3).

Formal environmental education

Cartoons can be (are?) a form of informal education through newspapers and other media, pointing out how some things don't make sense, and must be stopped. They can stimulate discussion in informal contexts, but also in formal education. We discussed how cartoons can be and are used in formal education contexts.

JH: Quite a number of my cartoons place environmental destruction into a political context – indifferent governments, greedy companies, capitalism, misplaced valuing of profit over the environment. These cartoons aren't simply about some aspect of the environment, they come with a worldview. If a teacher [were] to use them in a lesson, it would more likely be a history or politics lesson than a science one because you don't learn much about scientific principles from Mick Jagger saying, "It's a gas gas gas" (Figure 13.9), but you can start off meaningful discussion using cartoons such as Figures 13.3 and 13.5 in a history, geography or politics class. But that doesn't mean science teachers don't use my cartoons, or that they are humourless. My best teachers at school were all science teachers; admittedly that is a long time ago, but I remember they made the subject sing and engaged my interest in a way the history teachers did not. A good thing about being a cartoonist is that I don't have to separate things into disciplines, the cartoon is whatever is in my mind at the time.

AG: I would like to see your cartoons being used by science teachers as much as social science teachers. Science is too often seen and taught as a very dry and humourless subject. Yet it does need to be taught differently if we are to engage students in environmental issues and with STEM. So many schools are in desperate need of science teachers like the ones you had!

In the early development of the concept of environmental education science was seen as the appropriate home for it, but others, such as Arthur Lucas

(1980), expressed concern that "too many science educators seem to believe that their discipline is the vehicle for environmental education" (1), and Ian Robottom (1983) called science a limited vehicle for environmental education. The basis for their arguments was that science educators have difficulties with teaching the sociopolitical aspects of science topics. As feminist scientist Marion Namenwirth (1986), succinctly states, "Scientists firmly believe that as long as they are not *conscious* of any bias or political agenda, they are neutral and objective, when in fact they are only unconscious" (29, italics in original). Environmental issues are complex and require a multidisciplinary approach to resolve them, or even discuss them.

JH: Some cartoons take complex information and distil it, and there are obvious educational implications to this. I wanted to do a cartoon about the albedo effect,[2] and came up with one I call "Hysterical penguins . . ." (see Figure 13.10). The cartoon provides a simple description of the albedo effect, and gives it an emotional dimension. It's funny but it's also not funny.

AG: I would see this as a great way of introducing a lesson on the albedo effect. Just like I would use your Mick Jagger cartoon (Figure 13.9) in a science class to introduce a unit on gases (or make a point in an academic article (Gough 2011)). Chemistry can be boring for some students, so engaging them in a topic with a laugh is a good start.

JH: Mick Jagger and the song are from last century so would this cartoon still work in a high school classroom? The referents of cartoons can often date. On the other hand, I wish some of my cartoons had gone out of date more, such as the one I call "Global warming" – the cartoon shows a woman on the phone sitting on the roof of a grass-thatched hut with water rising around her (Figure 13.11). This cartoon is from the 1990s, but the only thing that has dated is the phone she is using; it has a curly cord attaching the handset.

AG: I have used this cartoon so often in presentations! And your message is so right – we are still waiting for our national government to give a damn about the environment.

JH: One thing cartoons can do, that this cartoon does, is turn a dry fact into something else. Maybe something more memorable, maybe something that makes you laugh and therefore relax and take in more information, maybe something that adds narrative to a dry scientific process.

AG: Exactly. Science can be dry facts so you have to engage students! I'm not sure I used cartoons when teaching high school students in the mid-1970s, which is consistent with John Banas et al.'s (2011) observation that "[c]ompared to college instructors, high- and intermediate-school teachers have been found to use slightly less humor in their classes" (121). But since finding your cartoons in the academic phase of my career, I have continued to use them over the years in the belief that using humour in teaching can have a positive effect on student enjoyment and

Figure 13.10 Hysterical Penguins (Horacek 2010, 7).

content retention (Garner 2006; Sambrani et al. 2014), and on student's emotional engagement (Hoad et al. 2013). I also use them to emphasise critical points. As Maxwell Boykoff and Beth Osnes (2019) argue, "comedy and humor potentially exert power to impact new ways of thinking/acting about anthropogenic climate change . . . [and] experiential, emotional, and aesthetic learning can inform scientific ways of knowing" (154).

I would certainly be using your cartoons about climate change, species extinction, and pollution, for example, if I were to teach in high schools nowadays – not only to get engagement through a laugh but to make key points about the effects of rising sea level, lack of political action and impacts on other living things to encourage curiosity

Figure 13.11 Global Warming (Horacek 2007, 7), first published in *The Australian Magazine* 12 May 2001.

and learning. Visual illustrations ("Use of pictures or items expected to promote humor") are one of 23 different forms of humour used in educational settings discussed by Banas et al. (2011, 123–124). They also include cartoons in the category of "Using funny props such as cartoons, water pistols, funny cards, etc." (124). But I think your cartoons are more than funny props.

JH: I think there is a difference between cartoons that are funny props and the cartoons that feel important to me to do. The Mick Jagger cartoon

I would consider a funny prop, the hysterical penguin cartoon I feel is something more. I'm also thinking that any teacher using water pistols might be seen by the students as trying too hard to be fun and likeable? It would depend on what the lesson was about, but it sounds almost in the wearing-a-fright-wig category.

I know a lot of teachers do use my cartoons in classes. And they have often been published in educational textbooks; I know because educational publishers do ask permission. Clearly the authors of the textbooks feel that they are of use for educational purposes, whether as discussion points or illustrative "texts". My cartoons have also been used on exam papers, along the lines of, "This cartoon seems to be saying blah blah. What do you think? Are there other ways of interpreting it?" I am also often told of my cartoons being stuck on doors and walls at universities, seen by friends and acquaintances.

AG: I am guilty of putting your cartoons on my university office door! Although I used your cartoons in conference presentations and teaching during the mid-1990s, I first asked for permission to use one of your cartoons in print for an article in this journal in 1999 (see Figure 13.12). Of course, in these pandemic times, bushwalks in the city have become the norm for us in Melbourne, which is something you would not have anticipated in 1998.

JH: From the beginning, I have always let people use my work in classes and talks for nothing. Because I am trying to make a living in the precarious world of freelancing, I try to charge for people publishing my work, but for those doing presentations, I simply ask if they can acknowledge that the work is mine. (Anecdotally I am pretty sure a lot of people use my work without asking too.) It is important to me that the work is out in the world, and if someone is wanting to use it to try to teach people, so much the better. If my cartoons can help someone teach about environmental issues, that is wonderful.

Although we do not have quantified evidence of how Judy's cartoons have been used in educational contexts, there have been many publishers and examination authorities who have sought formal permission to reproduce the cartoons in educational materials, particularly environment-related materials. Judy also has extensive anecdotal evidence that her cartoons are frequently used in classrooms and lecture theatres, such as in Annette's own environmental education practices, because of their perceived power in consciousness-raising, informing and encouraging action.

Conclusion

While cartoonists (and many educators) believe cartoons have power in consciousness-raising and as change agents, it is difficult to know whether

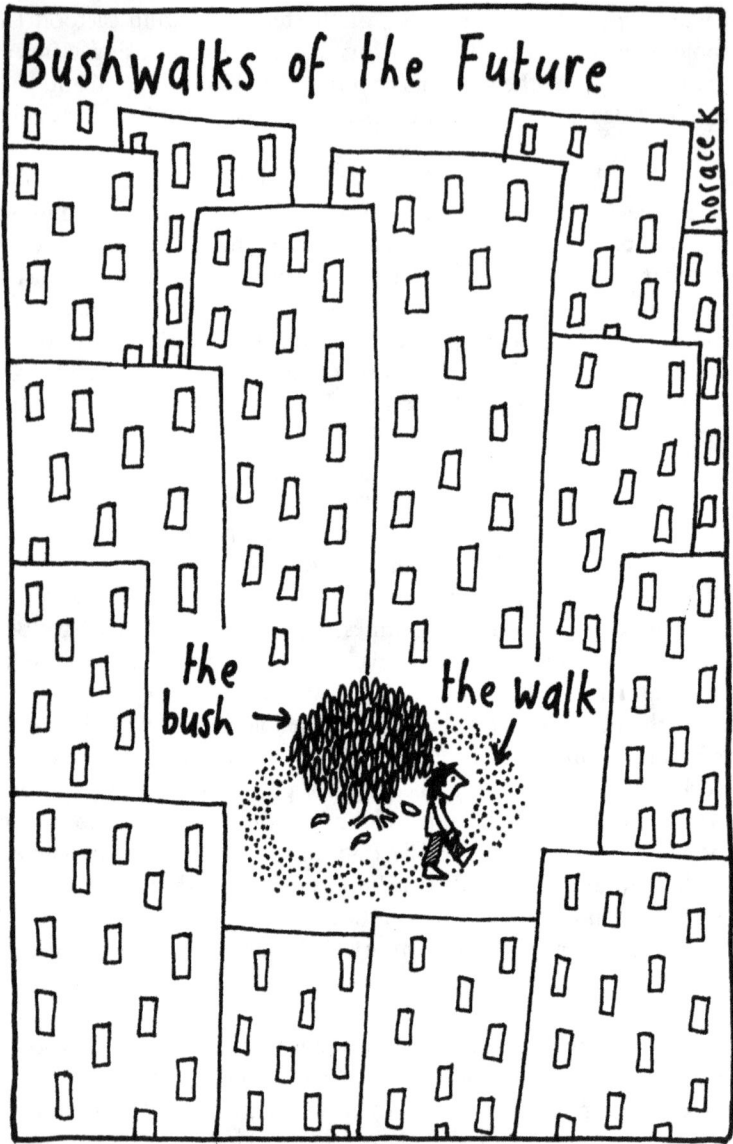

Figure 13.12 Bushwalks of the future (Horacek 1998, 82), first published in *The Age* 1980s.

they do have this power, and how we can tell. There is research that shows that using humour in teaching can have a positive effect on the facilitation of learning (Seidman and Brown 2013), student enjoyment and content retention (Garner 2006), and promoting more relaxed learning environments that support interactions among students (Tziatis 2012). This humour does not necessarily involve the visual, but research by Sambrani et al. (2014) found that visual material presented in a humorous manner is remembered better than equivalent visual non-humorous material.

Political cartoons are an established form of political satire and visual communication, but there has been little research about their societal and political impact. However, a recent study by Joop Van Holsteyn (2021) concluded that cartoons as a manifestation of political incivility do impact negatively on the evaluations of politicians; however, the effects are minor and more research is needed. Part of the problem is in finding an academic home for their study. Manning and Phiddian (2008, in van Holsteyn 2021) raise the question, "Do cartoons belong to art historians, political scientists, students of media, or interdisciplinary scholars of satire? It's not clear who 'owns' them, so there is no clear, pre-existing intellectual framework to shape debate" (219). Thus, van Holsteyn concludes that "knowledge of the impact of cartoons is predominantly based on speculation and anecdotal evidence" (219). This lack of an academic home is also reflected in what subjects in schools make use of political/ feminist/environmental cartoons.

Asserting a feminist environmentalist perspective through cartoons in educational contexts is central to our focus here. Male domination of the cartoon industry is longstanding, and males dominated the conceptualisations of environmental education (Gough 1999a). While women are now more involved in environmental education, the marginalisation of gender in the field continues (Gough 2021, Gough and Whitehouse 2019, 2020, Gough et al. 2017). Thus, Judy's cartoons perform a key role in drawing attention to both feminist and environmentalist perspectives for educational contexts.

Beyond this, our discussion has highlighted that there is a need for research into the impact of cartoons beyond speculation and anecdotal evidence. But the evidence that we have is that Judy's cartoons "are not only our friends today, but they still will be in twenty years' time. They're not ephemeral, they're universal" (Nicholson 1999, x). It is not every cartoon but a number of the cartoons in the book in which Peter Nicholson made that statement, and other cartoons from Judy's earlier books, are still being used by educators and activists who recognise these cartoons' educational and motivational power. The existing cartoons are put on all sorts of merchandise by Redbubble; there are greeting cards and calendars available through bookshops, and prints for sale through her website. The images are sought after and are generative.

Interpreting cartoons is a creative process between the cartoonist and readers, and cartoons can be used in educational contexts to provoke new thoughts and challenge ideas as part of unconventional approaches to engaging students in learning. Feminist environmental cartoons add another dimension by challenging prevailing viewpoints and encouraging other ways of thinking. This is their power and generativity for teaching and learning.

Disclosure statement

The authors report there are no competing interests to declare.

Acknowledgement

Judy and Annette acknowledge and pay respects to the Traditional Custodians of Melbourne/Naarm, the Wurundjeri and Boon Wurrung peoples of the Kulin nation, on whose unceded lands they live and work.

Biographies

Annette Gough is Professor Emerita of Science and Environmental Education in the School of Education at RMIT University, Melbourne, Australia. She has been an adjunct/visiting professor at universities in Canada, South Africa and Hong Kong, and is a Life Fellow of the Australian Association for Environmental Education (since 1992). She has led research and development projects funded by the Australian and Victorian governments and non-government bodies, worked with UNESCO, UNEP and UNESCO-UNEVOC, and has been co-editor of the *Australian Journal of Environmental Education*. She has over 150 publications and is an editorial board member for the *Australian Journal of Environmental Education, Environmental Education Research* and the *Journal of Biological Education* and co-editor of the Springer book series *International Explorations in Outdoor and Environmental Education*. Her research interests span environmental, sustainability and science education, research methodologies, posthuman and gender studies.

Facebook: www.facebook.com/profile.php?id=100009634599791

LinkedIn: www.linkedin.com/in/annette-gough-76b2431b/

Twitter: @Zyxo99

Judy Horacek is a freelance Australian cartoonist, writer and illustrator based in Melbourne, Australia. Her cartoons have been published widely in newspapers, journals and books both in Australia and overseas, including *The Age, The Australian, The Canberra Times* and *Ms Magazine*, and have appeared in hundreds of books and journals and campaigns. She has published ten collections of her cartoons and has twice been shortlisted for the Walkley Award for Best Cartoon. Judy also creates children's picture books, both on her own and with Mem Fox, including the incredibly successful *Where Is the Green Sheep?* She has an Honours Arts degree from Melbourne University in Fine Arts and English Literature, and an Arts Degree in Printmaking and Drawing from Australian National University. She regularly exhibits her watercolours and limited edition prints.

Facebook: www.facebook.com/judyhoracekofficial/

Facebook: www.facebook.com/judyhoracekcartoons

Instagram: @judyhoracek, www.instagram.com/judyhoracek/

Twitter: @judyhoracek

Website: https://twitter.com/judyhoracek

Notes

1 While this is not the place to make comparisons, we do note similarities between collaborative writing as a method of inquiry as used by Ken Gale, Jonathan Wyatt and others including Annette Gough and Noel Gough (see e.g. Gale and Wyatt 2008, 2009, 2013, 2017, 2021; Gough and Gough 2017; Wyatt and Gale 2011, 2018; Wyatt et al. 2010, 2011, 2014, 2017) and duoethnography as a methodology we are implementing here.
2 The albedo effect is an expression of the ability of surfaces to reflect sunlight (heat from the sun). Light-coloured surfaces return a large part of the sunrays back to the atmosphere (high albedo). Dark surfaces absorb the rays from the sun (low albedo). This has a significant impact on our climate. The lower the albedo, the more radiation from the sun that gets absorbed by the planet, and temperatures will rise. If the albedo is higher, and the Earth is more reflective, more of the radiation is returned to space, and the planet cools.

References

Archer, Carol, and Christopher Kelen. 2015. "Dialogic pedagogy in creative practice: a conversation in examples." *Pedagogy, Culture & Society* 23 (2): 175–202. doi:10.1 080/14681366.2014.932301

Banas, John A., Norah Dunbar, Dariela Rodriguez, and Shr-Jie Liu. 2011. "A review of humor in educational settings: four decades of research." *Communication Education* 60 (1): 115–144. doi:10.1080/03634523.2010.496867

Boykoff, Maxwell, and Beth Osnes. 2019. "A laughing matter? Confronting climate change through humor." *Political Geography* 68: 154–163. doi:10.1016/j.polgeo.2018.09.006

Chandler, Patrick, Beth Osnes, and Maxwell Boykoff. 2020. "Creative climate communications: teaching from the heart through the arts." In *Teaching Climate Change in the United States*, edited by Joseph Henderson, and Andrea Drewes, 172–185. Abingdon, UK: Routledge. doi:10.4324/9780367179496-12

Clarke, John. 2017. "Foreword." In *Random Life*, edited by Judy Horacek, 1–3. East Brunswick: Horacek Press.

Crabb, Annabel. (2021). *Ms Represented*. https://iview.abc.net.au/show/ms-represented-with-annabel-crabb

Di Chiro, Giovanna. 1987. "Environmental education and the question of gender: a feminist critique." In *Environmental Education: Practice and Possibility*, edited by Ian Robottom, 23–48. Geelong: Deakin University.

Fitzpatrick, Esther, and Sandy Farquhar. 2018. "Service and leadership in the university: Duoethnography as transformation." *Journal of Organizational Ethnography* 7 (3): 345–360. doi:10.1108/JOE-08-2017-0037

Gale, Ken, and Jonathan Wyatt. 2008. "Two men talking: a nomadic inquiry into collaborative writing." *International Review of Qualitative Research 1* (3): 361–380. doi:10.1525/irqr.2008.1.3.361

Gale, Ken, and Jonathan Wyatt. 2009. *Between the Two: A Nomadic Inquiry into Collaborative Writing and Subjectivity.* Newcastle: Cambridge Scholars Publishing.

Gale, Ken, and Jonathan Wyatt. 2013. "Assemblage/ethnography: troubling constructions of self in the play of materiality and representation." In *Contemporary British Autoethnography*, edited by Nigel P. Short, Lydia Turner, and Alec Grant, 139–155. Rotterdam: Sense. doi:10.1007/978-94-6209-410-9_9

Gale, Ken, and Jonathan Wyatt. 2017. "Working at the wonder: Collaborative writing as method of inquiry." *Qualitative Inquiry 23* (5): 355–364. doi:10.1177/1077800416659086

Gale, Ken, and Jonathan Wyatt. 2021, online. "Making trouble with ontogenesis: Collaborative writing, becoming, and concept forming as event." *Qualitative Inquiry*. 1–8. doi:10.1177/10778004211026898

Galleymore, Isabel. 2013. "A dark ecology of comedy: environmental cartoons, Jo Shapcott's Mad Cow poems and the motivational function of the comic mode." *Green Letters 17* (2): 151–163. doi:10.1080/14688417.2013.800336

Garner, R. L. 2006. "Humor in pedagogy: How ha-ha can lead to aha!" *College Teaching 54* (1): 177–180. doi:10.3200/CTCH.54.1.177-180

Gough, Annette. 1994. "Fathoming the fathers in environmental education: A feminist poststructuralist analysis." Unpublished PhD diss., Faculty of Education, Deakin University.

Gough, Annette. 1997. *Education and the Environment: Policy, Trends and the Problems of Marginalisation.* Melbourne, Victoria: Australian Council for Educational Research.

Gough, Annette. 1999a. "Recognising women in environmental education pedagogy and research: Toward an ecofeminist poststructuralist perspective." *Environmental Education Research 5* (2): 143–161. doi:10.1080/1350462990050202

Gough, Annette. 1999b. "Kids don't like wearing the same jeans as their mums and dads: So whose 'life' should be in significant life experiences research?" *Environmental Education Research 5* (4): 383–394. doi:10.1080/1350462990050404

Gough, Annette. 2011. "(W)rapping relationships between science education and globalization." *Cultural Studies of Science Education 6* (1): 77–88. doi:10.1007/s11422-010-9310-6

Gough, Annette. 2021. "Listening to voices from the margins: Transforming environmental education." In *Queer Ecopedagogies: Explorations in Nature, Sexuality, and Education*, edited by Joshua Russell, 161–181. Cham, Switzerland: Springer. doi:10.1007/978-3-030-65368-2_9

Gough, Annette, and Noel Gough. 2017. "Beyond cyborg subjectivities: Becoming-posthumanist educational researchers." *Educational Philosophy and Theory 49* (11): 1112–1124. doi:10.1080/00131857.2016.1174099

Gough, Annette, Connie Russell, and Hilary Whitehouse. 2017. "Introduction: Moving gender from margin to center in environmental education." *The Journal of Environmental Education*, 48 (1): 5–9.

Gough, Annette, and Hilary Whitehouse. 2019. Centering gender on the agenda for environmental education research. *The Journal of Environmental Education*, 50 (4–6): 332–347. doi:10.1080/00958964.2019.1703622

Gough, Annette, and Hilary Whitehouse. 2020. "Challenging amnesias: Recollecting feminist new materialism/ecofeminism/climate/education." *Environmental Education Research 26* (9–10): 1420–1434. doi:10.1080/13504622.2020.1727858

Greenall, Annette, and Jon C. Womersley, eds. 1977. *Development of Environmental Education in Australia – Key Issues*. Canberra, ACT: Curriculum Development Centre.

Greenall Gough, Annette. 1993. *Founders in Environmental Education*. Geelong: Deakin University Press.

Heilbrun, Carolyn G. 1999. *Women's Lives: The View from the Threshold*. Toronto, Canada: University of Toronto Press. doi:10.3138/9781442657557

Hoad, Colin, Craig Deed, and Alison Lugg. 2013. "The potential of humor as a trigger for emotional engagement in outdoor education." *Journal of Experiential Education 36* (1): 37–50. doi:10.1177/1053825913481583

hooks, bell. 2000. *Feminism is for Everybody: Passionate Politics*. Cambridge, MA: South End Press.

Horacek, Judy. 1992. *Life on the Edge*. North Melbourne: Spinifex.

Horacek, Judy. 1994. *Unrequited Love: Nos. 1–100*. Ringwood: McPhee Gribble.

Horacek, Judy. 1997. *Woman with Altitude*. South Melbourne: Hyland House.

Horacek, Judy. 1998. *Lost in Space*. St Leonards: Allen & Unwin.

Horacek, Judy. 1999. *If the Fruit Fits . . .* Sydney: Hodder.

Horacek, Judy. 2002. *I am Woman, Hear Me Draw*/Cartoons from the Pen of Judy Horacek. Canberra: National Museum of Australia.

Horacek, Judy. 2007. *Make Cakes Not War*. Carlton North: Scribe.

Horacek, Judy. 2010. If You Can't Stand the Heat. Carlton North: Scribe.

Horacek, Judy. 2017. *Random Life*. Brunswick East: Horacek Press.

Horacek, Judy. 2019. *Now or Never*. Brunswick East: Horacek Press.

Jenssen, Runa Hestad, and Ros Martin. 2021. "A tale of grappling: Performative duoethnography as expanded methodological thinking." *Reconceptualizing Education Research Methodology 12* (2): 59–82. doi:10.7577/rerm.4683

Kaltenbacher, Miriam, and Stefan Drews. 2020. "An inconvenient joke? A review of humor in climate change communication." *Environmental Communication 14* (6): 717–729. doi:10.1080/17524032.2020.1756888

Lather, Patti. 1991. *Getting Smart: Feminist Research and Pedagogy With/in the Postmodern*. New York, London: Routledge.

Lenz Taguchi, Hillevi. 2013. "Images of thinking in feminist materialisms: Ontological divergences and the production of researcher subjectivities." *International Journal of Qualitative Studies in Education 26* (6): 706–716. doi:10.1080/09518398.2013.788759

Lucas, Arthur M. 1980. "Science and environmental education: Pious hopes, self praise and disciplinary chauvinism." *Studies in Science Education 7*: 1–26. doi:10.1080/03057268008559874

Manning, Haydon, and Robert Phiddian. 2008. "Introduction: Controversial images." In *Comic Commentators: Contemporary Political Cartooning in Australia*, edited by Haydon Manning, and Robert Phiddian, 1–9. Perth: Network Books, pp. 1–9.

Manzo, Kate. 2012. "Earthworks: The geopolitical visions of climate change cartoons." *Political Geography 31*: 481–494. doi:10.1016/j.polgeo.2012.09.001

Martin, Angus. 1975. *The Last Generation: The End of Survival?* Glasgow, UK: Fontana.

Martinez, Deniss, and Ans Irfan. 2021, 4 November. "Colonialism, the climate crisis, and the need to center indigenous voices." *Environmental Health News.* www.ehn. org/indigenous-people-and-climate-change-2655479728.html

McDuff, Phil. 2019, 18 March. "Ending climate change requires the end of capitalism. Have we got the stomach for it?" *The Guardian.* www.theguardian.com/ commentisfree/2019/mar/18/ending-climate-change-end-capitalism

Murray, Robin L., and Joseph K. Heumann. 2007. "Environmental cartoons of the 1930s, '40s, and '50s: A critique of post-World War II progress?" *Interdisciplinary Studies in Literature and Environment* 14 (1): 20, 51–69. JSTOR, www.jstor.org/ stable/44086557. Accessed 19 July 2021. doi:10.1093/isle/14.1.51

Namenwirth, Marion. 1986. "Science seen through a feminist prism." In *Feminist Approaches to Science*, edited by Ruth Bleier, 18–41. New York: Pergamon.

Nicholson, Peter. 1999. "Introduction." In *If the Fruit Fits. . .*, edited by Judy Horacek, pp. ix–xi. Sydney, NSW: Hodder.

Norris, Joe. 2008. "Duoethnography." In *The SAGE encyclopedia of qualitative research methods*, Volume 1, edited by Lisa M. Given, 233–236. Los Angeles, CA: SAGE. doi:10.4135/9781412963909.n123

Norris, Joe, Richard D. Sawyer, and Darren Lund, eds. 2012. *Duoethnography: Dialogic Methods for Social, Health, and Educational Research.* New York: Routledge.

Osnes, Beth, Maxwell Boykoff, and Patrick Chandler. 2019. "Good-natured comedy to enrich climate communication." *Comedy Studies* 10 (2): 224–236. doi:10.1080/ 2040610X.2019.1623513

Palmieri, Sonia. (2021, 2 December). "Australia left behind on parliamentary gender equality reform." *The Sydney Morning Herald.* www.smh.com.au/national/australia-left-behind-on-parliamentary-gender-equality-reform-20211201-p59dt5.html

Robottom, Ian. 1983. "Science: A limited vehicle for environmental education." *Australian Science Teachers Journal* 29 (1): 27–31. doi:10.1017/S081406260000464X

Sambrani, Tanvi, Sneha Mani, Maureen Almeida, and Ewgeni Jakubovski. 2014. "The effect of humour on learning in an educational setting." *International Journal of Education and Psychological Research* 3 (3): 52–55.

Sawyer, Richard D., and Joe Norris. 2012. *Duoethnography.* Oxford, New York: Oxford University Press. doi:10.1093/acprof:osobl/9780199757404.001.0001

Seidman, Alan, and Stephen C. Brown. 2013. "College classroom humour: Even the pundits can benefit." *Education* 133 (3): 393–395.

Spender, Dale. 1990. *Man Made Language.* Second edition. London, UK: Pandora.

Spender, Dale. 1992. "Introduction." In *Life on the Edge*, edited by Judy Horacek, pp. v–ix. North Melbourne: Spinifex.

Starosielski, Nicole. 2011. "'Movements that are drawn': A history of environmental animation from The Lorax to FernGully to Avatar." *International Communication Gazette* 73 (1–2): 145–163. doi:10.1177/1748048510386746

Sund, Louise, and Karen Pashby. 2020. "Delinking global issues in northern Europe classrooms." *The Journal of Environmental Education* 51 (2): 156–170. doi:10.108 0/00958964.2020.1726264

Tlale, Lloyd D.N., and Norma R.A. Romm. 2019. "Duoethnographic storying around involvements in, and extension of the meanings of, engaged qualitative research." *Forum Qualitative Sozialforschung/Forum: Qualitative Social Research* 20 (1). doi:10.17169/fqs-20.1.3085.

Tziatis, Daniel. 2012. "Humour as an outdoor educator's tool." *Pathways: The Ontario Journal of Outdoor Education* 24 (2): 35–36.

Van Holsteyn, Joop. 2021. "Rude by nature? Political cartoons and political incivility." In *Political Incivility in the Parliamentary, Electoral and Media Arena: Crossing Boundaries*, edited by Annemarie S. Walter, 219–235. Abingdon, UK: Routledge. doi:10.4324/9781003029205-12

Weedon, Chris. 1987. *Feminist Practice and Poststructuralist Theory.* Oxford, UK: Blackwell.

Wyatt, Jonathan, and Ken Gale. 2011. "The textor, the nomads, and a labyrinth: A response to Graham Badley." *Qualitative Inquiry 17* (6): 493–497. doi:10.1177/1077800411409880

Wyatt, Jonathan, and Ken Gale. 2018. "Writing to it: Creative engagements with writing practice in and with the not yet known in today's academy." *International Journal of Qualitative Studies in Education 31* (2): 119–129. doi:10.1080/09518398.2017.1349957

Wyatt, Jonathan, Ken Gale, Susanne Gannon, and Bronwyn Davies. 2010. "Deleuzian thought and collaborative writing: a play in four acts." *Qualitative Inquiry 16* (9): 730–741. https://doi.org/10.1177/1077800410374299

Wyatt, Jonathan, Ken Gale, Susanne Gannon, and Bronwyn Davies. 2011. *Deleuze and Collaborative Writing: An Immanent Plane of Composition.* Peter Lang, New York.

Wyatt, Jonathan., Ken Gale, Susanne Gannon, and Bronwyn Davies. 2017. "Creating a space in between: Collaborative inquiries." In *Handbook of qualitative research*, edited by Norman K. Denzin and Yvonna S. Lincoln, 738–756. SAGE, London, England.

Wyatt, Jonathan., Ken Gale, Susanne Gannon, Bronwyn Davies, Norman K. Denzin, and Elizabeth A. St. Pierre. 2014. "Deleuze and collaborative writing: Responding to/with 'JKSB'." *Cultural Studies ↔ Critical Methodologies 14* (4): 407–416. https://doi.org/10.1177/1532708614530313

14 Cyborg subjectivities and liminal experiences

Annette Gough

Gough, A. "Body/mine: A chaos narrative of cyborg subjectivities and liminal experiences." *Women's Studies*, *34*(3–4): 249–264.2005. doi:10.1080/00497870590964147

Abstract

This is a story about how I came to construct my postoperatively scarred body as a mine site, my liminal experiences of reconstructively normalising its appearance as a cyborg, and what that means to me, conceptually and physically, as a feminist poststructuralist researcher interested in how the body of a theorist performs and is (re)presented within theoretical spaces. My story takes the form of a mitigated "chaos narrative" that questions linear modernist medical discourses of rehabilitation and restitution by discussing the chaotic dynamics of my cancer experiences, and it relates concepts of cyborg subjectivities and other feminist poststructuralist work to my corporeal body and the body of my research in environmental education.

A mastectomy does not put an end to an encounter with cancer. It does not guarantee a return to normality as in a restitution narrative, but, rather, it is a liminal experience that "announces the beginning of an even more frightening confrontation with life" (Bahar 2003, 1036). Carolyn Heilbrun eloquently describes this state of transition:

> Eve has been supposed to remark to Adam as they left the garden, my dear, we are in a state of transition, and of course they were. It is no coincidence that Eve delivers this line. While humanity in every era and stage in history has been marked by a strong sense of itself as being in a state of transition, women have always had a particularly close relationship to changeable terrain. In their quest for self knowledge, boundaries, and names, women have found themselves between varying cultural demands. In one view, perhaps the dominant one, the only way to gain positive status is to fit appropriately into approved categories: appropriately beautiful, appropriately young, appropriately thin, appropriately successful. In another view, women must abandon the appropriate and

DOI: 10.4324/9781003390930-17

seek out the liminal. The word limen means threshold. To be in a state of liminality is to be poised upon uncertain ground, on the brink of leaving one condition or country or self to enter upon another. When recognized, liminality offers women freedom to become themselves.

(Heilbrun, frontispiece)

In August 2001, I was in such a state of transition – poised on uncertain ground – and I needed to recognize the freedom that abandoning the appropriate and seeking out liminality might give me. But I was then still under the superficially benevolent colonialist administration of medical science. My body was part of its territory, and the disciplinary power of its gentle coercions towardsnormalization was very powerful. I simply felt colonized, oppressed, and powerless, and I accepted the medicalized discourses as if they were my own. As a "wounded storyteller" (Frank 1995), I was not yet in a position to respond to Arthur Frank's assertion that "illness is a call for stories: the body needs a voice which disease and illness take away" (quoted in Langellier 146), and reclaim the capacity to tell and hold on to my story. This narrative, a feminist poststructuralist critique of my experiences of body modification, is part of my reclamation process.

I have been an environmental educator for 30 years, and a recurring theme in my curriculum development work has been the impact of mining on the environment, and what "restoring the land" means after "Man" has removed a prized resource from various locations. It never occurred to me that my body could also become a mine site. But in August 2001, doctors detected a rare ore body, Paget's disease of the nipple,[1] on my living body and rushed to excavate the site. During this process I was reminded of, and began to identify with, a story told to me by Denise Spencer (1996), my former graduate student whose Master of Education thesis was a history of the Gove (North East Arnhem Land, Australia) bauxite mine from the Yolngu women's perspective. The Yolngu women told Spencer that the land is never the same after it has been mined, no matter how carefully the mining company "restores" it. This story tells how I came to terms with my postoperatively scarred body as a mine site that I saw as needing reconstruction to normalize my appearance, and what that entailed theoretically (as a feminist poststructuralist researcher) and physically (as an aging heterosexual female). Like Sarah Squire (55), I am interested in "how the body of the theorist is presented within theoretical spaces" when one is using a poststructuralist approach to research, and thus I respond to her question: "what is the place of personal experience in contemporary feminist writing on the body?" Like Squire, I come to this space through what Donna Haraway (1997, 12) calls a science-fictional wormhole, a "spatial anomaly that casts travellers into unexpected regions." My wormhole is my own illness experiences (Gough 2002), and now that I am here I want to explore "the possibilities of theorizing within stories instead of about them" (Bochner 141), and in so doing to rethink my own educational research work.

This is not a "woman as victim" or "survivor" story, nor is it an adventure story (or quest narrative), but closer to what Frank (1995) calls a "chaos narrative" that questions linear modernist medical discourses of rehabilitation and restitution by focusing on the chaotic dynamics of my cancer experiences. However, I recognize that chaos narratives can be difficult to read and understand, and I have thus produced a story that shuttles between "a linear chronology of recovery and progress" that seeks coherence and an invitation to "reflect on how experience with disease molds our narratives as well as on how narrative fashions and determines this experience" (Bahar 1032). In writing this story, I enact Donna Haraway's (1985, 1991) cyborg manifesto, relating her arguments and cyborg imagery to both my corporeal body and the body of my own research. As Danielle Devoss (835) notes, many researchers see a cyborg as a "border zone between social and body reality – a zone where postgenderedness is a possibility, a zone free of the boundaries of private and public," but my interest is to be one of the "few theorists [who] have approached the cyborg as a physical reality" as a result of my liminal experiences.

Although poststructuralism considers subjectivity to be nonunitary and discursively constituted, produced rather than already and unproblematically present, some feminist (including poststructuralist) and other researchers have argued for locating the personal experiences of the researcher in research stories (e.g., Probyn 1993; Squire 2002) and for the value of personal narratives in research (Bahar 2003; Bochner 2001; Crawley 2002; Frank 2000; Langellier 2001; MacLure 2000; Richardson 2001). For example, Elspeth Probyn (5) writes: "Underlying this crisis [of representation] there is for me a deeper evacuation of what 'experience' can be made to mean and how it may be put to work. After all it is difficult to speak one's self in a theoretical context that tends to exclude the positivity of experience." Langellier (147) similarly argues for "the theoretical significance of approaching identity as a performative struggle over the meanings of experiences as discourses navigate the body and the body anchors discourse," and Laurel Richardson (34) notes that we are always writing about our lives in our research stories.

As an educator I was interested to find that many of the writers who argue for more personal narratives in research are "wounded storytellers" (Frank 1995), for whom "[t]urning an experience into a story is a willful act to make one's afflictions meaningful; it is also a defense against suffering in silence" (Bochner 148). However, such storytelling is often as much for others as for the self: "the Other is a source of inspiration and purpose insofar as the wounded storyteller gains value and meaning from the expectation that the Other will learn, benefit and be guided by the tale that is told" (Bochner 149). In other words, storytelling has a pedagogical purpose too.

Constructing my body image as a mine site makes it "simultaneously an historical, natural, technical, discursive, material, entity" (Haraway 1997, 209) that provides me with an experience, not only of my own body but also of the ways in which my body is perceived by others. As Elizabeth Grosz (194–195) argues: "The subject's experience of the body is irreducibly bound up with the body's social status."

Haraway concludes her manifesto by asserting that she "would rather be a cyborg than a goddess" (p 181). In my work as a feminist environmental educator, this binary offers some generative possibilities, especially as I respond here to Judith Butler's (129) question: "How are the contours of the body clearly marked as the taken-for-granted ground or surface upon which gender significations are inscribed, a mere facticity devoid of value, prior to significance?"

In the subsequent sections of this narrative, I write of my corporeally and historically embodied experiences through the voice of a feminist poststructuralist researcher and environmental educator. In Sara Crawley's (72) terms, this writing "is both personal and theoretical, interweaving lived experience with how that experience, writ large, both is and can be represented as being." I present some vignettes of my breast cancer experiences and organize them to narrate a story of detection, surgery, and reconstruction using the metaphors of ore bodies and mine sites as an embodied display that locates my self (writes my body) in the practice of theorization. Of course, I present these vignettes as I interpret them now, recognizing that from a poststructuralist standpoint "there is no final knowledge; "the contingency and historical moment of all readings' means that, whatever the object of our gaze, it 'is contested, temporal and emergent'" (Lather 111).

Cyborg subjectivities

According to Haraway (1991, 149), "a cyborg is a cybernetic organism, a hybrid of machine and organism, a creature of social reality as well as a creature of fiction. . ... The cyborg is a matter of fiction and lived experience that changes what counts as women's experience in the late twentieth century." More recently, Rosi Braidotti (17) has argued that cyborg subjectivity is no longer science fiction – "the blurring of boundaries between humans and machines is socially enacted at all levels" – and that a cyborg is not a unitary subject position but "an embodied and socially embedded human subject that is structurally interconnected to technological elements or *apparati*." Cyborg subjectivities are dispositions to not only move among contestatory discourses and interrupt them but also dispose us to blur boundaries: "Cyborg writing is about the power to survive, not on the basis of original innocence, but on the basis of seizing the tools to mark the world that marked them as other" (Haraway 1991, 175). In exploring cyborg subjectivities, a challenge is thus "to combine the recognition of postmodern embodied subjects with resistance to power but also the rejection of relativism and cynicism" (Braidotti 245).

I deploy the notions of "embodied subject" and "embodied subjectivity" in this narrative from a poststructuralist position in which subjects are nonunitary, multiple, and complex; subjectivity is "an abstract or general principle that defies our separation into distinct selves and encourages us to imagine that, or simply helps us to understand why, our interior lives inevitably seem to involve other people, either as objects of need, desire and interest or as necessary sharers of common experience" (Mansfield 3).

This approach is consistent with Stanley's (2000, 11–12) view that in feminist methodology "knowledge is constructed from where the researcher/theoretician is situated," and that we need complex means of investigating and knowing the world which reject binary modes of thinking. My approach is also consistent with Stanley's interest in exploring how to link "methodological procedures" with "epistemological presuppositions." In making these links, my conceptualization of cyborg subjectivities is grounded in Harding's (1987, 3) understandings of methodology (as a theory and analysis of how research should proceed) and epistemology (as issues about an adequate theory of knowledge or justificatory strategy), and focused on "central questions such as who can be a 'knowed,' what can be known, what constitutes and validates knowledge, and what the relationship is or should be between knowing and being (that is between epistemology and ontology)" (Stanley & Wise 26).

This narrative of my body modification experiences also draws upon Foucault's (1975/1991) notion that power is corporeally revealed because events are "written" on the body: they shape the way we perform, or act out, our bodily selves. I have found Foucault's work relevant for feminist poststructuralist analysis because his emphasis on power relations in discourses offers feminists "a contextualization of experience and an analysis of its constitution and ideological power" (Weedon 125). Following Foucault, Weedon (113–114) argues: "Power is exercised within discourses in the ways in which they constitute and govern individual subjects. . ... Power also structures relations between different subjects within or across discourses." As I emerged from the fog of medicalization I came to see the connections between Foucault's (1975/1991) work on surveillance and power and my experiences of becoming an embodied mine site. These experiences eventually provided a theoretical scaffolding for me to begin writing my body, to think through "the place of the personal within theory" (Squire 56), to recognize these power relationships, and to experiment with going beyond them. As Foucault (1984/1988, 50) argues:

> The critical ontology of ourselves has to be considered not, certainly, as a theory, a doctrine, nor even as a permanent body of knowledge that is accumulating; it has to be conceived as an attitude, an ethos, a philosophical life in which the critique of what we are is at one and the same time the historical analysis of the limits that are imposed on us and an experiment with the possibility of going beyond them.

The gaze: exploring the ore body

From an early age my father taught me to "know the normal, look for the abnormal," so I knew that the small patch of reddened skin on the left side of my left nipple was "abnormal" long before I revealed it to my doctor. But the patch did not seem to be changing, and I was preoccupied with my mother's declining health (and then death), so I deferred seeking medical advice for several months.

I went to my family physician for the usual female cancer preventative testings in May 2001 and drew the patch to her attention. She immediately referred me to a breast clinic, where a mammogram showed nothing abnormal. I was given a cortisone cream to see if it would cure what might have been a rash. I mentioned Paget's disease, but the clinic's staff dismissed this as being highly unlikely given the mammogram result. If the patch got worse I was advised to come back immediately; otherwise, I had a follow-up appointment booked for 14 August, after my partner and I were due to return from a teaching commitment in Canada. The patch did not change, so we went to Canada without me visiting the doctor again.

On 31 July, the day after our return from Canada, I had an appointment with a dermatologist to remove some sun spots from my hand and nose. I showed him my patch. He immediately suspected Paget's disease, so he did a biopsy and sent the sample tissue to pathologists. He confirmed his diagnosis a week later, so for a week between receiving his advice and my 14 August appointment at the breast clinic, my family and I knew that I had Paget's disease.

Although some exploration of the ore body had already occurred through mammogram and biopsy, the medical gaze now became more intense and surveillant. At one level, I was subjected to the institutionally validated gaze that monitored me as a medical patient (Foucault 1976/1981) and sent me for ultrasound scans and discussions with surgeons. At another level I and other members of my family, though not telling each other this at the time, were putting my body under further surveillance by mining websites for information about Paget's disease – and becoming increasingly terrified because the web was full of worst-case scenarios, rather than the relatively good news story that I was likely to be among the 1% of the population who manifest Paget's disease of the nipple as a precancerous condition that is readily treated by surgery.

Disciplining the ill/ore body

I reacted to all of this surveillance by feeling alienated from my body and the world around me. I can now empathize with Jo Spence, who wrote "Property of Jo Spence?," on her breast the night before she had a mastectomy (see Figure 14.1) (Van Schaick 1998).[2]

Just as anti-mining protesters surround a site to prevent the machinery entering, I wanted to stop my body from having the surgery. I kept asking, "What if I didn't have the surgery? Is there an alternative?" I began to imagine myself as an animal living in a comfortable forest who hears the rumbling of machinery and the crashing of trees and wonders what is happening. The binaries of human/animal and human/machine were dissolving in me: I felt that I embodied Braidotti's (2002) Deleuzean notion of becoming-woman/animal/machine because after the surgery I would be living as a cyborg.

I tried to become a "scientist" again, a clinical observer of this subject/object who just happened to be me, distancing myself from my bodily experiences, controlling my subjectivity, a move that Luce Heshusius (1994) argues can result in an alienated consciousness which separates the knower from the known.

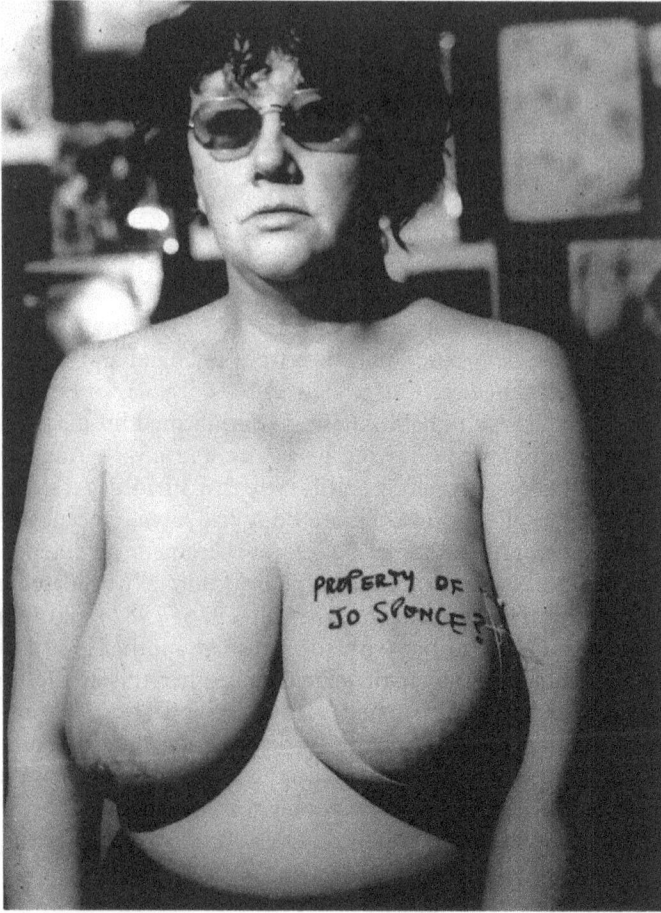

Figure 14.1 Jo Spence 1982.

Source: Photo credit Jo Spence/Terry Dennett. Reproduced with permission of Ryerson Image Centre.

Colonizing the mine site

The colonization of my body/mine site worked as efficiently as it probably could in the circumstances. The preparation for the mining process was exhausting, both mentally and emotionally, but the extraction itself was relatively quick, and, although painful, the pain was manageable. After much more surveillance of the ore body by assorted pathologists, my surgeon declared a successful extraction and invoked a restitution narrative. But the unresolved issue, both before and after surgery, was how to restore the terrain after the mine site had been abandoned?

An abandoned mine site is not a thing of beauty to any body, and it was no less so for being embodied in mine.

Normalization

Perhaps I set myself up for the gentle coercions toward postmastectomy normalization when, presurgery, I told the surgeon that I wanted to wake up with cleavage (and look like an externally "normal" female again as quickly as possible, rather than a "wounded storyteller"). Everything seemed to be moving so fast at that time – everyone seems to switch to full speed mode as soon as the cancer word is mentioned – and given that I did not have a rapidly growing malignant tumor, I wish there had been more time to think about what would/should happen after the surgery (and, although it was not realistic, about the surgery itself).

By saying that I wanted to wake up with cleavage I set the path I was going to travel. The surgeon gave me a booklet, *Breast Reconstruction: Your Choice* (Anti-Cancer Council of Victoria 1996), and we discussed the choices of breast implant and flap reconstruction. I was deterred by the hours of surgery involved in a flap reconstruction so I opted for an implant. No one mentioned that two-thirds of women who undergo mastectomies reject breast reconstruction altogether (Kasper 1995, in Langellier 2001).

I did not wake up from the mastectomy with cleavage, and its current approximation took three months as the surgeon gradually filled the tissue expander he had inserted during the mastectomy procedure and eventually replaced it with a silicon implant. This second operation occurred in December 2001. I had naively hoped that after this reconstruction surgery I would be normalized – able to wear my old clothes and not worry about necklines, able to wear my old bras, and able to perform stereotypical female normality to myself and others.

However, my breasts are no longer symmetrical; my implant is obvious to me in many necklines and there are other minor annoyances. That I did not feel normal after this surgery was a threshold. I remembered Denise Spencer's (1996) Yolngu women's story about the land never being the same after it has been mined no matter how careful the mining company is in the restoration process, but I also believed that the mining engineers/surgeons could work miracles on me.

At this time I wanted to be normalized, to be as close to resembling a "normal" female as possible, and not obviously marked as a cancer survivor with a "cyberboob." When I told my surgeon that my reconstruction did not feel or look "right" to me, he suggested that I might like to consult a psychiatrist, and/or a plastic surgeon. I took the latter option, and this opened up yet another new world of discourses of normalization, as this surgeon too handed me a book (Berger and Bostwick 1998). Here normalization was not just implants and flap reconstructions, there was a huge emphasis on nipple reconstructions and tattoos too. To me this is more than reconstructing the contours and soil layers of the land postmining; it is inscribing the fine details of gender significations with symbols of insignificance. A nipple serves erotic and functional purposes, but I could see no purpose for a "cybernipple."

Just how far can and does normalization go? Reading this book was empowering for me because it reminded me that I could research my own experiences and explore Foucault's (1984/1988, 26) technologies of the self, rediscovering that "being occupied with oneself and political activities are linked." Also, as Probyn (130) notes, Foucault's descriptions of the care of self are "a model of the self which operates at both ontological and epistemological levels in order to construct various modalities of the self."

Empowerment

After having mastectomies, women construct various modalities of the self. Some, like Rhea (Langellier 2001) and Deena Metzger (Van Schaick 1998), have tattoos on their scars. Rhea was inspired by Metzger's photograph[3] (see Figure 14.2), but Langellier does not indicate that Rhea knew the background to Metzger's story which Van Schaick (3) relates:

> The emphasis which Western culture has put on women's appearance leads Metzger to view her illness as ugliness, to feel shame, to apologize for the offensiveness of her appearance, to attempt to hide it. . . . The photograph represents a respite from previous periods of revulsion toward the breast cancer body, and Metzger chooses to emphasise her accomplishment.

Others seek the normalization offered by reconstruction surgery and, some, as evidenced by the 14 women's stories in Berger and Bostwick (1998), are happy with the results.

Figure 14.2 Deena Metzger, "I Am No Longer Afraid."
Source: Photograph by Hella Hammid. Reproduced with permission of Deena Metzger.

For me, empowerment came from being able to locate myself in the practice of theorization. As Squire (56), drawing on the work of Probyn (1993), notes: "Theory is not merely an abstraction, but is lived at the level of the experiential itself."

Foucault (1976/1981, 101) argues: "Discourse transmits and produces power; it reinforces it but it also undermines and exposes it, renders it fragile and makes it possible to thwart it." He also elaborates his concept of power, which he (1976/1981, 92) understands as "the multiplicity of force relations immanent in the sphere in which they operate and which constitute their own organisation." By focusing on the construction of my body as an embodied mine site, I can conceptualize my self as "a mode of holding together the epistemological and the ontological" (Probyn 4).

Conclusion

Being aware of tensions around writing the body in feminist poststructuralist research, I initially addressed these by writing of embodiment, and Paget's disease of the nipple as the specific aspect of bodily experience, because to recount a personal story "requires a willingness to try out styles of writing which make the writing of the theorist body possible" (Squire 58). Writing about a bulimic body Squire draws on Probyn's (1993) use of Foucault's notion of *le pli*, conceptualizing experience as doubled to read and write her lived experience of bulimia with a discursive analysis of that experience. I too have endeavored to use this approach here, combined with an attempt to write a feminist cyborg story, because "[f]eminist cyborg stories have the task of recoding communication and intelligence to subvert command and control" (Haraway 1991, 175).

According to Joanne Zylinska (2002, 216–217), corrective surgeries, such as mastectomies and breast reconstruction, have

> brought to the fore the instability of the relationship between nature and technics, but they have also foregrounded their mutual interdependence. These conceptual changes are transforming the ways in which we define identity, allowing emergence of less bounded and more connected models of human subjectivity. . . . Bodily experimentation challenges the possessive individualism that is characteristic of the capitalist model of selfhood, delineating instead the contours for what Celia Lury describes as "prosthetic culture."

The cyborg world that I now inhabit is one that Haraway (1991, 154) describes as being

> about lived social and bodily realities in which people are not afraid of their kinship with animals and machines, not afraid of permanently partial identities and contradictory standpoints. The political struggle is to see from both perspectives at once because each reveals both dominations and possibilities unimaginable from the other vantage point.

Breast cancer is one of the commonest forms of cancer in the Western world, and the most prevalent cancer among women. Once diagnosed with breast cancer, "even with the most optimistic prognosis" (Bahar 1026), a woman continues to live with her disease for the rest of her life: physically, psychologically, and culturally. Others who live with breast cancer have found their experiences generative for their feminist philosophical, political, and poetic work (see, for example, Hacker 2000; Lorde 1980; Sobchack 1995; Stacey, 1997). So have I: as Braidotti (2002) writes: "The embodied subject is thus a process of intersecting forces (affects) and spatio-temporal variables (connections)" (21), and "'embodied' accounts illuminate and transform our knowledge of ourselves and of the world" (12).

So what does this mean for my practice as an environmental educator? It has meant that I have come to better understand the fluidity of linguistic, health, and environmental boundaries. It has helped me to understand discourses of the Other better because I have experienced my "multiple, fragmentary, and unfinished" identity as "a performative struggle, always destabilized and deferred" (Langellier 147, 176). And it has helped me to better understand indigenous knowledge of the Australian environment: "There is a saying that Yolngu have that when you mine and strip someone's land it is like stripping them of their flesh and bones and leaving them bare to die" (from an interview with Rarriwuy in Spencer 139–140). I am not Yolngu and I do not claim their kinship, but my practice as an environmental educator has been changed through being an embodied subject in my own mine site story, stripped of my ore body. It is not a change I would have ever consciously sought, but scars and a "cyberboob" are now part of my body: "They are self-referential markers of the endless present/presence of . . . our 'physicality and our narrative,' and ultimately of the odd relation between 'corporeal and historical' existence" (Bahar 1049).

Acknowledgments

The intellectual, moral, physical, and sensual support of my partner, Noel Gough, is gratefully acknowledged, as are the comments of the anonymous reviewers; however, the final responsibility for the content of this story remains mine.

I also gratefully acknowledge the copyright holders for their permission to reproduce the images in this essay, specifically: the Jo Spence Archive London and Terry Dennett for the right to use Jo Spence's "Property of Jo Spence?";[4] and Deena Metzger for the right to use "I Am No Longer Afraid." The poster of "I Am No Longer Afraid" can be obtained from Donnelly/Colt, P.O. Box 188, Hampton, CT 06247, USA. Phone: 860–455–9621, Fax: 800–553–0006. Email: info@donnellycolt.com.

Notes

1 Paget's disease of the nipple is a rare form of breast cancer, affecting 1–4% of female patients presenting with breast cancer. It can be a precancerous condition or it may be the only indication of underlying breast cancer. The symptoms include a red, scaly

rash on the nipple, which is due to the presence of Paget's cells "that are themselves not cancerous, but are almost always associated with a cancer in the breast." www. oncolink.upenn.edu/experts/article.cfm?c=3&s=13&ss=22&id=1185)

2 Jo Spence, 1982. Photograph by Jo Spence/Terry Dennett. Reproduced with permission of Ryerson Image Centre.
3 Deena Metzger. "I Am No Longer Afraid." Photograph by Hella Hammid. Reproduced with permission of Deena Metzger.
4 In 2023 the copyright for this image is held by Ryerson Image Centre.

Works Cited

Anti-Cancer Council of Victoria. *Breast Reconstruction: Your Choice*. Carlton South: Anti-Cancer Council of Victoria. 1996.

Bahar, Sarah. "'If I'm one of the victims, who survives?': Marilyn Hacker's breast cancer texts." *Signs: Journal of Women in Culture and Society, 28*(4): 1025–1052. 2003.

Berger, Karen and Bostwick III, John. *A Woman's Decision: Breast Care, Treatment and Reconstruction*. Third Edition. New York: St Martin's Griffin. 1998.

Bochner, Arthur P. "Narrative's virtues." *Qualitative Inquiry, 7*(2): 131–157. 2001.

Braidotti, Rosi. *Metamorphoses: Towards a Materialist Theory of Becoming*. Cambridge: Polity Press. 2002.

Butler, Judith. *Gender Trouble: Feminism and the Subversion of Identity*. New York, London: Routledge. 1990.

Crawley, Sara L. "'They still don't understand why I hate wearing dresses!' An autoethnographic rant on dresses, boats and butchness." *Cultural Studies – Critical Methodologies, 2*(1): 69–92. 2002.

Devoss, Danielle. "Rereading cyborg(?) women: The visual rhetoric of images of cyborg (and cyber) bodies on the world wide web." *CyberPsychology & Behavior, 3*(5): 835–845. 2000.

Foucault, Michel. *Discipline and Punish: The Birth of the Prison*. Trans. Alan Sheridan. Harmondsworth: Pelican. 1975/1991.

Foucault, Michel. *The History of Sexuality, Vol. One, An Introduction*. Trans. Robert Hurley. Harmondsworth: Pelican. 1976/1981.

Foucault, Michel. *The Care of the Self: The History of Sexuality, Vol. Three*. Trans. Robert Hurley. New York: Vintage Books. 1984/1988.

Frank, Arthur W. *The Wounded Storyteller: Body, Illness, and Ethics*. Chicago: The University of Chicago Press. 1995.

Frank, Arthur W. "The standpoint of storyteller." *Qualitative Health Research, 10*(3): 354–365. 2000.

Gough, Annette. "I looked at the place where my breast used to be. . ." *WISENET Journal, 61*: 22–24. Accessed at www.wisenet-australia.org/issue60/annete%20 gough.htm. 2002.

Grosz, Elizabeth. *Space, Time, and Perversion: Essays on the Politics of Bodies*. New York, London: Routledge. 1995.

Hacker, Marilyn. *Squares and Courtyards*. New York: Norton. 2000.

Haraway, Donna. "A Cyborg Manifesto." In *Simians, Cyborgs, and Women: The Reinvention of Nature*. Ed. Donna Haraway. London: Free Association Books. 1991, 149–181. First published as A Manifesto for Cyborgs. *Socialist Review, 80*: 65–107. 1985.

Haraway, Donna. *Modest_Witness@Second_Millenium.FemaleMan©_Meets_Onco Mouse™*. New York: Routledge. 1997.

Harding, Sandra. "Introduction: Is there a feminist method?" In *Feminism and Methodology: Social Science Issues*. Ed. Sandra Harding. Bloomington: Indiana University Press and Open University Press. 1987, 1–14.

Harding, Sandra. "Introduction: Eurocentric scientific illiteracy – a challenge for the world community." In *The "Racial" Economy of Science: Toward a Democratic Future*. Ed. Sandra Harding. Bloomington: Indiana University Press. 1993, 1–29.

Heilbrun, Carolyn G. *Women's Lives: The View from the Threshold*. Toronto: University of Toronto Press. 1999.

Heshusius, Luce. "Freeing ourselves from objectivity: Managing subjectivity or turning toward a participatory mode of consciousness?" *Educational Researcher, 23*(3): 15–22.1994.

Kasper, A.S. (1995). "The social construction of breast loss and reconstruction." *Women's Health: Research on Gender, Behavior, and Policy, 1*: 197–219.

Langellier, Kristin M. " 'You're marked' breast cancer, tattoo, and the narrative performance of identity." In *Narrative and Identity: Studies in Autobiography, Self and Culture*. Eds. Jens Brockmeier and Donal Carbaugh. Amsterdam, Philadelphia: John Benjamins. 2001, 145–184.

Lather, Patti. *Getting Smart: Feminist Research and Pedagogy With/in the Postmodern*. New York, London: Routledge. 1991.

Lorde, Audre. *The Cancer Journals*. San Francisco: Spinsters/Aunt Lute. 1980.

MacLure, Maggie. "The repulsion of theory: Women writing research." In *Feminism and Educational Research Methodologies*. Ed. Heather Hodkinson. Manchester: The Manchester Metropolitan University. 2000, 61–77.

Mansfield, Nick. *Subjectivity: Theories of the Self from Freud to Haraway*. St Leonards, NSW: Allen & Unwin. 2000.

Probyn, Elspeth. *Sexing the Self: Gendered Positions in Cultural Studies*. New York: Routledge. 1993.

Richardson, Laurel. "Getting personal: Writing stories." *Qualitative Studies in Education, 14*(1): 33–38.2001.

Sobchack, Vivian. "Beating the meat/surviving the test or how to get out of this century alive." *Body & Society, I*(3–4): 209–214.1995.

Spencer, Denise. A history of the Gove Bauxite mine: From Yolngu women's perspective (North East Arnhem land). Unpublished Master of Education thesis. Geelong: Deakin University. 1996.

Squire, Sarah. "The personal and the political: Writing the theorist's body." *Australian Feminist Studies, 17*(37): 55–64.2002.

Stacey, Jackie. *Teratologies: A Cultural Study of Cancer*. London, New York: Routledge. 1997.

Stanley, Liz. "Children of our time: Politics, ethics and feminist research processes." In *Feminism and Educational Research Methodologies*. Ed. Heather Hodkinson. Manchester: The Manchester Metropolitan University. 2000, 5–35.

Stanley, Liz and Wise, Sue. "Method, methodology and epistemology in feminist research processes." In *Feminist Praxis*. Ed. Liz Stanley. London, New York: Routledge. 1990, 20–60.

Van Schaick, Elizabeth. "Palimpset of breast: Representation of breast cancer in the work of Deena Metzger and Jo Spence." *Schuylkill: A Creative and Critical Review from Temple University*. Fall 1998. Accessed at www.temple.edu/gradmag/fall98/schraick.htm.

Weedon, Chris. *Feminist Practice and Poststructuralist Theory*. Oxford: Blackwell. 1987
Zylinska, Joanna. "The future is monstrous: Prosthetics as ethics." In *The Cyborg Experiments: The Extensions of the Body in the Media Age*. Ed. Joanna Zylinska. London/
New York: Continuum. 2002, 214–236.

15 Gender, education, and the Anthropocene

Annette Gough

Gough, A. (2021). "Education in the Anthropocene." *Oxford Research Encyclopedia on Gender and Sexuality in Education.* doi:10.1093/acrefore/9780190264093.013.ORE_EDU-01391.R1
©2021 Reprinted by permission of Oxford Publishing Limited.

Summary

The term "Anthropocene" was coined in 2000 by Paul Crutzen and Eugene Stoermer to denote the present time interval as a new epoch of geological time dominated by human impact on the Earth. The starting date for the epoch is contentious – around the beginning of the Industrial Revolution (ca. 1800 CE), at the start of the nuclear age, or some other time, both earlier and later than these dates. The term itself is also contentious because of its humanist and human supremacy focus, and the way it hides troublesome differences between humans (including gender and cultural differences) and the intimate relationships among technology, humans, and other animals. Endeavours such as the United Nations Sustainable Development Goals aim to achieve gender equality by empowering women to participate in society. However, within this goal is the assumption that women and "other marginalized Others" can be assimilated within the dominant social paradigm rather than questioning the assumptions that maintain the subordination of these social groups. The goals also overlook the divergent impacts on women around the globe.

Education in an Anthropocene context necessitates a different pedagogy that provides opportunities for learning to live in and engage with the world and acknowledges that we live in a more-than-human world. It also requires learners to critique the Anthropocene as a concept and its associated themes to counter the humanist perspective, which fails to consider how the nonhuman and material worlds coshape our mutual worlds. In particular, education in the Anthropocene will need to be interdisciplinary, transdisciplinary, or cross-disciplinary; intersectional; ecofeminist or posthumanist; indigenous; and participatory.

Introduction

Until the early 2000s geologists classified the current geological period as the Holocene epoch (Lewis & Maslin, 2015). However, in 2000, Paul Crutzen

DOI: 10.4324/9781003390930-18

and Eugene Stoermer proposed that the epoch should be called the Anthropocene because humans are now exerting so much influence over planetary processes. This caused much discussion in the scientific community (Monastersky, 2015a, 2015b). The Anthropocene Working Group (AWG) of the International Commission on Stratigraphy was formed in 2008, and in 2019 the panel moved to finalize this renaming by submitting a formal proposal for the new name to the commission in 2021 (AWG, 2019; Subramanian, 2019). Nevertheless, there is still much indecision about when the Anthropocene started, if it exists, and whether it is an appropriate name. The Anthropocene is also contentious because of its humanist and human supremacy focus and the way it hides troublesome differences between humans (including gender and cultural differences) and the intimate relationships among technology, humans, and other animals. Does the Anthropocene end "the concept *nature*: a stable, nonhuman background to (human) history" (Morton, 2014, p. 258), or "is nature no longer separable from culture in this age of the Anthropocene" (Åsberg, 2017, p. 198)? And what does this mean for education? This chapter discusses these aspects in the context of the role of education in the Anthropocene, or whatever the present and future period is called.

Tensions around the "Anthropocene"

Before formally proposing the Anthropocene as a new geological epoch, the AWG needed to identify a definitive geological marker or global boundary stratotype section and point. Ideally this would be a single site with physical evidence in the sedimentary records that represents the start of the epoch (Lewis & Maslin, 2015). However, even this is contentious because some members of the AWG have argued that the Anthropocene is time-transgressive, with multiple beginnings, due to the progressive impacts of humans in the world since prehistoric agriculture (Subramanian, 2019). Simon Lewis and Mark Maslin (2015), for example, discuss suggestions for the beginning of the Anthropocene, including the impact of fire, preindustrial farming, sociometabolism, the meeting of Old and New World human populations, industrial technologies, and the atomic age. They conclude that there is currently not enough evidence to formally ratify a new Anthropocene epoch, and that "more widespread recognition that human actions are driving far-reaching changes to the life-supporting infrastructure of Earth may well have increasing philosophical, social, economic and political implications over the coming decades" (Lewis & Maslin, 2015, p. 178).

Some scientists see the Anthropocene as a reassertion of human dominion over the Earth:

If the Anthropocene is the epoch in which 'the human' itself has become a force of nature, then it only marks the full realization of what it has always implicitly been. . . . The practical consequence of these theories is the model of stewardship: their power nominates humans as guardians of the earth.

(Bajohr, 2019)

In this vein, Crutzen and Schwägerl (2011) write of "steer[ing] nature's course symbiotically instead of enslaving the formerly natural world" in a form of technological fix. Similarly, Bruno Latour (2014) believes that the Anthropocene is an era of negotiation in which humans must engage in a different kind of relationship with the rest of the members of our planet and acknowledge the need to leave behind ideas of human privilege.

While geologists and other scientists are still debating the existence of the Anthropocene, the concept has been taken up enthusiastically in the humanities. As Hannes Bajohr (2019) notes, "its attraction lies in its omnicompetent radiance: not only a geochronological coinage, it implies an ontology, a theory of history, and an anthropology," perhaps exploding "the classic separation between the history of nature and that of mankind [*sic*]." However, this enthusiasm is not necessarily supportive of the term. Indeed, the rise of the Anthropocene "has arisen at a most inconvenient moment" (Morton, 2014, p. 258) from a posthumanist position, such that as, Colebrook (2017) writes, "the notion that there is no such thing as the human (either by way of our difference from animals or because of intrahuman differences in culture and history) must give way to a sense of the human as defined by destructive impact" (p. 6). She then concludes,

> One effect of the Anthropocene has been a new form of difference: it now makes sense to talk of humans as such, both because of the damage "we" cause and because of the myopia that allowed us to think of the world as so much matter or "standing reserve." Humans are, now, different; and whatever the injustices and differences of history and colonization, "we" are now united in being threatened with nonexistence.
>
> (pp. 7–8)

These notions of Anthropocene are stimulating discussions about what it is to be human and our relationships with nonhumans. For example, Timothy Morton (2017) argues that our relationship with nonhumans decides the fate of our humanity, and humans need to develop a network of kindness and solidarity with nonhuman beings. Rosi Braidotti (2013) is troubled by the reassertion of humanism and what it hides:

> I am, however, seriously worried about the limitations of an uncritical reassertion of Humanism as the binding factor of this reactively assumed notion of a pan-human bond. I want to stress that the awareness of a new (negatively indexed) reconstruction of something we call humanity must not be allowed to flatten out or dismiss all the power differentials that are still enacted and operationalized through the axes of sexualization/

racialization/naturalization, just as they are being reshuffled by the spinning machine of advanced, bio-genetic capitalism.

(pp. 87–88)

While most agree that we do need new ways of thinking about our collective existence, "Anthropocene" is seen by many as the wrong descriptor because it surreptitiously purveys a human supremacy complex and the assertion of anthropocentrism (as distinct from an ecocentric worldview).

Women troubling the Anthropocene

Some scientists argue that adopting the Anthropocene makes "man" the center of the universe again, to the extent that "Man the taxonomic type" becomes "Man the brand" (Haraway, 1997, p. 74). And this "universal 'Man' is implicitly assumed to be masculine, white, urbanized, speaking a standard language, heterosexually inscribed in a reproductive unit and a full citizen of a recognized polity" (Braidotti, 2013, p. 65). This anthropocentrism ignores the global and gendered, classed and national disparities in human impacts on the planet that actually exist. As Stevens et al. (2018) argue, while Anthropocene "neatly evokes the contradiction between the human causes of environmental destruction and the human capacity to protect . . . the inherent social inequity of such drastic and rapid environmental change is not illuminated" (p. 2). They are also concerned that "the idea of the Anthropocene might even imply that all humanity is equally responsible" (p. 2). This is echoed by Jill Schneiderman (2017), who, among others, argues that "the Anthropocene does not acknowledge that some groups of human beings have had greater effects on the planet than others" (p. 184).

Women have been trying to get recognition of the impact of environmental degradation on their lives and livelihoods for decades. As Bella Abzug (1991) so clearly articulated in the lead-up to the 1992 Rio Earth Summit, "Women are half the world's population, yet we have almost no say in the environment and development policies that affect us, the lives of our families and the survival of this planet" (p. 2). Advocacy from groups such as the Women's Environment and Development Organisation resulted in women having their own chapter in Agenda 21, the outcomes document of the 1992 Earth Summit (United Nations [UN], 1993). Gender equality has remained on the UN sustainability agenda ever since, including as Sustainable Development Goal (SDG) 5 (UN, 2016). However, the UN Framework Convention on Climate Change (UNFCCC), which came into force in 1994, did not include any gender aspects. This led to the strategic actions to draw attention to the impact of climate change on women around the 2012 Conference of the Parties (COP 18), which produced the Gender Decision "promoting gender balance and improving the participation of women in UNFCCC negotiations and in the representation of Parties in bodies established pursuant to the Convention or the Kyoto Protocol" (UNFCCC, 2012). The impact of climate change on women is summarized by the Global Gender and Climate Alliance (2013):

- Women in developing countries are particularly vulnerable to climate change because they are highly dependent on natural resources for their livelihood.
- Women experience unequal access to resources and decision-making processes, with limited mobility in rural areas.
- Women make between 30 and 80% of what men earn annually.
- 103 out of 140 countries surveyed by the World Bank impose legal differences on the basis of gender that may hinder women's economic opportunities.
- Women make up half of the agricultural workforce in the least developed countries.
- In developing countries, women own between 10 and 20% of the land.
- Two-thirds of the world's illiterate adults are women.
- Socioeconomic norms can limit women from acquiring the information and skills necessary to escape or avoid hazards (e.g., swimming or climbing trees to escape rising water levels).
- Dress codes imposed on women can restrict their mobility in times of disaster, as can their responsibility for small children who cannot swim or run.
- A lack of sex-disaggregated data in all sectors often leads to an underestimation of women's roles and contributions, thus increasing gender-based vulnerability.

Women from the Third World, such as Vandana Shiva (1989, 1991, 2005), have been particularly strong in the environment movement, and there are many other individuals and organizations (such as UN WomenWatch) that are trying to get women's voices heard. Importantly, as Greta Gaard (2015) argues, the gendered environmental discourses that constructed them as individual "*victims* of environmental degradation in need of rescue" have shifted to now focus on "gender as a system structuring power relations" (pp. 21–22). This has been an important development in feminist responses to climate change.

While acknowledging that the term "Anthropocene" is contentious, here it is used for its capacity to do useful work as the term has been taken up within the humanities and by artists, social scientists, and scientists (Bajohr, 2019; Grusin, 2017). It is also increasingly being referenced in the education field, including arts education (e.g., Jagodzinski, 2013; Wallin, 2017), environmental education (e.g., Crex Crex Collective, 2018; Greenwood, 2014; Malone, 2017; Malone & Truong, 2017; Taylor, 2017; Thorne & Whitehouse, 2018), science education (de Freitas & Truman, 2020; Gilbert, 2015; Wagler, 2011), early childhood education (e.g., Nxumalo, 2020; Somerville & Powell, 2019), educational research (e.g., Charteris et al., 2018; Lloro-Bidart, 2015; Somerville, 2017), teacher education (e.g., Brennan, 2017), and higher education (e.g., Carstens, 2016; Decuypere et al., 2019). So what does the Anthropocene mean for education?

Changing roles for education

Education has a role in socializing people to live in societies. Educational institutions socialize individuals by passing on the social and cultural values and knowledge of the group as a form of social reproduction. In many ways, society "wants to keep and continue itself by reproducing as it is" (Kurt, 2015, p. 224). Education has had a function in social reproduction from the time of the Industrial Revolution and the introduction of free, compulsory, and non-religious education for all children. Legislation on compulsory education was first introduced in France in 1841, followed by England, Canada, Germany, and Australia in the 1870s, the United States in the 1920s, and eventually in China in 1986. As Michael Apple (2004) argues: "Educational institutions provide one of the major mechanisms through which power is maintained and challenged" (p. vii). While education gives people access to jobs and a "better" life (and access to universal education remains one of the SDGs for 2030 [UN, 2016]), the knowledge and values implicit in the current dominant education process remain contentious, and many critique the social reproduction role of education and how the education establishment goes about achieving this. For example, bell hooks (1989) discusses the ways in which racism, sexism, and class exploitation work in the lives of black females and how they are dehumanized; Sandy Grande (2004) explores the intersection between dominant modes of critical educational theory and the sociopolitical landscape of American Indian education; and Antonia Darder and Rodolfo Torres (2013) discuss the link between educational practice and the larger socioeconomic and structural dimensions that shape Latinos' lives. Shirley Steinberg (2012) draws on cultural studies and extends our notions of cultural pedagogy to focus attention on the complex interactions of power, knowledge, identity, and politics, and "the methods by which cultural differences along the lines of race, class, gender, national origin, religion, and geographical place are encoded in consciousness and processed by individuals" (p. 233).

Educational institutions achieve social reproduction through controlling the ways people access economic and cultural resources and power, deciding "whose knowledge is 'official' and about who has the right to decide both what is to be taught and how teaching and learning are to be evaluated" (Apple, 2004, p. vii). This role was questioned by social reconstruction educators such as Schiro (2007), who saw society as unhealthy and believed that education could provide the means to reconstruct society. Some of the roots of social reconstructionism can be traced to John Dewey (1916), who described education as "that reconstruction or reorganization of experience which adds to the meaning of experience, and which increases ability to direct the course of subsequent experience" (p. 76). However, in the mid-20th century, the growth of critical theory led educators to embrace social reconstructionism as they acknowledged the oppression of women in the dominant forms of education:

> These forms of critical theory revolted against traditional ways of viewing and conceptualizing our world; against powerful (oppressive,

exploitative, and/or dominant) social groups who made economic, cultural, and educational decisions affecting the lives of those less powerful; and against rationalist, Eurocentric cultural traditions that privileged those who were white, educated, rich, and male in comparison to those who were nonwhite, uneducated, poor, or female. They focused on the subjective and social construction of knowledge rather than on objective knowledge.

(Schiro, 2007, p. 156)

Kemmis et al. (1983) developed the notion of the "socially critical school" as a way of opening up possibilities for social change and reconstruction, along with improved or enhanced curriculum and schooling. This was very much consistent with critical theory and the views of Michael Apple (*Ideology and Curriculum* was first published in 2004). Within the socially critical school model, knowledge is viewed as "constructed through social interaction and thus as historically, culturally, politically and economically located" (p. 11), and is linked to action, emancipation, and social critique.

Several scholars (e.g., Greenall Gough & Robottom, 1993; Huckle, 1991) discuss how the field of environmental education, and subsequently education for sustainability and for sustainable development, needs to be concerned about the social construction of knowledge, arguing that it "should adopt a critical approach to encourage careful awareness of the various factors involved in the situation" (UNESCO, 1980, p. 27). Greenall Gough and Robottom (1993) provide a case study of how a school was adopting a socially critical curriculum and students were taking action for the environment.

Sadly, the flourishing of socially critical approaches to education were short-lived, and with the rise of neoconservative and neoliberal agendas the revolution went backwards. Since Apple (2004) wrote of the desire to return to a supposed Eden that was also "a politics of cultural control that marginalized the lives, dreams, and experiences of identifiable people," society has returned to

shallow understandings of science [witness the climate change deniers], the search for technical solutions based on this (mis)understanding of science, a new managerialism that relies on the massiveness of the resurgent regime of "measuring anything that moves in classrooms," the reduction of education to workplace skills and the culture of the powerful.

(Apple, 2004, p. xii)

This is, of course, the exact opposite of what is needed for education in the Anthropocene: "we now live on a bio-physically different planet than the one in which modern civilization developed and in which our common assumptions about education were formed" (Greenwood, 2014, p. 281).

There are some exceptions to this desire for the maintenance of cultural capital and a return to Eden, and there are those who see education as having a role in preparing critical thinkers and agents of change. David de Carvalho, the chief executive officer of the Australian Curriculum, Assessment and Reporting Authority, writes that education serves "in ascending order, social, cultural and personal needs and values . . . [thus] paradoxically serv[ing] simultaneously the purposes of social and cultural continuity, and social and cultural change" (de Carvalho, 2019).

More significantly, there has been some movement, particularly at an international political level, that recognizes and advocates for a very different role for education; one that is directed at social reconstruction and transformation. This view is encapsulated in the SDGs (UN, 2016), but the origins, in an environmental context, can be traced to the 1972 UN Conference on the Human Environment (UN, 1973) and the subsequent UN Conference on Environment and Development (UN, 1993), World Summit on Sustainable Development (UN, 2002), and the Rio+20 UN Conference on Sustainable Development (UN, 2012). These origins are important, as they mark the long-standing international concern about the impact of human activity on the state of the environment and increasing concern about gender, race, and class issues.

Prioritizing concern for the environment and society

The declaration from the 1972 UN Conference on the Human Environment held in Stockholm provides a vision and set of common principles focused on preserving and enhancing the human environment: "The protection and improvement of the human environment is a major issue which affects the well-being of peoples and economic development throughout the world. It is the urgent desire of the peoples of the whole world and the duty of all governments" (UN, 1973, p. 3).

The Common Vision in the outcomes document from the Rio+20 conference, *The Future We Want* (UN, 2012), is grounded in a very different orientation. It opens with a commitment to ensuring "the promotion of an economically, socially and environmentally sustainable future for our planet and for present and future generations" (p. 1), and continues with a focus on mainstreaming sustainable development, "integrating economic, social and environmental aspects and recognizing their interlinkages, so as to achieve sustainable development in all its dimensions" (p. 2), which is a very different focus from the Stockholm Declaration's concern for the protection and improvement of the human environment. Here, economic development is much more foregrounded.

Another change is in the prioritizing of the human condition over that of the environment: "Eradicating poverty is the greatest global challenge facing the world today and an indispensable requirement for sustainable development" (UN, 2012, p. 1). This is also the first SDG (UN, 2016). Like the concerns of the Anthropocene, these goals are very human centered, with the focus of the first five being human issues: eradicating poverty, removing hunger, human health and well-being, education for all, and achieving gender equality. It is not really until Goals 9, 11, 12, 14, and 15 that there are concerns about the state of the environment and reducing human impact on it through reducing resource consumption and protecting and conserving life in the water and on land. What these goals add to considerations of the Anthropocene is recognition of gender, age, sex, disability, race, ethnicity, origin, religion, economic or other status as human factors related to achieving sustainable development. However, these goals are not perfect. For example, UN Women (2012) called for a "new gender responsive global development framework," explicitly positioning women as "agents of change, innovators and decision makers" (p. 11). Kate Wilkinson (2016) reviews the Rio+20 outcomes document (UN, 2012) and the SDGs and concludes that they reaffirm a liberal understanding of gender equality because "there are still many barriers to full and equal participation by women in environmental issues" related to "ideological and structural assumptions that inform the dominant social paradigm within the concept of sustainable development" (Wilkinson, 2016). Wilkinson also points out that the analysis of green economy introduced in the outcomes document "maintains separation and distance between humanity and nonhuman nature" (p. 558).

One response to this agenda being so human centered has been the development of concerns for the more-than-human (and posthumanism, which recognizes that there is no one unified cohesive "human," Seaman, 2007), as well as animal studies and "the green plant-human relationships that undergird human cultures as well as the darkly petroleum-fueled industrialization, mass species extinctions, and strange new ecosystems in the Anthropocene" (Sullivan, 2019, p. 152). Morton (2017) adds that there is a need to negotiate the politics of humanity in order to reclaim the upper scales of ecological coexistence and to resist corporations that would rob humans of kinship with nonhuman beings. Donna Haraway (2018) in many ways sums up these positions when she argues that "there can be no environmental justice or ecological reworlding without multispecies environmental justice and that means nurturing and inventing enduring multispecies – human and nonhuman – kindreds" (p. 102).

Others are not so hopeful. For Roy Scranton (2015), humanity's task is "learning to die in the Anthropocene," so "we need a new vision of who 'we' are. We need a new humanism – a newly philosophical humanism, undergirded by renewed attention to the humanities" (p. 19).

Rethinking education in/for the Anthropocene

Many educators have enthusiastically embraced the notion of the Anthropocene. For example, Jane Gilbert (2015) sees the Anthropocene as possibly "the 'crisis to end all crises,' the catalyst to provoke real change" (p. 188) in science education, and Reinhold Leinfelder (2013) asserts, "The Anthropocene concept appears particularly useful also for educational purposes, since it uses metaphors, integrates disciplinary knowledge, promotes integrative thinking, and focuses on the long-term perspective and with it our responsibility for the future" (p. 26).

Others, however, argue that we need to interrogate the contested nature of the term itself and its political and cultural implications, "rather than unwaveringly accepting the 'age of humans'" (Lloro-Bidart, 2015, p. 132).

Background

The concerns about human impact on the planet that underlie the naming of the Anthropocene have been around for some time – hence the convening of the 1972 UN Conference on the Human Environment in Stockholm and subsequent conferences. And there has been an environmental education movement since the late 1960s, albeit mainly on the fringes of formal education in many places, even though it has been seen as serving "as a catalyst or common denominator in the renewal of contemporary education" (UNESCO, 1978, p. 20). Thus, perhaps the time for environmental education has finally arrived as the most suitable form of education for the Anthropocene.

One of the earliest international agreements on environmental education was the Belgrade Charter Framework for Environmental Education (UNESCO, 1975), which states:

> Millions of individuals will themselves need to adjust their own priorities and assume a "personal and individualised global ethic" – and reflect in all of their behaviour a commitment to the improvement of the quality of the environment and of life for all the world's people . . .
>
> The reform of educational processes and systems is central to the building of this new development ethic and world economic order . . .
>
> This new environmental education must be broad based and strongly related to the basic principles outlined in the United Nations Declaration on the *New Economic Order*.
>
> (pp. 1–2)

A similar statement could be written about the need for education in the Anthropocene.

Indeed, Target 4.7 of SDG 4 on Education calls on countries to ensure that all learners are provided with the knowledge and skills to promote sustainable development, including, among others, through education for sustainable development and sustainable lifestyles, human rights, gender equality, promotion of a culture of peace and nonviolence, global citizenship, and appreciation of cultural diversity and of culture's contribution to sustainable development (UN, 2016).

Addressing the Anthropocene through education

Teresa Lloro-Bidart (2015) identifies three overarching conceptual and/or practical shifts that need to be engaged in education in/for the Anthropocene:

- interdisciplinarity, transdisciplinarity and cross disciplinarity;
- community- and/or participatory-based approaches in the natural sciences; and

• alternative modes of thought, including "mobile lives", "post-carbon social theory", Indigenous, ecofeminist/posthumanist and connectivity to *oikos* perspectives. (p. 133)

Many of these are part of (environmental) education for sustainability discussions and have been for some time. Environmental education has always been thought of as being interdisciplinary: the word "interdisciplinary" is mentioned multiple times in the report from the 1977 UNESCO-UNEP *Intergovernmental Conference on Environmental Education*. For example, one of the guiding principles for environmental education is that it should "be interdisciplinary in its approach, drawing on the specific content of each discipline in making possible a holistic and balanced perspective" (UNESCO, 1978, p. 27). Indeed, it is its interdisciplinary character that has made it difficult for environmental education to find a place in the school curriculum – is it a separate subject or a cross-disciplinary theme (Gough, 1997)?

Environmental education also has a participatory orientation: one of the five categories of objectives included in the Tbilisi Declaration was "Participation: to provide social groups and individuals with an opportunity to be actively involved at all levels in working toward resolution of environmental problems" (UNESCO, 1978, p. 27). For some this was a contentious aspect – while it is fine to be educated *about* and *in* the environment, acting *for* the environment, as in a socially reconstructionist or transformative approach to education, was not acceptable (Gough, 1997). Such concerns led David Greenwood (2014) to question: "are schools relevant to the complex realities of a changing planet? Or, do they mainly serve an outdated vision of an industrial society that is turning rapidly into a complex mix of decline and transformation?" (p. 279). Education in the Anthropocene requires participatory approaches because people need to learn to work together and live with climate change and the other environmental crises, as well as work across cultures and genders in addressing environmental issues.

Including alternative modes of thought is probably where the traditional conceptions of environmental education are extended. Although there has been encouragement for valuing Indigenous knowledge in environmental education for some time (e.g., Lowan-Trudeau, 2015; Shava, 2013; Simpson, 2002; Tuck et al., 2014), it is still not a common consideration. Similarly, the need for connections with nonhuman nature and the more-than-human have been part of ecofeminist and posthumanist discourses but have not been widely taken up in environmental education (Bell & Russell, 2000; Fawcett, 2013; Gough & Whitehouse, 2003, 2018; Lloro-Bidart, 2017; Russell & Bell, 1996; Snaza et al., 2014). Rather, the emphasis in environmental education has been more humanist – people know what is good for the environment and how to protect it. Such an approach is found in the UNESCO (2019) *Framework for the Implementation of Education for Sustainable Development Beyond 2019*, which recognizes that "climate change is a real and rapidly-evolving threat for humanity" (p. 2) but does not provide a great deal of guidance for dealing with this threat, except to "encourage learners to undertake transformative actions for sustainability" (annex II, p. 4). The framework is silent on connecting with nonhuman nature and the more-than-human.

Social and environmental justice are additional considerations for education in the Anthropocene. For example, Huey-Li Li (2017) argues that "schooling should embrace and engage ecological and human vulnerability. In this way, education might better assume ethical responsibility for mitigating the ongoing ecological decline" (p. 435). She sees the vulnerability of communities as growing because of human activities. These vulnerabilities include increased poverty, increased violence (due to firearms and drugs as well as family violence), greater urban density, environmental degradation, and climate change, all of which are included in the SDG. Li argues that people need to be resilient to minimize or overcome their vulnerabilities. She then questions why schools ignore the "irrefutable glocal ecological devastation that renders students vulnerable" (p. 442) and proposes ecologizing education as a solution. This dismisses the arbitrary distinction between natural and human-induced ecological disasters (because both render people vulnerable) and commences with an inclusive recognition of human embodiment in the biophysical environment and the coterminal coexistence of human and ecological vulnerability "to cultivate an active and responsive citizenry that is capable of and committed to promoting and implementing ecologically congenial cultural and social transformation" (p. 450) and addresses the interrelated social and environmental issues.

Richard Kahn (2008) has argued for ecopedagogy to bring about liberation for animals, nature, and the oppressed people of the Earth. Martusewicz et al. (2015) envisage ecojustice education as working toward diverse, democratic, and sustainable communities. Randy Haluza-DeLay (2013) also argues for educating for environmental justice and recognition that poor, racialized communities are disproportionately impacted by environmental degradation. Lloro-Bidart and Finewood (2018) and Maina-Okori et al. (2018) encompass these concerns within intersectionality. An intersectional approach to education in and for the Anthropocene would enable consideration of social and environmental justice issues.

Constance Russell (2019) implements "an intersectional approach to learning and teaching about humans and other animals in educational contexts," which includes a social justice dimension. Joshua Russell (2019) argues for critical pedagogies that espouse feminist, posthumanist, queer, and Indigenous methodologies; decenter the traditional human/adult/Western perspective; and emphasize the materiality and knowledges that emerge from careful attunement to place, nonhuman animals as agents, decolonizing Indigenous perspectives, and the perceptual worlds of children.

Thus, Education in the Anthropocene needs to be socially reconstructive and transformative. Business-as-usual and social reproduction in a neoliberal and neoconservative agenda will not work as society, and our environment have been drastically altered. Interventions such as socially critical schools and environmental education have brought elements of this education to the surface, but we need these and more to confront the future.

Gendered and global dimensions of educating in the Anthropocene

While an intersectional approach to education that takes into account gender, age, sex, disability, race, ethnicity, origin, religion, and economic status is important in the Anthropocene, and a gender-based approach is particularly important because women are so impacted by the events that comprise much of what constitutes the Anthropocene (see the section on "Women Troubling the Anthropocene"), *The Global Gender Gap Report 2017* (World Economic Forum, 2017) benchmarked gender parity in 144 countries and came to the conclusion that, at present rates of social change, global gender equality was still more than 200 years away. The UN Girls' Education Initiative's (2019) *Situational Analysis of SDG 4 with a Gender Lens* similarly found huge deficiencies in girls' engagement with education and found that gender equality is often underrepresented in ESD curricula. This is not a good sign for a humankind facing a planetary accounting in the next decade because the absence of gender parity is analyzed as being largely responsible for the climate crisis. Sandra Harding (2015) admonishes:

> The failure to address women's issues directly in development contexts not only damages women's chances for flourishing and for equality; it also renders it impossible to achieve the eradication of poverty and advance of other, noneconomic kinds of flourishing that supposedly have been the goals of development projects for more than six decades.
>
> (p. 79)

The concept of climate justice is gaining international traction through nongovernmental organizations such as The Climate Reality Project and the UNFCCC. A Gender and Climate Change Gender Action Plan (GAP) was proposed at COP 23 in Bonn, Germany, in 2017, noting that "a gender-responsive climate policy continues to require further strengthening in all activities concerning adaptation, mitigation and related means of implementation . . . as well as decision making on the implementation of climate policies" (UNFCCC, 2017, p. 1). The GAP aims to increase the number of women who are active in climate decision-making, and train policy makers on bringing gender equality into climate funding programs, and engage grassroots women's organizations for local and global climate action. These actions are consistent with the targets and indicators of the SDGs (UN, 2016) and the associated education for sustainable development. Indeed, the *Framework for the Implementation of Education for Sustainable Development Beyond 2019* specifically mentions the vulnerability of women to hazards induced by climate change and their need for access to ESD:

> ESD is an instrument to achieve all the SDGs, and each of the SDGs comes with specific gendered challenges. ESD takes on a cross-disciplinary and systemic approach that enables the question of gender equality to be linked to the various issues of sustainable development. There is, for example, a gendered facet of vulnerability to hazards induced by climate change. When disasters occur, more women die than men because social rules of conduct mean that, for example in the case of flooding, women often have not learned to swim, and have behavioural restrictions that limit their mobility in the face of risk. It should therefore become a priority to provide women with access to ESD. In this regard, ESD actively promotes gender equality, and creates conditions and strategies that empower women.
>
> (annex II, p. 3)

The marginalization of women in environmental education research and practice has been noted for some time (e.g., Di Chiro, 1987; Gough, 1999a, 1999b, 2013; Gough & Whitehouse, 2003, 2018; Gray, 2018; Mitten et al., 2018; Piersol & Timmerman, 2017; Russell & Fawcett, 2013), but this is slowly changing. Bringing feminist perspectives into education in/for the Anthropocene is to recognize the complexity of human roles and relationships with respect to environments and realize that there are multiple subjectivities and multiple ways of knowing and interacting with environments that cannot be encapsulated within the notion of universalized subjects. Each feminist approach has something unique to offer, and when taken together, they could be particularly potent. The anticolonial methodologies that comes from Black, Chicana, and Indigenous researchers, such as Dolores Calderon (2014) and Fikile Nxumalo and Stacia Cedillo (2017), provide interdisciplinary frameworks that can be used to examine the way multiple colonialisms (post, settler, internal, etc.) operate insidiously in educational contexts across the globe. Ecofeminist, posthumanist, and intersectional researchers, such as Chattopadhyay (2019), Gough and Whitehouse (2018, 2020), Lloro-Bidart (2017), Lloro-Bidart and Michael Finewood (2018), and Maina-Okori et al. (2018), argue for the dismantling of nature-culture binaries, the disruption of anthropocentric views, new ways of encountering more-than-human worlds, and assertion of more political standpoints. These different perspectives and methodologies are concerned with the development of pedagogies that are sensitive to relationships between humans, nonhuman nature, and the more-than-human, which can help transform education in the Anthropocene for the better.

Conclusion

This chapter has focused on understandings of and contestations around the concept of the Anthropocene, the changing role of education in society, and what education in the Anthropocene could include, particularly from multiple feminist approaches and perspectives because women are so affected by climate change and other environmental crises. Education in the Anthropocene needs to be very different from the education currently being practiced in schools. It necessitates a different pedagogy and a different curriculum. Are we ready?

Further Reading

Cutter-Mackenzie, A., Malone, K., & Barratt Hacking, E. (Eds.). (2020). *Research handbook on childhoodnature: Assemblages of childhood and nature*. Cham, Switzerland: Springer.

Darder, A. (1991). *Culture and power in the classroom: A critical foundation for bicultural education*. New York, NY: Bergin & Garvey.

Gough, A. (2018, September). Working with/in/against more-than-human environmental sustainability education. *On Education: Journal for Research and Debate, 2*. https://doi.org/10.17899/on_ed.2018.2.3

Gough, A., & Whitehouse, H. (2019). Centering gender on the agenda for environmental education research. *Journal of Environmental Education, 50*(4–6), 332–347. http://doi.org/10.1080/00958964.2019.1703622

Gray, T., & Mitten, D. (Eds.). (2018). *The Palgrave international handbook of women and outdoor learning*. Cham, Switzerland: Palgrave Macmillan.

Haraway, D. J. (2016). *Staying with the trouble: Making kin in the Chthulucene*. Durham, NC: Duke University Press.

Jagodzinski, J. (Ed.). (2018). *Interrogating the anthropocene: Ecology, aesthetics, pedagogy, and the future in question*. Cham, Switzerland: Palgrave Macmillan.

Jickling, B., Blenkinsop, S., Timmerman, N., & Sitka-Sage, M. D. D. (Eds.). (2018). *Wild pedagogies: Touchstones for renegotiating education and the environment in the Anthropocene*. Cham, Switzerland: Palgrave Macmillan.

Lloro-Bidart, T., & Banschbach, V. S. (Eds.). (2019). *Animals in environmental education: Interdisciplinary approaches to curriculum and pedagogy*. Cham, Switzerland: Palgrave Macmillan.

Malone, K., Truong, S., & Gray, T. (Eds.). (2017). *Reimagining sustainability in precarious times*. Singapore: Springer.

Reyes, V. C., Charteris, J., Nye, A., & Mavropoulou, S. (Eds.). (2018). *Educational research in the age of the Anthropocene*. Hershey, PA: IGI Global.

Stevens, L., Tait, P., & Varney, D. (Eds.). (2018). *Feminist ecologies: Changing environments in the Anthropocene* (pp. 1–22). Cham, Switzerland: Palgrave Macmillan.

Stevenson, R. B., Brody, M., Dillon, J., & Wals, A. (Eds.). (2013). *International handbook of research on environmental education*. New York, NY: Routledge.

United Nations Educational, Scientific and Cultural Organization. (2020). *Education for sustainable development: A roadmap*. Paris, France: UNESCO. https://unesdoc.unesco.org/ark:/48223/pf0000230514

References

Abzug, B. (1991). Women want an equal say in UNCED. *The Network 92, 9*, 2.

Anthropocene Working Group. (2019). *Working group on the Anthropocene*. http://quaternary.stratigraphy.org/working-groups/anthropocene/.

Apple, M. W. (2004). *Ideology and curriculum* (3rd ed.). New York, NY: Routledge.

Åsberg, C. (2017). Feminist posthumanities in the Anthropocene: Forays into the postnatural. *Journal of Posthuman Studies, 1*(2), 185–204. www.jstor.org/stable/10.5325/jpoststud.1.2.0185

Bajohr, H. (2019). *Anthropocene and negative anthropology: Return of man.* https://publicseminar.org/essays/anthropocene-and-negative-anthropology/

Bell, A., & Russell, C. (2000). Beyond human, beyond words: Anthropocentrism, critical pedagogy, and the poststructuralist turn. *Canadian Journal of Education, 25*(3), 188–203.

Braidotti, R. (2013). *The posthuman.* Cambridge, UK: Polity.

Brennan, M. (2017). Struggles for teacher education in the age of the Anthropocene. *Journal of Education, 69.* https://doi.org/10.17159/2520-9868/i69a02

Calderon, D. (2014). Anticolonial methodologies in education: Embodying land and Indigeneity in Chicana feminisms. *Journal of Latino/Latin American Studies, 6*(2), 81–96.

Carstens, D. (2016). The Anthropocene crisis and higher education: A fundamental shift. *South African Journal of Higher Education, 30*(3), 255–273. http://doi.org/10.20853/30-3-650

Charteris, J., Nye, A., Mavropoulou, S., & Reyes, V. (2018). Preface: Positioning educational research in the age of the Anthropocene – a "wicked" epoch. In V. C. Reyes, J. Charteris, A. Nye, & S. Mavropoulou (Eds.), *Educational research in the age of the Anthropocene* (pp. xv–xxx). Hershey, PA: IGI Global.

Chattopadhyay, S. (2019). Infiltrating the academy through (anarcha-)ecofeminist pedagogies. *Capitalism Nature Socialism, 30*(1), 31–49.

Colebrook, C. (2017). We have always been post-Anthropocene: The Anthropocentric counterfactual. In R. Grusin (Ed.), *Anthropocene feminism* (pp. 1–20). Minneapolis: University of Minnesota Press.

Crex Crex Collective. (2018). On the Anthropocene. In B. Jickling, S. Blenkinsop, N. Timmerman, & M. D. D. Sitka-Sage (Eds.), *Wild pedagogies: Touchstones for renegotiating education and the environment in the Anthropocene* (pp. 51–61). Cham, Switzerland: Palgrave Macmillan.

Crutzen, P. J., & Schwägerl, C. (2011, January 24). Living in the Anthropocene: Toward a new global ethos. *Yale Environment 360.* https://e360.yale.edu/features/living_in_the_anthropocene_toward_a_new_global_ethos

Crutzen, P. J., & Stoermer, E. F. (2000). The "Anthropocene." *Global Change Newsletter, 41,* 17–18.

Darder, A., & Torres, R. D. (Eds.). (2013). *Latinos and education: A critical reader.* New York, NY: Routledge.

de Carvalho, D. (2019). ACARA: The role of education. *Education Matters Primary.* https://educationmattersmag.com.au/acara-the-role-of-education/

de Freitas, E., & Truman, S. E. (2020). New empiricisms in the Anthropocene: Thinking with speculative fiction about science and social inquiry. *Qualitative Inquiry.* http://doi.org/10.1177/1077800420943643

Decuypere, M., Hoet, H., & Vandenabeele, J. (2019). Learning to navigate (in) the Anthropocene. *Sustainability, 11,* 547. https://doi.org/10.3390/su11020547

Dewey, J. (1916). *Democracy and education: An introduction to the philosophy of education.* New York, NY: Macmillan.

Di Chiro, G. (1987). Applying a feminist critique to environmental education. *Australian Journal of Environmental Education, 3*(1), 10–17.

Fawcett, L. (2013). Three degrees of separation: Accounting for naturecultures in environmental education research. In R. B. Stevenson, M. Brody, J. Dillon, & A. Wals (Eds.), *International handbook of research on environmental education* (pp. 409–417). New York, NY: Routledge.

Gaard, G. (2015). Ecofeminism and climate change. *Women's Studies International Forum, 49*, 20–33.

Gilbert, J. (2015). Transforming science education for the Anthropocene: Is it possible? *Research in Science Education, 46*, 187–201. https://doi.org/10.1007/s11165-015-9498-2

Global Gender and Climate Alliance. (2013). *Overview of linkages between gender and climate change.* https://issuu.com/undp/docs/pb1_asiapacific_capacity_final

Gough, A. (1997). *Education and the environment: Policy, trends and the problems of marginalisation.* Melbourne, Australia: Australian Council for Educational Research.

Gough, A. (1999a). The power and promise of feminist research in environmental education. *Southern African Journal of Environmental Education, 19*, 28–39.

Gough, A. (1999b). Recognising women in environmental education pedagogy and research: Toward an ecofeminist poststructuralist perspective. *Environmental Education Research, 5*(2), 143–161. https://doi.org/10.1080/1350462990050202

Gough, A. (2013). Researching differently: Generating a gender agenda for research in environmental education. In R. B. Stevenson, M. Brody, J. Dillon, & A. Wals (Eds.), *International handbook of research on environmental education* (pp. 375–383). New York, NY: Routledge.

Gough, A., & Whitehouse, H. (2003). The "nature" of environmental education from a feminist poststructuralist viewpoint. *Canadian Journal of Environmental Education, 8*, 31–43.

Gough, A., & Whitehouse, H. (2018). New vintages and new bottles: The "nature" of environmental education from new material feminist and ecofeminist viewpoints. *The Journal of Environmental Education, 49*(4), 336–349. https://doi.org/10.1080/00958964.2017.1409186

Gough, A., & Whitehouse, H. (2020, online). Challenging amnesias: Feminist new materialism/ecofeminism/women/climate/education. *Environmental Education Research, 26*(9–10), 1420–1434. https://doi.org/10.1080/13504622.2020.1727785

Grande, S. (2004). *Red pedagogy: Native American social and political thought.* Lanham, MD: Rowman & Littlefield.

Gray, T. (2018). Thirty years on, and has the gendered landscape changed in outdoor learning? In T. Gray & D. Mitten (Eds.), *The Palgrave international handbook of women and outdoor learning* (pp. 35–54). Cham, Switzerland: Palgrave Macmillan.

Greenall Gough, A., & Robottom, I. (1993). Towards a socially critical environmental education: Water quality studies in a coastal school. *Journal of Curriculum Studies, 25*(4), 301–316. https://doi.org/10.1080/0022027930250401

Greenwood, D. (2014). Culture, environment, and education in the Anthropocene. In M. P. Mueller et al. (Eds.), *Assessing schools for generation R (responsibility)* (pp. 279–292). Cham, Switzerland: Springer.

Grusin, R. (Ed.). (2017). *Anthropocene feminism.* Minneapolis: University of Minnesota Press.

Haluza-DeLay, R. (2013). Educating for environmental justice. In R. B. Stevenson, M. Brody, J. Dillon, & A. Wals (Eds.), *International handbook of research on environmental education* (pp. 394–403). New York, NY: Routledge.

Haraway, D. (1997). *Modest_Witness@Second_Millennium.FemaleMan_Meets_Onco-Mouse: Feminism and technoscience*. London, UK: Routledge.

Haraway, D. (2018). Staying with the trouble for multispecies environmental justice. *Dialogues in Human Geography, 8*(1), 102–105.

Harding, S. (2015). *Objectivity and diversity: Another logic of scientific research*. Chicago, IL: University of Chicago Press.

hooks, b. (1989). *Talking back: Thinking feminist, thinking black*. Cambridge, MA: South End Press.

Huckle, J. (1991). Education for sustainability: Assessing pathways to the future. *Australian Journal of Environmental Education, 7*, 43–62.

Jagodzinski, J. (2013). Art and its education in the Anthropocene: The need for an avant-garde without authority. *Journal of Curriculum and Pedagogy, 10*(1), 31–34. https://doi.org/10.1080/15505170.2013.790000

Kahn, R. (2008). Towards ecopedagogy: Weaving a broad-based pedagogy of liberation for animals, nature, and the oppressed people of the earth. In A. Darder, R. Torres, & M. Baltodano (Eds.), *The critical pedagogy reader* (2nd ed., pp. 522–540). New York, NY: Routledge.

Kemmis, S., Cole, P., & Suggett, D. (1983). *Orientations to curriculum and transition: Towards the socially critical school*. Melbourne, Australia: Victorian Institute of Secondary Education.

Kurt, I. (2015). Education and social reproduction in schools. *European Journal of Social Sciences Education and Research, 2*(4), 223–226. http://doi.org/10.26417/ejser.v5i1.p223–226

Latour, B. (2014). Agency at the time of the Anthropocene. *New Literary History, 45*(1), 1–18.

Leinfelder, R. (2013). Assuming responsibility for the Anthropocene: Challenges and opportunities in education. *RCC Perspectives, 2*, 9–28.

Lewis, S. L., & Maslin, M. A. (2015). Defining the Anthropocene. *Nature, 519*, 171–180. https://doi.org/10.1038/nature14258

Li, H.-L. (2017). Rethinking vulnerability in the age of the Anthropocene: Toward ecologizing education. *Educational Theory, 67*(4), 435–451. http://doi.org/10.1111/edth.12264

Lloro-Bidart, T. (2015). A political ecology of education in/for the Anthropocene. *Environment and Society: Advances in Research, 6*, 128–148. http://doi.org/10.3167/ares.2015.060108

Lloro-Bidart, T. (2017). A feminist posthumanist political ecology of education for theorizing human-animal relations/relationships. *Environmental Education Research, 23*(1), 111–130.

Lloro-Bidart, T., & Finewood, M. H. (2018). Intersectional feminism for the environmental studies and sciences: Looking inward and outward. *Journal of EnvironmentalStudiesandSciences,8*,142–151.https://doi.org/10.1007/s13412-018-0468-7

Lowan-Trudeau, G. (2015). *From bricolage to metissage: Rethinking intercultural approaches to Indigenous environmental education and research*. Bern, Switzerland: Peter Lang.

Maina-Okori, N. M., Koushik, J. R., & Wilson, A. (2018). Reimagining intersectionality in environmental and sustainability education: A critical literature review. *The Journal of Environmental Education, 49*(4), 286–296. http://doi.org/10.1080/00958964.2017.1364215

Malone, K. (2017). Ecological posthumanist theorizing: Grappling with child-dog-bodies. In K. Malone, S. Truong, & T. Gray (Eds.), *Reimagining sustainability in precarious times* (pp. 161–172). Singapore: Springer.

Malone, K., & Truong, S. (2017). Sustainability, education, and Anthropocentric precarity. In K. Malone, S. Truong, & T. Gray (Eds.), *Reimagining sustainability in precarious times* (pp. 13–16). Singapore: Springer.

Martusewicz, R. A., Edmundson, J., & Lupinacci, J. (2015). *Ecojustice education: Toward diverse, democratic, and sustainable communities* (2nd ed.). New York, NY: Routledge.

Mitten, D., Gray, T., Allen-Craig, S., Loeffler, T. A., & Carpenter, C. (2018). The invisibility cloak: Women's contributions to outdoor and environmental education. *The Journal of Environmental Education, 49*(4), 318–327.

Monastersky, R. (2015a). Anthropocene: The human age. *Nature, 519,* 144–147. http://doi.org/10.1038/519144a

Monastersky, R. (2015b, January 16). First atomic blast proposed as start of Anthropocene. *Nature.* http://doi.org/10.1038/nature.2015.16739

Morton, T. (2014). How I learned to stop worrying and love the term Anthropocene. *Cambridge Journal of Postcolonial Literary Inquiry, 1*(2), 257–264. http://doi.org/10.1017/pli.2014.15

Morton, T. (2017). *Humankind: Solidarity with nonhuman people.* London, UK: Verso.

Nxumalo, F. (2020). Situating Indigenous and Black childhoods in the Anthropocene. In A. Cutter-Mackenzie, K. Malone, & E. Barratt Hacking (Eds.), *Research handbook on childhoodnature: Assemblages of childhood and nature* (pp. 535–556). Cham, Switzerland: Springer.

Nxumalo, F., & Cedillo, S. (2017). Decolonizing "place" in early childhood studies: Thinking with Indigenous onto-epistemologies and Black feminist geographies. *Global Studies of Childhood, 7*(2), 99–112.

Piersol, L., & Timmerman, N. (2017). Reimagining environmental education within academia: Storytelling and dialogue as lived ecofeminist politics. *The Journal of Environmental Education, 48*(1), 10–17.

Russell, C. (2019). An intersectional approach to learning and teaching about humans and other animals in educational contexts. In T. Lloro-Bidart & V. S. Banschbach (Eds.), *Animals in environmental education: Interdisciplinary approaches to curriculum and pedagogy* (pp. 35–52). Cham, Switzerland: Palgrave Macmillan.

Russell, C., & Bell, A. (1996). A politicized ethic of care: Environmental education from an ecofeminist perspective. In K. Warren (Ed.), *Women's voices in experiential education* (pp. 172–181). Dubuque, IA: Kendall Hunt.

Russell, C., & Fawcett, L. (2013). Moving margins in environmental education. In R. B. Stevenson, M. Brody, J. Dillon, & A. Wals (Eds.), *International handbook of research on environmental education* (pp. 365–374). New York, NY: Routledge.

Russell, J. (2019). Attending to nonhuman animals in pedagogical relationships and encounters. In T. Lloro-Bidart & V. S. Banschbach (Eds.), *Animals in environmental education: Interdisciplinary approaches to curriculum and pedagogy* (pp. 117–137). Cham, Switzerland: Palgrave Macmillan.

Schiro, M. S. (2007). *Curriculum theory: Conflicting visions and enduring concerns.* Thousand Oaks, CA: SAGE.

Schneiderman, J. S. (2017). The Anthropocene controversy. In R. Grusin (Ed.), *Anthropocene feminism* (pp. 169–195). Minneapolis: University of Minnesota Press.

Scranton, R. (2015). *Learning to die in the Anthropocene: Reflections on the end of a civilization*. San Francisco, CA: City Light Publishers.

Seaman, M. J. (2007). Becoming more (than) human: Affective posthumanisms, past and future. *Journal of Narrative Theory, 37*(2), 246–275.

Shava, S. (2013). The representation of Indigenous knowledges. In R. B. Stevenson, M. Brody, J. Dillon, & A. Wals (Eds.), *International handbook of research on environmental education* (pp. 384–393). New York, NY: Routledge.

Shiva, V. (1989). *Staying alive: Women, ecology and development*. London, UK: Zed Books.

Shiva, V. (1991). Biodiversity, biotechnology and profits. In V. Shiva (Ed.), *Biodiversity: Social and ecological perspectives* (pp. 43–58). London, UK: Zed Books/World Rainforest Movement.

Shiva, V. (2005). *Earth democracy: Justice, sustainability, and peace*. Boston, MA: South End Press.

Simpson, L. (2002). Indigenous environmental education for cultural survival. *Canadian Journal of Environmental Education, 7*(1), 13–25.

Snaza, N., Appelbaum, P., Bayne, S., Carlson, D., Morris, M., . . . Weaver, J. A. (2014). Toward a posthuman education. *Journal of Curriculum Theorizing, 30*(2), 39–55.

Somerville, M. (2017). The Anthropocene's call to educational research. In K. Malone, S. Truong, & T. Gray (Eds.), *Reimagining sustainability in precarious times* (pp. 17–27). Singapore: Springer.

Somerville, M., & Powell, S. J. (2019). Thinking posthuman with mud: And children of the Anthropocene. *Educational Philosophy and Theory, 51*(8), 829–840. http://doi.org/10.1080/00131857.2018.1516138

Steinberg, S. (2012). Critical pedagogy and cultural studies research: Bricolage in action. *Counterpoints, 422*, 230–254. www.jstor.org/stable/42981761

Stevens, L., Tait, P., & Varney, D. (2018). Introduction: "Street-fighters and philosophers" – traversing ecofeminisms. In L. Stevens, P. Tait, & D. Varney (Eds.), *Feminist ecologies: Changing environments in the Anthropocene* (pp. 1–22). Cham, Switzerland: Palgrave Macmillan.

Subramanian, M. (2019, May 21). Anthropocene now: Influential panel votes to recognize Earth's new epoch. *Nature*. http://doi.org/10.1038/d41586-019-01641-5

Sullivan, H. I. (2019). Petro-texts, plants, and people in the Anthropocene: The dark green. *Green Letters: Studies in Ecocriticism, 23*(2), 152–167. http://doi.org/10.10 80/14688417.2019.1650663

Taylor, A. (2017). Beyond stewardship: Common world pedagogies for the Anthropocene. *Environmental Education Research, 23*(10), 1448–1461.

Thorne, M., & Whitehouse, H. (2018). Environmental stewardship education in the Anthropocene (part two): A learning for environmental citizenship conceptual framework. *The Social Educator, 36*(1), 17–28.

Tuck, E., McKenzie, M., & McCoy, K. (2014). Land education: Indigenous, postcolonial, and decolonizing perspectives on place and environmental education research. *Environmental Education Research, 20*(1), 1–23.

UN Women. (2012). *The future women want: A vision of a sustainable development for all*. New York, NY: United Nations Entity for Gender Equality and the Empowerment of Women.

United Nations. (1973). *Report of the United Nations conference on the human environment. Stockholm, 5–16 June 1972*. New York, NY: United Nations. https://digitallibrary.un.org/record/523249?ln=en

United Nations. (1993). *Agenda 21: Earth summit – the United Nations programme of action from Rio.* New York, NY: United Nations. http://doi.org/10.18356/a9d63da6-en

United Nations. (2002). *Report of the world summit on sustainable development, Johannesburg, South Africa, 26 August-4 September 2002.* New York, NY: United Nations. https://digitallibrary.un.org/record/478154?ln=en

United Nations. (2012). *The future we want: Outcomes document adopted at Rio + 20.* https://sustainabledevelopment.un.org/content/documents/733FutureWeWant.pdf

United Nations. (2016). *Sustainable development goals: 17 goals to transform our world.* www.un.org/sustainabledevelopment/sustainable-development-goals/

United Nations Educational, Scientific and Cultural Organization. (1975). *The Belgrade charter: A global framework for environmental education.* https://unesdoc.unesco.org/ark:/48223/pf0000017772

United Nations Educational, Scientific and Cultural Organization. (1978). *Intergovernmental conference on environmental education, Tbilisi, USSR, October 14–26, 1977. Final report.* Paris, France: UNESCO.

United Nations Educational, Scientific and Cultural Organization. (1980). *Environmental education in the light of the Tbilisi conference.* Paris, France: UNESCO.

United Nations Educational, Scientific and Cultural Organization. (2019). *Framework for the implementation of education for sustainable development beyond 2030.* Paris, France: UNESCO. https://unesdoc.unesco.org/ark:/48223/pf0000370215.page=7

United Nations Framework Convention on Climate Change. (2012). *Promoting gender balance and improving the participation of women in UNFCCC negotiations and in the representation of parties in bodies established pursuant to the convention or the Kyoto Protocol.* https://digitallibrary.un.org/record/739142?ln=en#record-files-collapse-header

United Nations Framework Convention on Climate Change. (2017). *Gender and climate change: Establishment of a gender action plan.* https://unfccc.int/resource/docs/2017/sbi/eng/l29.pdf

United Nations Girls Education Initiative. (2019). *Situation analysis of SDG 4 with a gender lens.* https://bangkok.unesco.org/content/situation-analysis-sdg4-gender-lens

Wagler, R. (2011). The Anthropocene mass extinction: An emerging curriculum theme for science educators. *American Biology Teacher, 73*(2), 78–83.

Wallin, J. J. (2017). Pedagogy at the brink of the post-Anthropocene. *Educational Philosophy and Theory, 49*(11), 1099–1111. http://doi.org/10.1080/00131857.2016.1163246

Wilkinson, K. (2016). Is this the future we want? An ecofeminist comment on the UN conference on sustainable development outcome document. In K. Rubenstein & K. G. Young (Eds.), *The public law of gender: From the local to the global* (pp. 538–560). Cambridge, UK: Cambridge University Press.

World Economic Forum. (2017). *The global gender gap report 2017.* www.weforum.org/reports/the-global-gender-gap-report-2017

Section IV

Conclusion

And so, to the end: looking forward. After 30 years of researching and writing around feminism and environmental education it is hard not to be pessimistic about how we seem to have taken three steps forward and two back. This has particularly been the case since the COVID-19 pandemic, but the seeds were there before then. In 2000 the United Nations agreed on Millennium Development Goals for 2015. In 2015 their report "acknowledges uneven achievements and shortfalls in many areas. The work is not complete, and it must continue in the new development era" (United Nations 2015a, 4), and so we have the Sustainable Development Goals (SDG) (United Nations 2015b). As discussed in Chapter 16, progress on over 50 per cent of the SDG targets "is weak or insufficient, and on 30 per cent, it has stalled or gone into reverse" (United Nations 2023, 2). The 2023 report on the SDG concludes, "Unless we act now, the 2030 Agenda could become an epitaph for a world that might have been" (2). This fate became particularly apparent in July 2023 as the world had its hottest month on record. And there is much evidence that women and other vulnerable groups are most affected by climate change (UN Women 2022).

There have been successes. There are more women taking on leadership roles in environment-related organisations, and women now seem to dominate the membership of these organisations. But given patriarchy's usual approach to things that are seen as female dominated (such as nursing and teaching), has this meant that the environment has been marginalised?

The insidiousness of neoliberalism and its influence was something we did not see until it was too late, as is well indicated by the authors in *Neoliberalism and Environmental Education* (Henderson et al. 2017). This has flowed into arguments about the "Capitalocene," which I see as a preferable term to "Anthropocene" (see Chapter 16).

Getting environmental education, and particularly gendered perspectives on environmental education, into curricula is an ongoing challenge, and, as discussed in Chapter 16, there is a need for a different, socially transformative, pedagogy too.

Finally, we need to look at future generations and how they might respond. The New Nihilists, which are seen as including people like Greta Thunberg

DOI: 10.4324/9781003390930-19

(Syfret 2023), are our future – but how will they respond? All we are left with is hope.

References

"Explainer: How gender inequality and climate change are interconnected". *UN Women*, 28 February 2022. Accessed 18 November 2023. www.unwomen.org/en/news-stories/explainer/2022/02/explainer-how-gender-inequality-and-climate-change-are-interconnected

Henderson, Joseph, David Hursh, and David Greenwood, eds. 2017. *Neoliberalism and Environmental Education*. Abingdon/New York: Routledge.

Syfret, Wendy. 2023. "Rise of a new generation of nihilists". *The Age*, 23 October. www.theage.com.au/lifestyle/life-and-relationships/new-nihilism-how-gen-z-is-embracing-a-life-of-futility-and-meaninglessness-20231016-p5ecra.html

United Nations. 2015a. *The Millennium Development Goals Report 2015*. New York: United Nations.

United Nations. 2015b. *Transforming our world: The 2030 agenda for sustainable development*. Accessed 10 November 2023. https://sdgs.un.org/publications/transforming-our-world-2030-agenda-sustainable-development-17981

United Nations. 2023. *The sustainable development goals report 2023: Special edition*. Accessed 12 November 2023. https://unstats.un.org/sdgs/report/2023/

16 Where to now for gendered environmental education research?

Annette Gough

Abstract

This chapter discusses future directions for educational research from feminist and other gendered perspectives in light of the current sense of urgency about climate change and other ecological crises. Starting with a discussion of the global context, particularly from United Nations and UNESCO documents, the chapter then focuses on the Anthropocene/Capitalocene, changing conceptions of humans and nature, listening to new voices and the need to change both school curricula and pedagogy to take into account feminist and other gendered perspectives as well as changing conceptions of humans and nature. The chapter concludes with a discussion of generational differences and the possibilities of nihilistic hope.

Introduction

In an ideal world this chapter would be focused on the progress we have made towards gender equality, environmental protection and addressing climate change in the past 20-plus years and building on that. But this is not an ideal world. Instead, as Rosi Braidotti (2022, 57) eloquently summarises, we are at crisis point:

> The sense of emergency arising from the posthuman convergence intro-
> duces new patterns of discrimination upon older modes of oppression.
> It exposes the inner power structures of advanced capitalism and the
> sexualised, racialised and naturalised political economies of exploitable
> labour and dispensable bodies that support it. Importantly, it com-
> bines this analysis with a sense of urgency arising from the spectre of
> the Sixth Extinction, the ecological crisis and climate change. Environ-
> mental scarcity clashes with technological abundance within the fast
> flows of capital, triggering a short-circuiting of planetary dimensions.

A sense of urgency was made very clear in July 2023 when the United Nations secretary general Antonio Guterres proclaimed, "Climate change is here. It is

DOI: 10.4324/9781003390930-20

terrifying. . ... The era of global warming has ended. The era of global boiling has arrived" (UN Climate Change News 2023). Earlier chapters discussed how climate change most affects the more vulnerable in society such as women, children and marginalised groups, and progress on the other ecological crises included in the Sustainable Development Goals (United Nations 2015) are generally worsening (United Nations 2023).

This chapter discusses some key themes for environmental education research from feminist and other gendered perspectives going forward. These include the global context, the Anthropocene/Capitalocene, changing conceptions of humans and nature, listening to new voices and changing the curriculum and pedagogy.

Researching global context

In 2015 the United Nations agreed on a set of Sustainable Development Goals (SDGs) to transform the world by 2030. Always ambitious, the COVID-19 pandemic had a major impact on progress towards achieving the goals. In the early days of the pandemic the United Nations (2020) reflected on the global challenges it posed to each of the SDGs and developed *A UN Framework for the Immediate Socio-economic Response to COVID-19*. This document outlined five streams of work, to be supported by the United Nations Development system, "connected by a strong environmental sustainability and gender equality imperative to build back better" (1). The framework seemed to be a pivot point, foregrounding the importance of the environment rather than just humans, and referring to "mutually beneficial symbiotic relation between humans and their surrounding ecosystems":

> Once the health crisis is over, we cannot have business-as-usual practices that increase emissions and other environmental externalities like pressure on wildlife and biodiversity. The performance and resilience of our socio-economic systems depend on the state of the natural environment and ecosystems. A mutually beneficial symbiotic relation between humans and their surrounding ecosystems is inter alia the answer to more resilient economies and societies. Securing the global environmental commons requires living within planetary boundaries, conserving and sustainably managing globally shared resources and ecosystems, as well as their shared vulnerabilities and risks to promote human wellbeing.
>
> (4)

Sadly, this framework and its focus on a different relationship between humans and their environments were usurped by a business-as-usual neoliberal agenda, and since 2020 the climate crisis has continued, and gender equality and biodiversity have gone backwards. The 2021 SDG progress report (United

Nations Department of Economic and Social Affairs 2021) noted, among other impacts, that

- The economic slowdown in 2020 had done little to slow the climate crisis. Concentrations of major greenhouse gases continued to increase, while the global average temperature was about 1.2°C above pre-industrial levels, dangerously close to the 1.5°C limit established in the Paris Agreement.
- The COVID-19 pandemic has adversely affected progress towards gender equality: violence against women and girls has intensified; child marriage is expected to increase; and women have suffered a disproportionate share of job losses and increased care work at home.
- The world fell short on 2020 targets to halt biodiversity loss and 10 million hectares of forest being lost each year between 2015–2020.

The 2023 progress report (United Nations, 2) does not paint a prettier picture:

> Progress on more than 50 per cent of targets of the SDGs is weak and insufficient; on 30 per cent, it has stalled or gone into reverse. These include key targets on poverty, hunger and climate. Unless we act now, the 2030 Agenda could become an epitaph for a world that might have been.

Before the pandemic, Elspeth Probyn (2016, 12) commented, "gender and sexuality, as well as ethnicity and class, have been squeezed out of current debates on the Anthropocene, climate change, and the more-than-human". This seems to be even more the case now, despite women often facing "higher risks and greater burdens from the impacts of climate change in situations of poverty and due to existing roles, responsibilities and cultural norms" (UN Climate Change News 2023).

We need to continue to work to keep gender and the environment, including climate change, on the environmental education research agenda. For much too long environmental educators have allowed education for sustainability (or sustainable development) to take over the space. The SDGs reduced environmental education to education for sustainable development (ESD) and buried it in a list of knowledge and skills needed to promote sustainable development:

> By 2030, ensure that all learners acquire the knowledge and skills needed to promote sustainable development, including, among others, through education for sustainable development and sustainable lifestyles, human rights, gender equality, promotion of a culture of peace and non-violence, global citizenship and appreciation of cultural diversity and of culture's contribution to sustainable development.
>
> (United Nations 2015, Target 4.7)

Recently there have been some glimmers of hope that the environment is again being recognised in the ESD discourses, together with gender. The UNESCO *Education for Sustainable Development: A Roadmap* (2020) recognises the need to address environmental sustainability crises:

> The current climate emergency and other environmental sustainabil-
> ity crises are the product of human behaviour. . .. This means not only
> addressing environmental challenges but also revisiting the complex mix
> of social and economic issues such as inequality that are intertwined with
> the cause and impact of these problems. . . . We must urgently learn to
> live differently.
>
> (UNESCO 2020, 6)

The roadmap sees ESD as empowering learners with knowledge, skills, values and attitudes to take informed decisions and make responsible actions for environmental integrity, economic viability and a just society empowering people of all genders, for present and future generations, while respecting cultural diversity (UNESCO 2020, 8), that is, living differently.

In addition, the *Berlin Declaration on Education for Sustainable Development* (UNESCO 2021a) re-prioritised the environment on the education agenda, particularly relationships with nature:

> We are convinced that urgent action is needed to address the dramatic
> interrelated challenges the world is facing, in particular, the climate cri-
> sis, mass loss of biodiversity, pollution, pandemic diseases, extreme pov-
> erty and inequalities, violent conflicts, and other environmental, social
> and economic crises that endanger life on our planet. We believe that
> the urgency of these challenges, exacerbated by the Covid-19 pandemic,
> requires a fundamental transformation that sets us on the path of sus-
> tainable development based on more just, inclusive, caring and peaceful
> relationships with each other and with nature.
>
> (UNESCO 2021a, 1)

It also included a commitment by signatories to "[e]mphasize gender equality and non-discrimination in access to knowledge and skills, and ensure gender mainstreaming in ESD which enables a more profound and holistic understanding of sustainability challenges and potential solutions" (UNESCO 2021a, 3).

This Declaration prompted *UNESCO News* (2021) to proclaim, "UNE-SCO declares environmental education must be a core curriculum component by 2025". This is a very different position on the environment compared with the Decade of Education for Sustainable Development (DESD) implementation scheme (UNESCO 2005) which combined the Millennium Development Goals (MDG) process, the Education for All (EFA) movement and the United Nations Literacy Decade (UNLD) with ESD and buried environmental concerns within ESD. The Declaration also included a commitment to "[r]ecognize climate change as a priority area of ESD of particular importance to Small Island Developing States (SIDS), as they require special attention in terms of ESD implementation due to their increasing vulnerability to climate change and natural hazards" (UNESCO 2021a, 3). These statements firmly put climate change on the education agenda. However, we are yet to see any concerted action on these commitments, particularly in Australia.

Researching in the Anthropocene/Capitalocene

As discussed in Chapter 15, while the term "Anthropocene" is receiving increasingly frequent usage, prominent feminists argue that using Anthropocene makes "man" the centre of the universe again, to the extent that "Man the taxonomic type" becomes "Man the brand" (Haraway 1997, 74). And this "universal 'Man' is implicitly assumed to be masculine, white, urbanized, speaking a standard language, heterosexually inscribed in a reproductive unit and a full citizen of a recognized polity" (Braidotti 2013, 65). Teresa Lloro-Bidart (2015, 132) urges that we need to interrogate the contested nature of the term itself and its political and cultural implications, "rather than unwaveringly accepting the 'age of humans'". The anthropocentrism of the Anthropocene ignores the global and gendered, classed and national disparities in human impacts on the planet that actually exist, and there are better alternatives for naming the causes of the planetary crisis.

In addition to Anthropocene, other names proposed for describing these troubled times, many of which are still very human-centred but which name the causes, are: Capitalocene (Demos 2016; Haraway 2015); Corporatcene, Plasticene and Petrolcene (Schneiderman 2017); Econocene and Gynocene (Norgaard 2013); Plantationocene and Chthulucene (Haraway 2015); and Pyrocene (Pyne 2019). The arguments for the Capitalocene – the age of capital – are strong:

> [I]t has the advantage of naming the culprit, locating climate change not merely in fossil fuels, but within the complex and interrelated processes of global-scale economic-political organization stretched over histories of enclosures, colonialisms, industrializations, and globalizations, which have both evolved within nature's web of life as well as brought ecological transformations to it.
>
> (Demos 2016, n.p.)

Donna Haraway (2015, 2016) sees the name Anthropocene as more of a boundary event than an epoch, as a marker of severe discontinuities because "what comes after will not be like what came before" (2015, 160), and Capitalocene as a useful word, but argues for Chthulucene as a more inclusive name "for the dynamic ongoing sym-chthonic forces and powers of which people are a part, within which ongoingness is at stake" (2015, 160) for the past, present and future. However, she does not discard Anthropocene and Capitalocene: "Despite its problems, the term Anthropocene was and is embraced because it collects up many matters of fact, concern, and care; and I hope Capitalocene will roll off myriad tongues soon" (2016, 184).

Although "Anthropocene" is increasingly being referenced in the education field, including environmental education, Capitalocene has a lot going for it in terms of naming the culprits behind our environmental crises, and it warrants focus in environmental education research, particularly given how much women are impacted and how it undermines the Nature/Culture binary. Jason Moore (2018) argues, "Popular Anthropocene is but the latest of a long series of environmental concepts that deny the multi-species violence and inequality of capitalism and assert that the devastation created by capital is the responsibility of all humans" (239), whereas "[t]he Capitalocene *is* an argument about thinking ecological crisis" (239), recognising that the rise of capitalism "was tightly linked to the expulsion of women from Society, and their forcible relocation into the realm of Cheap – and cheapened – Natures" (270) which are outside the circuit of capital but essential to its operation (i.e. "the unpaid work of 'women, nature, and colonies' (Mies 1986, 77)") (242). Moore argues that by addressing the Capitalocene we can transcend the Nature/Culture binary, address the re/production of gendered and racialised domination and displace it with a value-relational ontology.

Thus, environmental education research should be thinking with Capitalocene rather than Anthropocene.

Research changing conceptions of humans and nature

Moore's (2018) discussion of the Capitalocene highlights how the association of nature with women has connections with the rise of capitalism, which seems a good reason for rejecting the binary. This binary can also invoke an essentialism that led to the marginalisation of ecofeminism in the 1990s (Gaard 2011).

In this century there have increasingly been arguments for replacing an essentialist association between women and nature with various versions that are associated with versions of materialist feminist environmentalism (Gaard 2011) and a partnership ethic (Merchant 2016). As Stacy Alaimo (2015) argues, given the current state of the planet, feminist theory needs to contend with "nature" in ways that are not only part of a social justice agenda but also to the survival of innumerable species, ecosystems and life forms. She proposed a concept of transcorporeality (Kuznetski and Alaimo 2020, 139), which she argues is

> the opposite of distancing or dividing the human from external nature. It implies that we're literally enmeshed in the physical material world, so environmentalism cannot be an externalized and optional kind of pursuit, but is always present, always at hand. It's not about other places, because everything that we do, within global capitalism, has far-reaching planetary impacts.

Other changing conceptions of nature have drawn attention to the entanglement of human and more-than-human life. As discussed in several earlier chapters, these have particularly been associated with new materialism. As Karen Barad (2007, 170) argues:

> Bodies do not simply take their places in the world. They are not simply situated in, or located in, particular environments. Rather "environments" and "bodies" are intra-actively co-constituted. Bodies ("human", "environmental," or otherwise) are integral "parts" of, or dynamic reconfigurings of, what is.

This entanglement of the world and everything in it leads Barad to argue that there are no separate "objects" with boundaries in nature, but there are identifiable "phenomena", which are the "*ontological* inseparability of agentially intra-acting components" and "basic units of reality", where " 'intra-action' *signifies the mutual constitution of entangled agencies*" (p. 33, emphasis in original).

Haraway (2018, 102) makes a similar argument that "there can be no environmental justice or ecological reworlding without multispecies environmental justice and that means nurturing and inventing enduring multispecies – human and nonhuman – kindreds". For Haraway, "Kin making requires taking the risk of becoming-with new kinds of person-making, generative and experimental categories of kindred, other sorts of 'we', other sorts of 'selves', and unexpected kinds of sympoetic, symchthonic human and nonhuman critters" (2018, 102). Briony Towers, Blanche Verlie and I have already explored this notion of making kin in our chapter about fire as our unruly kin and humanity's entanglement with fire, foregrounding the agency of human and more-than-human materiality (i.e., "nature") and their entanglement (Gough et al. 2022).

Another important area is in human-animal relations/relationships (Lloro-Bidart 2017), particularly from feminist and other gendered perspectives as we continue to break down the binaries.

What is clear from these deliberations is that we need to be researching differently – no longer focused on human dominion over other living things but focused on how humans are entangled with their more-than-human materiality.

Changing voices

While feminist research in environmental education has continued with a few publications each year since the mid-1990s, as demonstrated in Figure 16.1, the last decade has seen a significant rise in the number of publications being reported as being related to feminist research in environmental education.

Starting about ten years ago, Connie Russell, Teresa Lloro and Marcia McKenzie started compiling a bibliography of feminist research in environmental education, now expanded to also include Gender and EE references. This was subsequently expanded by the editors and authors associated with the special issues on gender and environmental education of *The Journal of*

Feminist research in EE bibliography

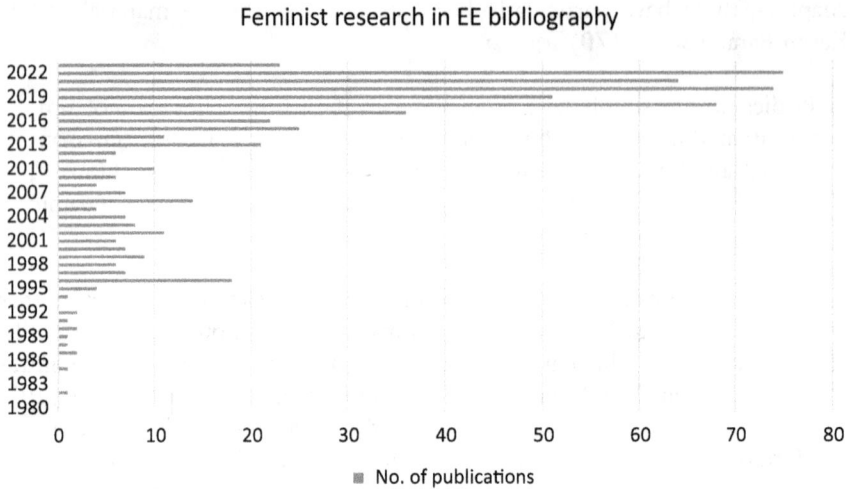

Figure 16.1 Date distribution of publications in *Feminist Research in EE Bibliography* (Russell 2023).

Environmental Education (in 2017 and 2018), and it continues to be expanded on at least an annual basis in a process described by Russell (2023, 1):

> This bibliography is focused on environmental education literature that makes critical connections to gender and/or explicitly makes use of feminist analyses. **Please add to this bibliography** by sending APA-formatted references (your own that have been missed or that are hot off the press or others that you have come across in your work) to Connie Russell at crussell@lakeheadu.ca who will periodically share updated versions with the EE Intersectional Feminist Caucus on our Facebook site and to any others who express interest. Please feel free to share.

No value judgements are made on inclusions in the bibliography, and some may query how some of the entries meet the criteria, but what is clear is that there has been a significant increase in the number of publications being reported: the July 2016 version of the Bibliography had 91 entries, the July 2023 version had 606 entries.

The publications until around 2006 were mainly focused on arguments around putting women into the environmental and outdoor education agenda, identifying gender differences, girls' experiences of programmes and identifying the absence of women's interests – and they tended to be written by Western white middle-class females. The entries since 2007 are increasingly embracing new ways of thinking and researching – drawing on post-colonialism, intersectionality, assemblages and entanglements, queer theory, new materialism, and ethical and political discourses. This is definitely a case of let a thousand flowers bloom as more researchers become interested in gender and new materialist perspectives, such as in the two special issues of *The Journal of Environmental Education* on gender and environmental education research (2017 and 2018), the edited collection *Queer Ecopedagogies: Explorations in Nature, Sexuality, and Education* (Russell 2021), and in articles such as Shiva Zarabadi's (2022), and Fikile Nxumalo and Marleen Tepeyolotl Villaneuva's (2020).

Going forward we need to be listening to these previously Othered voices and to provide opportunities for these voices to be heard, recognising the importance of intersectionality, assemblages and entanglements.

Challenging curricula and pedagogy

Finding a place for environmental education in formal education curricula has been a challenge since the 1970s, and even where there seems to be a place – such as "sustainability" being a cross-curriculum priority in the Australian Curriculum (ACARA 2023) – it is not mandatory nor is it examinable. Such curriculum statements are also deficient from an indigenous perspective (Blecher 2023). Neither do they allow for an intersectional approach to teaching and learning about humans and other animals (see Russell 2019, for example).

The findings from a relatively recent survey of the extent to which environmental issues are integrated in primary and secondary education policies and curricula across 46 UNESCO Member States (UNESCO 2021b, 1) are therefore not surprising:

> Over half of education policies and curricula studied made no mention of climate change. Only 19 per cent made reference to biodiversity. Countries have made progress: 83 per cent of education policies and curricula studied addressed the environment at least once, and 69 per cent mentioned sustainability – but it is clear that more needs to be done to prepare learners with the knowledge, skills, values and attitudes to act for our planet.

Environmental education researchers need to continue to argue for more environmental content in curricula at all levels of education, and they also need to draw attention to the silences in the content of these curricula. There is generally no recognition of the gendered, racialised and classist nature of the impact of environmental problems on humans as well as nonhuman and material worlds, and how the nonhuman and material worlds co-shape our mutual worlds. These aspects are generally neglected in the dominant humanist-focused curriculum. As David Chandler (2023, 1) argues,

> [T]he modern ontology (of the human as self-determined subject separate from a world composed of other-determined objects) is a problematic abstraction, failing to capture the complexity of real life. This failure means that we do not take into full account the exploitation of the natural environment.

Going forward we also need a different pedagogy – one that provides opportunities for learning to live in and engage with the world and acknowledges that we live in a more-than-human world (as discussed in Chapter 15). The different pedagogy is not very different from the socially critical pedagogy that was seen as consistent with good environmental education in the 1990s (see, for example, Greenall Gough and Robottom 1993), but it is very different from the more traditional vocational neo-classical or liberal progressive approaches to curriculum and pedagogy (Kemmis et al. 1983) that are in general use in educational institutions. We need a socially transformative rather than socially reproductive approach to learning. Acknowledging the more-than-human world expands notions of the total environment (natural and built, technological and social) that were in the Tbilisi Declaration (UNESCO 1978) to include consideration of the nonhuman and material worlds that co-shape our mutual worlds. Such considerations also mean a different approach to the content of environmental sustainability education, reinforcing that environmental issues are cross-disciplinary/interdisciplinary/multidisciplinary/transdisciplinary, not the provenance of a single discipline.

Conclusion

In an article from last century (Gough 1999), I referenced Ted Mooney's (1982, 80) statement, "here we are getting older, and there they are getting different", to draw attention to just how different the (significant life) experiences of the youth of today are from those which were significant for the baby boomers (those born between 1946 and 1964). I think it is important to keep Mooney's notion in mind as we look to the future in environmental education research.

Recent discussions of "a new generation of nihilists" (Syfret 2023) – that is, Gen Z, born between 1997 and 2012 – and "nihilist hope" (Chandler 2023) made me think again of the Mooney quote. There are many indications that the new generation is getting different. According to WGSN (2023), The New Nihilists

> are overwhelmed by global problems and they have lost faith in the ability of governments or institutions to fix them, so they are seeking solace by stepping back from the world. It's not that they have given up caring – they are simply finding that caring less is an effective coping mechanism. And while nihilism is typically seen as a negative sentiment (file it next to cynicism and pessimism), The New Nihilists are discovering that relinquishing responsibility can be a source of joy, giving them the freedom to live by their own rules, envision new realities, and set their own metrics of success and happiness outside of societal expectations.

Describing the New Nihilists, Wendy Syfret (2023, 23) argues that "they've taken the opportunity to challenge the status quo altogether . . . before you can be truly motivated to reimagine a different world, you have to stop believing in the old one".

In a similar vein, David Chandler (2023, 12) writes of nihilist hope which he sees as seeking

> to move beyond the affirmative approaches which seek to find hope in forces already in existence, and therefore potentially accessible through work on freeing the mind or upon embodied affordances and relations. There is no desire to affirm the world as it exists. It is not just the modernist construction of the 'human' but the ontology of "world" itself that is problematized in the nihilist grounding in the injustice and constitutive violence of modernity.

This group of Gen Z – "the outsiders, the independent thinkers, the rule-breakers and the navel-gazers" (WGSN 2023) – offers hope for change. If they are reimagining a different world, one can hope that it will be a world which is more tolerant of difference and willing to problematise and address injustices. The baby boomers, and previous generations, opened Pandora's box and created our environmental and climate crises – all we have is the hope that was left in the box.

References

Alaimo, Stacy. 2015. "Nature." In *The Oxford Handbook of Feminist Theory*, edited by Lisa Disch, and Mary Hawkesworth, 530–550. New York: Oxford University Press.

Australian Curriculum Assessment and Reporting Authority (ACARA). 2023. *Australian Curriculum: Sustainability*. Accessed 18 November 2023. https://v9.australiancurriculum.edu.au/teacher-resources/understand-this-cross-curriculum-priority/sustainability

Barad, Karen. 2007. *Meeting the Universe Halfway: Quantum Physics and the Entanglement of Matter and Meaning*. London: Duke University Press.

Blecher, Fi. 2023, online. "Settler futurity in the local and global: Problematising education for sustainable development in the Australian curriculum." *Policy Futures in Education*. Advance online publication. doi:10.1177/14782103231209551

Braidotti, Rosi. 2013. *The Posthuman*. Cambridge: Polity.

Braidotti, Rosi. 2022. *Posthuman Feminism*. Cambridge: Polity.

Chandler, David. 2023. "The politics of the unseen: Speculative, pragmatic and nihilist hope in the anthropocene." *Distinktion: Journal of Social Theory*. Advance online publication. doi:10.1080/1600910X.2023.2235916

Demos, T.J. 2016. "Anthropocene, Capitalocene, Gynocene: The many names of resistance." *Frontiers of Solitude*, 11 September. http://frontiers-of-solitude.org/blog/442

Gaard, Greta. 2011. "Ecofeminism revisited: Rejecting essentialism and re-placing species in a material feminist environmentalism." *Feminist Formations*, 23(2), 26–53.

Gough, Annette. 1999. "Kids don't like wearing the same jeans as their mums and dads: So whose 'life' should be in significant life experiences research?" *Environmental Education Research*, 5(4), 383–394.

Gough, Annette, Briony Towers, and Blanche Verlie. 2022. "Fire as unruly kin: Curriculum silences and human responses." In *Reimagining Science Education in the*

Anthropocene, edited by Maria Wallace, Jesse Bazzul, Marc Higgins, and Sara Tolbert, 91–106. Cham: Palgrave MacMillan.

Greenall Gough, Annette, and Ian Robottom. 1993. "Towards a socially critical environmental education: Water quality studies in a coastal school." *Journal of Curriculum Studies*, 25(4), 301–316.

Haraway, Donna. 1997. *Modest_Witness@Second_Millennium.FemaleMan_Meets_OncoMouse: Feminism and technoscience*. London, UK: Routledge.

Haraway, Donna. 2015. "Anthropocene, Capitalocene, Plantationocene, Chthulucene: Making Kin." *Environmental Humanities*, 6, 159–165.

Haraway Donna. 2016. *Staying with the trouble: Making kin in the Chthulucene*. Durham, NC: Duke University Press.

Haraway, Donna. 2018. "Staying with the trouble for multispecies environmental justice." *Dialogues in Human Geography*, 8(1), 102–105.

Kemmis, Stephen, Peter Cole, and Dahle Suggett. 1983. *Orientations to Curriculum and Transition: Towards the Socially-Critical School*. Melbourne: Victorian Institute of Secondary Education.

Kuznetski, Julia, and Stacy Alaimo. 2020. "Transcorporeality: An interview with Stacy Alaimo." *Ecozon: European Journal of Literature Culture and Environment*, 11(2), 137–146. doi:10.37536/ECOZONA.2020.11.2.3478

Lloro-Bidart, Teresa. 2015. "A political ecology of education in/for the Anthropocene." *Environment and Society: Advances in Research*, 6, 128–148.

Lloro-Bidart, Teresa. 2017. "A feminist posthumanist political ecology of education for theorizing human-animal relations/relationships." *Environmental Education Research*, 23(1), 111–130.

Merchant, Carolyn. 2016. *Autonomous Nature: Problems of Prediction and Control from Ancient Times to the Scientific Revolution*. New York, NY: Routledge.

Mooney, Ted. 1982. *Easy Travel to Other Planets*. London: Jonathan Cape.

Moore, Jason W. 2018. "The capitalocene part II: Accumulation by appropriation and the centrality of unpaid work/energy." *The Journal of Peasant Studies*, 45(2), 237–279.

Norgaard, Richard B. 2013. "The Econocene and the Delta." *San Francisco Estuary and Watershed Science*, 11(3), 1–5. http://escholarship.org/uc/item/4h98t2m0

Nxumalo, Fikile, and Marleen Tepeyolotl Villanueva. 2020. "Listening to water: Situated dialogues between Black, Indigenous and Black-Indigenous feminisms." In *Transdisciplinary Feminist Research Practices*, edited by Carol A. Taylor, Christina Hughes, and Jasmine B. Ulmer, 59–75. Abingdon/New York: Routledge.

Probyn, Elspeth. 2016. *Eating the Ocean*. Durham: Duke University Press.

Pyne, Steve. 2019. "Winter isn't coming. Prepare for the Pyrocene." *History News Network*, August 25. https://historynewsnetwork.org/article/172842.

Russell, Connie. 2019. "An intersectional approach to teaching and learning about humans and other animals in educational contexts" In *Animals in Environmental Education: Interdisciplinary Approaches to Curriculum and Pedagogy*, edited by Teresa Lloro-Bidart, and Valerie S. Banschbach, 35–52. Cham: Palgrave Macmillan.

Russell, Connie. Compiler. 2023. "Gender and EE/feminist research in EE Bibliography." Unpublished document.

Russell, Joshua, ed. 2021. *Queer Ecopedagogies: Explorations in Nature, Sexuality, and Education*. Cham: Springer.

Schneiderman, Jill S. 2017. "The Anthropocene controversy." In *Anthropocene Feminism*, edited by Richard Grusin, 169–195. Minneapolis: University of Minnesota Press.

Syfret, Wendy. 2023. "Rise of a new generation of nihilists". *The Age*, 23 October. www.theage.com.au/lifestyle/life-and-relationships/new-nihilism-how-gen-z-is-embracing-a-life-of-futility-and-meaninglessness-20231016-p5ecra.html

UNESCO. 1978. *Intergovernmental Conference on Environmental Education: Tbilisi (USSR), 14–26 October 1977: Final Report.* Paris: UNESCO.

UNESCO. 2005. *United Nations Decade of Education for Sustainable Development (2005–2014): International Implementation Scheme. ED/DESD/2005/PI/01.* Paris: UNESCO.

UNESCO. 2020. *Education for Sustainable Development: A Roadmap.* Paris: UNESCO.

UNESCO. 2021a. *Berlin Declaration on Education for Sustainable Development.* UNESCO World Conference on Education for Sustainable Development, 17–19 May. https://en.unesco.org/sites/default/files/esdfor2030-berlin-declaration-en.pdf

UNESCO. 2021b. *Learn for Our Planet: A Global Review of How Environmental Issues are Integrated in Education.* Paris: UNESCO.

"UNESCO declares environmental education must be a core curriculum component by 2025." *UNESCO News*, 20 May 2021, updated 20 April 2023. Accessed 12 November 2023. www.unesco.org/en/articles/unesco-declares-environmental-education-must-be-core-curriculum-component-2025?TSPD_101_R0=080713870fab20006ad b122333a8fe93d50c54608e0ca4f69cb261ba7db10c5731fa2d65df53a6310872c70d d4143000a16f6712bbea63c0c26218718a135a61c815cd655439e0b4144695522 fad7c7889b408890a3a39e542036e95aceb2afa#:~:text=UNESCO%20has%20 called%20for%20Education,order%20to%20preserve%20the%20planet.

United Nations. 2015. *Transforming Our World: The 2030 Agenda for Sustainable Development.* Accessed 10 November 2023. https://sdgs.un.org/publications/ transforming-our-world-2030-agenda-sustainable-development-17981

United Nations. 2020. *A UN Framework for the Immediate Socio-Economic Response to COVID-19.* New York: United Nations. Accessed 12 November 2023. https:// unsdg.un.org/sites/default/files/2020-04/UN-framework-for-the-immediate-socio-economic-response-to-COVID-19.pdf

United Nations. 2023. *The Sustainable Development Goals Report 2023: Special Edition.* Accessed 12 November 2023. https://unstats.un.org/sdgs/report/2023/

United Nations Climate Change News. 2023. *Five Reasons Why Climate Action Needs Women.* Accessed 12 November 2023. https://unfccc.int/news/five-reasons-why-climate-action-needs-women#:~:text=Women%20often%20face%20higher%20 risks,for%20the%20young%20and%20elderly.

United Nations Department of Economic and Social Affairs (DESA). 2021. Sustainable development report shows devastating impact of COVID, ahead of 'critical' new phase. *Africa Renewal*, 6 July. Accessed 12 November 2023.

WGSN. 2023. *Executive Summary – Future Consumer 2025.* Accessed 18 November 2023. https://createtomorrowwgsn.com/4151918-wp-fc-2025/2/?aliId= eyJpIjoiSGdHb1RYZEhlQWFBY0JYVCIsInQiOiJLdDV4VkIyUGJ6WVdIcGllK2 RpaFwvUT09In0%253D

Zarabadi, Shiva. 2022. "Watery assemblages: The affective and material swimming-becomings of a Muslim girl's queer body with nature." *Australian Journal of Environmental Education*, *38*(3–4), 451–461.

Index

For Product Safety Concerns and Information please contact our EU
representative GPSR@taylorandfrancis.com
Taylor & Francis Verlag GmbH, Kaufingerstraße 24, 80331 München, Germany

www.ingramcontent.com/pod-product-compliance
Lightning Source LLC
Chambersburg PA
CBHW052118230326
41598CB00080B/3845

* 9 7 8 1 0 3 2 4 8 8 2 1 9 *